HENRY HOYLE HOWORTH

THE GLACIAL NIGHTMARE AND THE FLOOD

A SECOND APPEAL TO COMMON SENSE FROM THE EXTRAVAGANCE OF SOME RECENT GEOLOGY

VOLUME I

Elibron Classics
www.elibron.com

Elibron Classics series.

ISBN 1-4212-3848-9 (paperback)
ISBN 1-4212-3847-0 (hardcover)

This Elibron Classics Replica Edition is an unabridged facsimile of the edition published in 1893 by Sampson Low, Marston & Co., Ltd., London.

THE GLACIAL NIGHTMARE
AND THE FLOOD.

VOL. I.

THE

GLACIAL NIGHTMARE

AND THE FLOOD

A SECOND APPEAL TO COMMON SENSE
FROM THE EXTRAVAGANCE OF
SOME RECENT GEOLOGY

BY

SIR HENRY H. HOWORTH, K.C.I.E., M.P., F.G.S., ETC.

AUTHOR OF "THE MAMMOTH AND THE FLOOD," "THE HISTORY OF
THE MONGOLS," "CHINGHIZ KHAN AND HIS ANCESTORS,"
ETC., ETC.

IN TWO VOLUMES

VOL. I.

LONDON
SAMPSON LOW, MARSTON & COMPANY
Limited
St. Dunstan's House
FETTER LANE, FLEET STREET, E.C.
1893

To the Memory of

CHARPENTIER, FORBES AND MURCHISON,

THREE TEACHERS FROM WHOM I HAVE LEARNT

MORE THAN I CAN TELL.

I wish it were possible for me to emulate their gifts. Their industry, patience and accuracy, their sober and mature judgment, their scientific courage, and above all, the possession of that eagle eye which enabled them to overlook a great horizon without having their view distorted by the smaller details of the landscape. Their thought and reasoning were pursued by no mediæval ghosts in the shape of scholastic formulæ, and they were not afraid to attribute great events to corresponding causes, however much the induction might jar with the shibboleths of the orthodox science of the day. If I cannot emulate their gifts I trust I may have profited by their teaching and example, and that what follows would not have been deemed by them an unworthy tribute from their disciple and follower,

THE AUTHOR.

PREFACE.

It is a singular and a notable fact, that while most other branches of science have emancipated themselves from the trammels of metaphysical reasoning, the science of geology still remains imprisoned in "*à priori*" theories.

It was many years ago that one of the teachers to whom I am under great obligations, Adam Sedgwick, protested with the gravity of a true philosopher against this method of reasoning. "The study of the great physical mutations on the surface of the earth," he said, "is the business of geology. But who can define the limits of these mutations? They have been drawn by the hand of Nature, and may be studied in the record of her works, but they never have been, and never will be, fixed by any guesses of our own, or by any trains of *à priori* reasoning based upon hypothetical analysis. We must banish all *à priori* reasoning from the threshold of our argument; and the language of theory can never fall from our lips with any grace or fitness, unless it appear as the simple enunciation of those general facts, with which, by observation alone, we have at length become acquainted. . . . An hypothesis is indeed (when we are all agreed on receiving it) an admirable means of marshalling scattered facts together. But by those who differ from us, an hypothesis will ever be regarded with just suspicion; for it too often becomes, even in spite of our best efforts, like a false horizon in astronomy, and vitiates all the results of our observations, however varied, or many times repeated." [1]

With these words I most cordially agree, but if they were wise and just in 1831, they are equally wise and just now, when our science has become more professional and when authority therefore has become more paramount. "That the

[1] Address to the Geol. Soc.

primary laws of Nature are immutable—that all we now see is subordinate to those immutable laws—and that we can only judge of effects which are past by the effects we behold in progress" are truisms of all science. When this sound doctrine is perverted, however, to teaching men that "the physical operations now going on are not only the type, but the measure of intensity of the physical powers acting on the earth at all anterior periods. This is to assume an unwarrantable hypothesis with no *à priori* probability." This is the language of another great geologist, a distinguished observer and generalizer.

The words were originally directed against the development of the views of Hutton and Playfair, which we owe to the brilliant and fascinating pen of Lyell. He had explained with great wealth of illustration and a wonderful command of limpid and picturesque language, the necessity of studying Nature at work, if we are to unravel the history of Nature from the ruins of her handiwork. He taught, as it had never been taught before, that rain, snow, and frost, that river and sea, that volcano and earthquake, that these and similar instruments which Nature is continuously employing now, are capable, if we allow a sufficient draft upon time, of producing the largest effects. So far no one quarrels with him, and so far he was following modern methods in science, those methods which are alone inductive and empirical. When he went beyond this he travelled beyond the true limits of induction, and against the journey Sedgwick, Murchison, Hopkins and others protested. The protest was, however, largely addressed to an obstinate and perverse generation, which in its rebellion against the old teaching whereby every difficulty was solved by an appeal to the direct interposition of the Almighty, refused to listen to any other voice than that of Uniformity. Not uniformity in the sense that Nature has worked with similar tools and with similar methods in all ages, but in the sense that she has always worked with the same vigour and intensity, or, as Mr. Conybeare translated the argument as applied to geology, because a child grows two inches every year, therefore, that is the normal growth of a human being during the three score years and ten which are his allotted pilgrimage.

To uniformity in its former sense we all adhere, against uniformity in its latter and transcendental and metaphysical sense the following work is meant to be a protest as its predecessor was.

Misleading as I deem the arguments of Lyell and his scholars to have been when applied to the older beds, they were much more so when applied to explain the superficial mantle of gravel, clay, sand, etc., which covers the ragged and ruined surface of the older rocks, and gives to the earth its generally smooth and undulating outline. The former deposits are for the most part arranged in regularly stratified beds, with a regular succession which can be studied in many places in an undisturbed condition. It is very different with the superficial soft beds, which are so incongruous and heterogeneous in structure, which mantle the country irrespective of its contour, which often contain blocks of stone that have travelled hundreds of miles from home, and which afford so many puzzles to us all. Yet if there be a geological horizon, which it is important that we should study on sound principles, it is assuredly this one, for it enshrines the last completed chapter in the history of the world, and, among many other interesting riddles, contains the explanation of the mysterious problem of the origin of our race.

To explain these beds has been the anxious effort of several generations of geologists, and the difficulty of the task may be measured by the divergent theories which have been forthcoming about them, and by the further fact that they still remain in many respects the despair of geology.

Forbes, in his chapter on the Geological agency of Glaciers, writes as follows: "The occurrence of vast masses of primitive rocks, apparently without any great wear and tear of travelling, upon secondary or alluvial surfaces at great distances from their origin, has been one of the numerous opprobria of geology. It is peculiarly so because a thousand circumstances demonstrate that the deposition of these masses has taken place at the very last period of the earth's history. No considerable changes of surface have occurred since. These blocks are superficial, naked, deposited upon bare rock, which has received no coating of soil since, and are often placed in positions of such ticklish equilibrium that any considerable convulsion of nature, whether by earthquake or debacle, must inevitably have displaced them. A geologist might therefore justly be asked, "If you cannot account for these very latest and plainest phenomena of change and transport on the earth's surface, whose various revolutions you pretend to explain, how shall we follow you, when you tell us of the metamorphoses of slates and the throes of granite?"[2]

This was written in 1843. In 1866 M. d'Archiac, one of the most accomplished and trustworthy observers of recent times, could still write: "One of the epochs of Nature about which most has been written in recent times, that which immediately preceded our own and of which the duration was apparently not very long is, nevertheless, that about which we know the least, and which has given rise to the greatest number of hypotheses. The slight development of marine and lacustrine deposits, compared with the great predominance if not universal distribution of the erratic depôts, their small thickness over wide extents of country, the entanglement of one part, and the regular succession of the other, the absence of regularity, symmetry and continuity in the disposition of the beds, have caused the comparisons and relations which it has been attempted to evolve from them to be incomplete and uncertain."[3]

I could easily quote passages from the geological writings of men in the forefront of our science on both sides of the Atlantic, to show that in the year 1892 we are still face to face with the unsolved problem of rationally explaining the manifold and perverse difficulties presented by the latest geological deposits.

I have thought it right in the following pages to begin with an account at considerable length of the various theories which have been propounded to explain a large series of these beds. I have done so more especially, because my own conclusion largely involves a return to the views of those older geologists, who wrote before the world was dazzled by the extravagant development of the Glacial theory, and because their careful labours have been recently ignored.

At one time most men agreed that these beds were assorted and distributed by water, and the great conflict between the Huttonians and their opponents, was as to whether the water

Forbes' Travels, 40 and 41. [3] Geologie et Paleontologie 693.

acted in a normal way, that is in the form of rivers, or the sea, or in an abnormal way, that is by means of catastrophes or *debacles*, as it was the fashion to call them. It must be said that the Huttonians had found it hard to keep the field, for it seemed impossible to realize that water acting in any normal way could transport great blocks of stone for hundreds of miles, and spread vast sheets of gravel and clay over whole continents.

Hence the stretching of the theory of Uniformity almost to the breaking point by Lyell and Darwin, who in order to meet the facts, postulated the submergence of whole continents, and the floating over them from some unknown Arctic sea of fleets of icebergs which were supposed to have sailed in and out of the Alpine valleys, across Germany and Russia, etc. etc., and to have dropped their burdens as they went. The sea bottom is then supposed to have risen again, and to have disclosed what we now see.

The impossibility of maintaining this view in presence of the facts, will, I think, be found to be satisfactorily shown in the penultimate chapter of the following work; although this was scarcely needed, since the view itself was ultimately abandoned by its principal champions.

At this stage Charpentier and Agassiz came to the rescue of the sorely tried theory of Uniformity. The Swiss geologists, who had studied the *modus operandi* and results of glacier work, had also noticed signs and evidences in Switzerland that the modern glaciers are only shrunken relics of much larger glaciers which at one time existed there, and that the unmistakable work of these old glaciers could be traced far down the valleys in the shape of polished and striated rocks, transported *débris*, etc., etc. This led to inquiry being made elsewhere, and it soon became patent that, unless we are to ignore the clearest evidence, we must allow that in the last period of the earth's history there was a development of glaciers on a large scale in nearly all latitudes where high land existed.

This view, which was first propounded and elaborately argued by Charpentier, has stood the test of criticism, and seems to be established beyond question.

When ice was shown to have had a greater extension, and consequently a greater potency in former times than now, it opened up possibilities as to its capacity to explain a great deal more than had been dreamt of. The position was at once seized by Schimper and Agassiz, and their arguments and rhetoric took the world by storm.

It was plausible to argue that the same force which could carry great masses of granite and other rock from the upper Alpine valleys into the lower country, was competent when working on a larger scale to scratch, furrow and smooth wide extents of country many leagues away from any mountain chain, or any sloping ground; to remove masses of gravel, boulders and clay, mixed with unweathered stone, over many degrees of latitude and longtitude, and to explain all the varied and widespread phenomena of the drift. Hence the invocation of a Glacial Period to explain what were deemed to be plainly glacial phenomena.

The Glacial theory as thus propounded by Schimper and Agassiz, was naturally very welcome to the champions of Uniformity, for it seemed to offer a way of escape from a great difficulty without appealing to other than every-day causes.

It is true that the magnitude and wide geographical distribution of the facts to be explained involved unusual courage on the part of those who held this view, and some of his followers were startled and confounded by the lengths to which Agassiz was prepared to go. For he not only invoked a time when a portentous winding sheet of ice covered the temperate zone, but he also took it into the tropics, and argued that a similar icy mantle once filled the valley of the Amazon in Brazil.

While this last extension of the theory met with considerable opposition, the main conclusion for which he fought, namely, that the drift was deposited by continental ice-sheets conterminous with the drift itself, became the predominant creed of a large and powerful school of geologists.

Some of Agassiz's eager scholars soon saw that the new theory needed external support. If, as they argued, the doctrine of Uniformity is conciliated by the prospect of a great ice age and widespread ice-sheets, it also requires that such an ice age should be explained without a breach in the continuity of Nature's laws. Hence the fervid appeals by the Glacial school of geologists to astronomy and meteorology to furnish a satisfactory basis on which their theory might stand, and which should enable them to say that an ice age was not a mere accidental occurrence, whose explanation is uncertain, but a part of the great scheme by which the Universe is governed, and the natural result of the ordinary operation of Nature's laws.

The challenge was not neglected, and many elaborate theories have been evolved by sharp-witted and ingenious men, in order to satisfy the champions of the apparently triumphant and unanswerable Glacial theory. A criticism of these various theories will be found in the following pages, where it is attempted to be shown that they have all completely failed, either to meet the facts of geology, or to be consistent with the laws of physics, and it seems to be as plain as anything can be that the dictum of the great French geologist, Charles Martins, is still true, namely, "that, *if the Glacial theory be sound,* its explanation is a secret which the future alone can disclose."

One of the chief objects of the following book, however, is to show that the Glacial theory, as usually taught, is not sound, but that it ignores, and is at issue with, the laws which govern the movements of ice, while the geological phenomena to be explained refuse to be equated with it. This is partially acknowledged by the principal apostles of the ice theory. They admit that ice as we know it in the laboratory, or ice as we know it in glaciers, acts quite differently to the ice they postulate, and produces different effects, but we are bidden to put aside our puny experiments which can be tested, and to turn from the glaciers which can be explored and examined, to the vast potentiality of ice in the shape of portentous ice-sheets beyond the reach of empirical tests, and which we are told acted quite differently to ordinary ice. That is to say, they appeal from sub-lunary experiments to *à priori* arguments drawn from a transcendental world. Assuredly this is a curious position for the champions of Uniformity to occupy. In regard to it I will quote a fable already utilized by Hugh Miller. 'A wolf,' says Plutarch, "peeping into a hut where a company of shepherds were assembled, saw them regaling themselves with a joint of mutton. 'Ye gods!' he exclaimed, 'what a clamour they would have raised if they had caught *me* at such a banquet.'"

I hold that the Glacial theory, as ordinarily taught, is based not upon induction, but upon hypotheses, some of which are incapable of verification, while others can be shown to be false, and it has all the infirmity of the science of the Middle Ages. This is why I have called it a Glacial Nightmare. Holding it to be false, I hold further that no theory of modern times has had a more disastrously mischievous effect upon the progress of Natural Science. It is not merely in the domain of geology that its baneful influence has been felt. We cannot take up a text book, in which the profounder problems of Biology are treated, problems like the distribution of animals and plants, the pedigree of life, the origin, and beginnings of the human race, without being impressed with its influence as a factor.

In all these and many other inquiries, the postulate of an ice age forms a necessary element of current theories. The length of the following argument is in fact the homage paid to the widespread influence of the theory, which made it incumbent in disputing it to do so with such care and completeness as I can command, amid manifold duties, and indifferent health.

What, then, is shortly the burden of the following argument?

I admit completely, that the position maintained by Charpentier in his work on Glaciers is unassailable, first, because it makes no appeal to any occult and hidden forces underlying the movements of ice, but proves the existence of greater glaciers formerly by comparing and equating the ruins they have left with the ruins made by existing glaciers. Secondly, because it is consistent with all the geological facts that we can summon to test it by. On the other hand, I not only disbelieve in, but I utterly deny the possibility of ice having moved over hundreds of miles of level country, such as we see in Poland and Russia, and the prairies of North America, and distributed the drift as we find it there. I further deny its capacity to mount long slopes, or to traverse uneven ground except when under the impulse of gravity. I similarly deny to it the excavating and denuding power which has been attributed to it by those who claim it as the excavator of lakes and valleys, and I altogether question the legitimacy of arguments based upon a supposed physical capacity which cannot be tested by experiment, and which is entirely based upon hypothesis. This means that I utterly question the prime postulate of the Glacial theory itself.

The purport of the following pages, however, is not wholly critical and destructive. If we sweep away the Glacial theory in its extravagant form, we have not swept away the difficulties which it was introduced to explain. It is our duty, it seems to me, in view of the importance of the issue, to meet these difficulties. How then are they to be met? The answer to this question I have tried to give in the last chapters of this book. The answer is not one consistent with the modern developments of the theory of Uniformity, and for those who hold with Ramsay that Nature has not varied either in the rapidity, the intensity, or the quality of her work since she began to mould the world, there is nothing in those chapters which can prove acceptable. My masters, Sedgwick and Murchison, taught me a very different lesson which I have seen no reason whatever to unlearn. They taught me that no plainer witness is to be found of any physical fact than that Nature has at times worked with enormous energy and rapidity, and at others much more evenly and quietly, and that the rocky strata teem with evidence of violent and sudden dislocations on a great scale. That these catastrophes were aimless and lawless, I do not believe. On the contrary, they were the result of law, but of a law whose tendency we have not yet perhaps duly measured, and whose key we might perhaps have discovered if we had not been pursuing the fantastic shadows of metaphysical reasoning for so many years.

This conclusion is, in effect, the same as that arrived at by some of the best writers on the subject. Let me quote from one of them.

"No one appreciates more highly than I do," says Mr. Babbage, "the labours of those older geologists," that is, Lyell and his school, "who first taught an earlier generation to estimate more truly the power and efficiency of certain forces which act very slowly but continuously during long periods of time. The error in scientific speculation against which they fought was the error of laying exclusive or exaggerated stress upon forces of a particular kind—forces the existence of which they did not deny, but which had worked only at comparatively distant intervals and in alliance always with other forces, the operation of which is ceaseless. It is precisely the same error in principle, though exhibited in a different form, which is now exhibited by those geologists who attribute almost everything to running water and to scraping ice. They are simply catastrophists in a new dress. They attribute extravagant power and stupendous effects to one form of force instead of to another. There is, indeed, one difference, and it is a difference in favour of the older school of catastrophists rather than of the younger—that whereas there never could be any doubt of the adequacy of subterranean force to produce the effects ascribed to it, there is the greatest doubt of the adequacy of rain and ice to effect in any time, however long, the stupendous changes ascribed to them by Mr. Geikie. On the other, there is really nothing stupendous about these effects when they are regarded in connection with the known and visible effects of subterranean force. The highest ranges of mountains we have are, relatively to the circumference of the earth's crust, infinitely smaller than the puckers on an orange-skin." [4]

Again, the same writer says: "Magnitude is all relative. The store of Time and the store of Force may be regarded as both unlimited. But it does not follow that in accounting for any given effect we are entitled to draw to an unlimited extent either upon the one or upon the other. Extravagant demands may as easily be made upon the one or upon the other. The inventions and imaginations to which the extreme glacialists resort are, beyond all comparison, more violent than those which were common with the old convulsionists. Whole continents are built up upon the top of the existing mountains, which there is no proof whatever ever existed; and then these continents are all ground down by ice or washed away by ordinary surf, and yet so that not a fragment shall be left behind. I venture to believe that I shall have some support from the great leaders of geological science, who, in power of intellect, are still young among us, when I record my dissent from the extravagant theories of the younger glacialists." [5]

With this reasoning I quite agree. The earth seems to me

[4] Babbage, Journ. Geol. Soc. xxiv. 272 and 273. [5] Id. xxiv. 273.

to be full of evidence of intermittent violence and repose. In facing the solution of the Drift problem I must be taken therefore to postulate, not merely the possibility of catastrophes, but to maintain that they have occurred frequently in the world's history.

Secondly, I would claim that while the Glacial theory makes demands upon the powers of ice which are inconsistent with its proved qualities and cannot be made to fit in with the facts which have to be explained, the power to which I appeal makes no demands whatever upon any force but that of which we can establish the competency, both by direct experiment and by theoretical calculation, and that, so far as we know, it explains all the facts.

Thirdly, this explanation is one which was deemed satisfactory and complete by the Fathers of Geology, men who were quite as keen observers and quite as keen critics as their descendants, and who were also more independent, and less dominated by official orthodoxy in Science. For a long time some of the most brilliant masters of our science were advocates of the Diluvial theory as an adequate explanation of the facts. It is true that some of them attempted to explain too much by this cause, and that until Charpentier published his famous work the operations of land ice, on a much larger scale than that in which it works now, had been overlooked.

It is true, also, that the views they maintained were in some cases sophisticated by an appeal to untenable postulates, but in the main the theory, of which I am an advocate, is a return to that so strenuously supported by Hall and Conybeare, Von Buch, Sedgwick, Murchison, D'Archiac, Phillips, and many others.

Lastly, I claim to have already established the necessity of this appeal on entirely different grounds. In my previous work on the "Mammoth and the Flood," I collected a great mass of evidence which went to show that the Mammoth and some of its companions, including so-called palæolithic man, were swept away in wide areas by a great flood of waters which drowned them, and then covered them with continuous layers of loam and gravel. The facts and the arguments adduced in that book have been referred to many times by many writers for their manifold character and wide-reaching and cumulative effect, and I am pleased to think that some of the first authorities in this country and elsewhere in the realms of Physics of Natural History and Archæology, have assured me that they considered the argument unanswerable in so far as the particular facts referred to are concerned, and that what is needed is that it should be supplemented by a similar argument based on geological grounds. On the other hand, in the many reviews of that book which have appeared here in France and Germany and in America, I know of none in which its arguments have been met or traversed, while the great majority give a verdict of suspense dependent on the production of adequate geological evidence.

In the following pages, then, I claim to have shown that a widespread flood, which seems to be a necessary postulate, if we are to adequately grapple with the extinction of the Pleistocene fauna, is an equally necessary postulate if the geological facts are to be duly explained. At all events, I claim for my arguments that they deserve adequate criticism. I am not aware that I have omitted to notice a single fact or difficulty which needs explanation, and I have tried to make the book a fairly complete monograph on the controversies with which it deals. That it contains many faults, both of commission and omission, I know well, and those who have tried to wade through the vast multitude of publications dealing with the problems here discussed, and who realize also the many and critical and necessary side issues which have had to be sifted and criticized in the process of exhausting the problem, will be tender and considerate to my method and my plan. For the facts, arguments, and conclusions adduced, I ask for no quarter. If they can be shown to be false and misleading, they had better be swept away mercilessly. If they are true, as I believe them to be, they cannot be ignored. Meanwhile, I am ready to enter the arena and to fight hard for them.

The book does not exhaust the problem, nor does it profess to do so. When it was begun I hoped to have brought together all the geological evidence bearing upon it, but the necessity of entering at much greater length than I intended into the history of the Glacial theory if I was to do justice to the predecessors, upon whose shoulders I am privileged to stand, and the necessity also of discussing at adequate length the many theories of Ice motion and the various astronomical and meteorological explanations of the Ice age, have made it impossible to condense all my arguments in one book. The work is, in fact, limited to the so-called drift beds. That the same conclusion follows, from a consideration of the distribution of the valley gravels, the brick earths and loams, the loess, chernozem and Pampas mud, etc., I have tried to show in papers printed in the *Geological Magazine*. The facts and arguments contained in those papers I hope to enlarge and deal with in another volume, in which the geological evidence forthcoming from countries where the true drift does not occur, will have a part, and in which it may be possible perhaps to discuss the extent and the cause of the Pleistocene Flood, and the question of whether it was one of a recurrent series of similar catastrophes, as I am disposed to think it was, or an exceptional and unique event.

At present I merely claim to have made good my promise to produce adequate geological evidence in support of the theory I advanced in my former work. This theory seems to me to be overwhelmingly established, and to be an absolutely necessary postulate if we are to treat our science as an Inductive science at all. Being so, I would go a step further and not only return to older opinions, but also return to older and clearer and better nomenclature. The Pleistocene Flood, though far from being universal, was certainly one of the most widespread catastrophes which the world has seen. It forms a great dividing line in the superficial deposits as was maintained long ago, and as such, it is a very useful landmark which ought to appear in our nomenclature, and I do not know of any better terms than ante-diluvian and post-diluvian to mark the two great divisions of the post-Pliocene beds.

Ante-diluvian times, so far as we can see, were marked in the temperate latitudes of both hemispheres by accumulations of ice in the shape of large glaciers on the high lands (then probably much higher) of Western Europe, of North America, of Australia, New Zealand, and perhaps South Africa. Alongside of these glaciers, and in contact with them in all these latitudes, were wide champaign and wooded districts in which the Mammoth and the woolly rhinoceros were the most prominent animals in North Asia and Europe, the Mammoth and the Mastodon in North America, the Mastodon and the great Sloths in South America, the various species of gigantic Kangaroos and Wombats in Australia, the great wingless birds in New Zealand. They lived and thrived in the near neighbourhood of the ante-diluvian glaciers, just as the apteryx now thrives in the luxuriant forests near the great glaciers in New Zealand, and just as the tiger and the rhododendron thrive close to the Himalayan glaciers.

Meanwhile, in Northern Asia and Western Europe, and in North and South America certainly, and probably also in Australia, ante-diluvian man lived alongside of and hunted the ante-diluvian animals, and if we are to clear up his pedigree we must find the true interpretation of these ante-diluvian beds.

Presently came a tremendous catastrophe, the cause of which, as I have tried to show in the *Geological Magazine*, was the rapid and perhaps sudden upheaval of some of the largest mountain chains in the world, accompanied probably by great subsidences of land elsewhere. The breaking up of the earth's crust at this time, of which the evidence seems to be overwhelming, necessarily caused great waves of translation to traverse wide continental areas, as Scott Russell, Hopkins, Whewell, and Murchison argued they would, and these waves of translation as necessarily drowned the great beasts and their companions, including palæolithic man, and covered them with continuous mantles of loam, clay, gravel and sand, as we find them drowned and covered. They also necessarily took up the great blocks which the glaciers had fashioned, and transported them to a certain distance and distributed them, and the Drift associated with them, as we find them distributed. This Induction, whose details are contained in the following pages, seems to me to be complete, not only because it adequately explains the facts, but because it is the only theory that does so, and I know nothing against it, but the almost pathetic devotion of a large school of thinkers to the Religion founded by Hutton, whose High Priest was Lyell, and which in essence is based on *à priori* arguments like those which dominated mediæval scholasticism and made it so barren.

I must now say a few words in justification of the mode in which this work has been put together.

A casual reader, on turning over its pages, will be apt to say: Here is a compilation, a mere collection of opinions and of facts gleaned from other sources. Perhaps if he will take the pains to read a little more closely, he will find that this would be a very inadequate and unfair view of the position. Of course, like every other man who has written a scientific book, I am dependent for a great deal on what others have written and what others have discovered. If it had not been for the scientific decalogue which prescribes, *inter alia*, that the man who first makes an induction is entitled to the credit of it, I could very easily have written the whole book without giving a reference and without quoting a passage. This also explains the laborious traversing of a vast number of books dating from the earlier days of geological writing, which are most of them forgotten, if justice was to be done to the first rational explanations of the Drift phenomena. Hence also the marshalling of many successive views in illustration of the progress of a complete theory on the subject. The survey is, I know, not exhaustive, for the subject is vast beyond conception, but I trust that justice has been done to some men at least, who lived before Agamemnon and whose keen eyes and whose sound judgment it has been the fashion to decry, while much of what they have written has been treated as obsolete.

This is my explanation of the large number of opinions quoted, and of references made which perhaps unfairly disguise the fact that I have tried, in view of what follows, to examine the drift phenomena in many countries, and have frequently worked out conclusions for myself for which I have quoted others, because others had already been over the same ground and reached the same results, and were therefore entitled to the credit of what they first described. The general argument and the general theory, with a great deal of illustration, is entirely my own. The men of water had been long ago dispersed and scattered, the men who would dare to appeal to cataclysms and catastrophes are few indeed. Those who would venture to jeopardize their character for scientific sobriety by reviving and extending the views of the geologists of fifty years ago on great diluvial movements are not to be found anywhere. Whether right or wrong in his conclusion, it is perhaps well that, among the erratic heretics who are careless of prestige and indifferent to conventional opinion, one should occasionally be found to challenge the dominant creed by assailing its foundations.

I hope I may have escaped one penalty of a good deal of polemical writing, and that nothing contained in the following pages will be found offensive to the living or unfair to the dead. If I cannot agree with some of the conclusions of Lyell, of the Geikies, and of Ramsay, I am well aware of the gap that separates us in reputation and in knowledge, and of the debt I owe them all, and no one has a profounder regard and appreciation of what they have done than myself. They have taught me a great deal of what I know.

I must now turn momentarily to protect myself against misconstruction in another direction. This work, like its predecessor, is written in the interests of no school of thought, and is meant to be merely an inductive argument from the facts. It may be used by some to support a position which is not mine, and which, as I have said elsewhere, is not a conclusion of Science, but a dogmatic opinion, partly based upon the primitive traditions of mankind and partly on ingenuous speculations in cosmology made before the birth of science. It has no other value.

No doubt if the catastrophe I have described really occurred in the human period, as my contention is, the fact must have impressed the imagination of the survivors, and it is not unnatural to find it reflected in many scattered traditions.

This is a very different thing, however, from the conclusion that the views here established support a narrative whose dogmatic value depends upon its details, or give any countenance to the notion that the postulated flood was universal, that it destroyed all life save what was sheltered in one vantage, or that the human race began its genealogy again from one fortunate family.

The story in the Bible is interesting as an early example of a widespread tradition, and nothing more. Whether directly derived from Babylonia, or through some intermediate source, we cannot with our present knowledge suppose that the narrative was incorporated in the Pentateuch before the tenth and perhaps not before the seventh century B.C., and that is the measure of its value as *a Jewish* record. To suppose that religion or morals are helped or furthered by trying to give some greater authority to the story by factitious appeals to geology, must end in disappointment and in failure, and it is right to say plainly that it involves both a dangerous logic and a misleading hope to continually try and equate notions which are incommensurable, while it must inevitably mislead those who ought not to be misled. The grave and wise words of Sir Francis Bacon ought assuredly to be ever present to us when we embark on the impossible task of equating science and faith. "Tanto magis haec vanitas inhibenda venit et coercenda quia ex divinorum et humanorum male—sana admixtione, non solum eduntur philosophia phantastica sed etiam religio hæretica." These wise words are thus paraphrased by a great geologist, who was also a church dignitary, my master, Sedgwick. "This vanity merits castigation and reproof the more, as, from the mischievous admixture of divine and human things, there is compounded at once a fantastical philosophy and an heretical religion." [6]

It is not part of my subject, but I cannot help parenthetically saying that Divines would do well to separate rather more sharply than hitherto the annalistic and ritualistic from the essentially religious books of the Bible. It is not Moses and Samuel who among the ancient sears are very helpful to us in these days in the realm of morals, or in presenting high ideals and standards for our guidance. It is rather Isaiah, Jeremiah and Job.

I have to thank my old friend Dr. Woodward for allowing me to discuss some of the issues contained in the following pages in the *Geological Magazine.* I have further to thank the Council of the Manchester Literary and Philosophical Society for permission to reproduce a long paper which I published in its Transactions, dealing with the various theories of ice motion, and I have also to thank a large number of geologists

[6] Sedgwick, Address, Geol. Soc., 22.

from whose works I have derived infinite pleasure and profit, and whose labours I have read and criticized.

Much of what is of value in the following work belongs to them. In regard to the mistakes and blemishes which it contains, and which are largely my own, I have to plead that the book has had to be written under manifold disabilities and especially the double burden of time being heavily mortgaged by other duties, and health and eyesight having been equally uncertain. This has prevented the necessary revision which the book would otherwise have had. I know well that in many cases its style is slovenly and that many verbal slips occur throughout it. These accidents will, I trust, not entirely monopolize the attention of my critics, who will perhaps do some justice to the long and conscientious labours it enshrines and the important issue it tries to raise. I shall feel indebted to any one who is disposed to comment on what I have said, if he will let me have an opportunity of profiting by the criticism, and will send me what he may write.

In introducing another work to the frigid welcome of a world overloaded with books, I cannot forget the patience of those who have been the uncomplaining partners of my long toil at home. In conclusion, it is with genuine sympathy and appreciation that I appropriate the words of the weary monk, who, after long days in the Scriptorium wrote, "Explicit, expliceat; ludere Scriptor eat."

The Athenæum,
November, 1892.

CONTENTS.

CHAPTER I.

THE CHAMPIONS OF WATER. I.

CHAPTER II.

THE CHAMPIONS OF WATER. II.

CHAPTER III.

THE CHAMPIONS OF WATER. III.

CHAPTER VIII.

ON THE ALLEGED RECURRENCE OF GLACIAL EPOCHS, AND ON SUPPOSED INTER-GLACIAL BEDS.

CHAPTER IX.

APPEALS TO TRANSCENDENTAL PHYSICS AND ASTRONOMY.

LIST OF ERRATA.

Page 19, line 37, *insert* and *before* La Sarra.
,, 24, ,, 19, *for* toisés *read* toises.
,, 42, ,, 14, *for* geog. *read* geol.
,, 45, ,, 39, *after* ante *insert* 14.
,, 47, ,, 13, *for* st. *read* à.
,, 58, ,, 11, *erase* comma *after* covering.
,, 64, ,, 19, *for* is *read* an.
,, 81, ,, 17, *for* Betws *read* Bettws.
,, 83, ,, 24, *insert* with *after* intervene.
,, 84, ,, 22, *for* to *read* from.
,, 87, ,, 4, *insert* and *after* north.
,, 111, ,, 25, *for* Gimone *read* Gimont.
,, 117, ,, 4, *insert* hyphens *between* porphyry like *and* rock.
,, 118, ,, 16, *for* state *read* State.
,, ,, ,, 26, *for* Klaadu *read* Kcaadn.
,, 120, ,, 19, *for* les *read* le.
,, 122, ,, 24, *for* Silicia *read* Silesia.
,, 127, ,, 22, *insert* of *after* south.
,, 151, ,, 30, *for* Sedgewick *read* Sedgwick.
,, 178, ,, 24, *after* he *insert* concluded that.
,, 189, ,, 4, *for* Hills. While *read* Hills, while.
,, 192, ,, 25, *for* glacial *read* glacier.
,, 217, ,, 3, *for* glacé *read* glace.
,, 232, ,, 36. *for* deposited *read* caused.
,, 233, ,, 21, *insert* and *after* Norway.
,, 294, ,, 38, *erase* comma *after* surface.
,, 297, ,, 18, *for* Hall *read* Hull.
,, 300, ,, 37, *for* Mantel *read* Mantell.
,, 309, ,, 25, *for* Ischer *read* Escher.
,, 316, ,, 9, *for* whole *read* while.
,, 319, ,, 31, *for* Faxo *read* Faxö.
,, 320, ,, 27, *for* E *read* A.
,, 339, ,, 28, *for* 239 F *read* —239 F.
,, 384, ,, 24, *insert* that is *after* and.
,, 401, ,, 11, *erase* that.
,, 423, ,, 3, *erase* comma *after* and.
,, 425, ,, 12, *for* an astronomical *read* a meteorological.
,, 426, ,, 25, *for* Somervell *read* Somervail.
,, 432, ,, 20, 21. *erase* in the arctic regions.
,, 433, ,, 24, *for* their *read* these.
,, 461, ,, 18, *for* have left *read* bear.
,, ,, ,, 24, *for* north *read* south.
,, 462, ,, 37, *for* is *read* are.
,, 468, ,, 20, *for* Neu *read* Neue.
,, 469, ,, 4, *for* Innspruck *read* Innsbruck.
,, 494, ,, 18, 21. *erase* the inverted commas *before* In *and insert them before* variegated.
,, 495, ,, 10, *for* Gereas *read* Geraes.

Page 498, line 39, *for* Schomberg *read* Schomburgk.
,, 516, ,, 36, *insert* the account of *after* in.
,, 529, ,, 19, *for* casue *read* cause.
,, 540, ,, 7, 8, *erase the words from* The *in line* 7 *to* that *in line* 8, *and instead of* the scratching *insert* The *in line* 8.
,, ,, ,, 14, *for* it *read* a glacier.
,, 551, ,, 37, *for* stripes *read* strips.
,, 556, ,, 20, *for* Feriet *read* Ferret.
,, 560, ,, 8, *for* immediate *read* intermediate.
,, ,, ,, 15, *for* glaciers *read* crevasses.
,, 562, ,, 3, *for* state of the crevasses *read* curves.
,, 605, ,, 24, *insert* Memoirs of *before* U. S.
,, 610, ,, 18, *for* Bell *read* Ball.
,, 612, ,, 1, *insert* in *after* rule.
,, 626, ,, 10, *for* its *read* the glacier's.
,, 628, ,, 4, *for* Dranse *read* Drance.
,, 660, ,, 32, *for* Moteratsch *read* Morteratsch.
,, 662, ,, 18, *for* caused *read* causes.
,, 663, ,, 12, *for* inseparable *read* insuperable.
,, 671, ,, 21, *for* luxurious *read* luxuriant.
,, 677, ,, 34, *for* deep *read* steep.
,, 681, ,, 8, *for* green *read* granite.
,, ,, ,, 18, *for* close *read* else.
,, ,, ,, 30, *for* erratic *read* erratics.
,, 684, ,, 32, *for* it *read* them.
,, 698, ,, 15, *for* asäs *read* asär.
,, 739, ,, 43, *for* Id. *insert* Q. J. G. S.
,, 757, ,, 6, *for* St. *read* South.
,, 776, ,, 27, *for* ärar *read* äsar.
,, 779, ,, 7, *for* burst *read* bursts.
,, 783, ,, 10, *insert* earth *after* brick.
,, ,, ,, 11, *for* earth *read* land.
,, 786, ,, 37, *for* were *read* are.
,, 832, ,, 18, *for* Bell *read* Belt.
,, 846, ,, 30, *for* , as *read* of.
,, 848, ,, 21, *erase* comma *after* Lowestoft.
,, 856, ,, 25, *for* Holmforth *read* Holmfirth.
,, 857, ,, 35, *for* Kmmeredge *read* Kimmeridge.
,, 860, ,, 37, *for* were *read* was.
,, 864, ,, 40, *insert* of the chan *after* depth.
,, 865, ,, 7, *insert* comma *after* solitary.
,, 887, ,, 30, *for* Schomburgh *read* Schomburgk.

THE GLACIAL NIGHTMARE.

CHAPTER I.

THE PHENOMENA OF THE DRIFT.

The Champions of Water (Part I.). Writers of the Eighteenth Century.

"Frangitur enim occurrentibus saxis, et per angusta eluctatus ubicumque vincit, aut vincitur, fluctuat."—SENECA, "Natur. Quæs," lib. iv. ch. 2.

Abundance of boulders in North Europe—Some remarkable boulders—Views of the unscientific about them—The Swedish writers—Swedenborg, 1719-22—Tilas, 1740-42—Linnæus, 1744—Bergmann, 1769—Writers on the Alps, &c.—Bourguet, 1729—Ehrhart, 1735—Arduino, 1759—Ferber, 1773—Writers on the erratics of Germany, Russia, &c.—Guettard, 1762—Guldenstadt, 1768—Pallas, 1768—J. A. De Luc and his explosion theory, 1779, &c.—Silberschlag, 1780—De Luc the younger—Larivière, 1829, and Dassen, 1835—De Saussure, 1779, &c.—Dolomieu and the theory of fluvial erosion, 1793—Hutton, 1795—Playfair, 1802—Ebel, 1808—Conybeare—The fluvial erosion theory answered by De Luc—Sir James Hall and Charpentier—The views of Mierotto, 1790—De la Metherie, 1795—Bertrand, 1797—Aikin, 1797—De Lasterye, 1798-9.

EVERY one who has travelled in the North of Europe, in Sweden, Finland or North Russia, must have had his attention arrested by the innumerable boulders and great rounded masses of stone which strew the whole country.

In Wraxall's Travels in the Baltic countries we find him stating, in a letter written from Wyborg in June, 1774, "that he had seen no part of Sweden, except East Gothland, so free from those vast stones which Nature in her anger scattered over these kingdoms," and in describing the neighbourhood of

Louisa, in Finland, he says: "It is difficult to depicture the aspect of the country in which it stands. For the space of at least a league before I arrived there, and for more than two leagues after I quitted it, the earth may almost be said to have disappeared from my view, so completely was it covered with stones; or rather rocks: for many of them, from their magnitude, may well merit that appellation. It seems as if they had fallen from the sky; and Ovid, had he been acquainted with this portion of the globe, might have here placed 'the Campi Phlegraei' where Jupiter overcame the Titans. The road, compelled to respect these formidable impediments, performs a thousand tortuous evolutions, in order to avoid them, and serpentines beautifully for many miles. Neither cultivation nor population can possibly exist among such a wilderness of stones."

De Luc, speaking of North Germany, says "the blocks are there accumulated in such numbers that it might be supposed they had been brought together for the purpose of building some large town."

Dr. Clarke says of these ruinous-looking landscapes: "The soil, if it may bear the name of soil, is altogether incorrigible: it consists of enormous loose fragments of bare granite, piled together till they become mountains, and form steep precipices. Upon these boulders there appears hardly a trace of any vegetable earth or even of any kind of covering; yet they are thickly planted with forests of tall *pines, birch* and *juniper* trees which, in a marvellous manner, have found nourishment for their roots in the interstices between the boulders." [1]

These great stones, shaggy with grey mosses and weather-beaten, the ruins of a world that has passed away, are a perpetual source of glamour and of romance as we drive up and down the forest roads of the north in the twilight, and they form a fitting background to the adventures of giants and dwarfs with which the imagination of the old-world folk peopled them.

Nor are such stones confined to the North. As we move further south, if they become more unfrequent, their strange shapes and unexpected positions are even more obtrusive, and

[1] Travels, part 3, sect. 2, 95.

especially is this so when they consist of some kind of rock foreign to the country.

In the cultivated valleys few now remain; they are not convenient neighbours to the plough and harrow, and have been largely broken up to make the roads or used to build walls. It is otherwise with the wastes, especially the upland wastes and commons.

It will not be uninteresting to call attention to a few specimens of these stones, especially remarkable for their size. The most famous of all, perhaps, is that called *Pierre-à-Bôt*, a boulder of fine-grained granite, planted on the Jura mountains near Neuchatel, and which must have come from Martigny in the Valais. It is 50 feet long, 40 feet high and 20 wide, thus having a cubic area of 40,000 feet. To reach its present home from its original one it had to travel 60 to 70 geographical miles, and to cross the lake of Neuchatel. Another one, near Zurich, is called the *Pflugstein*, and can be traced to the Alps of Glarus, whence it had to cross Lake Zurich. It rises to 60 feet above the ground, and its cubic area, according to Heer, is 72,000 feet with a weight of 4500 tons. The *Pierre des Marmottes*, a granite boulder near Monthey in the lower Valais, can be traced to the valley of Ferret, a distance of 30 to 40 miles, and measures about 60 feet in length, 30 in height and 33 in width, and has a cubic area of about 55,000 feet. The *Pierre-du-trésor*, a granite boulder near Orsiéres, measures over 90,000 cubic feet. *Bloe-Monstre*, near Devent, a boulder of limestone from the valley of Avanzon, measures about 55 feet in length, 43 in width and 60 in height, with an area of about 130,000 cubic feet.[2]

A famous boulder, called the stone of the Markgraf, on the Rauhe Alp, weighed 560,000 lbs., and out of it has been carved a vase nearly 23 feet in diameter.[3] The gigantic basin before the New Museum at Berlin was wrought out of one half of an erratic block of Swedish red granite, the other half of which is still lying near Fürstenwalde.[4]

A block in the diocese of Flade, in the island of Mors, is about 23 feet long, 18 wide, and of unknown depth.[5] Forchhammer describes a great pyramidal block, buried in the

[2] Braun, Recreations in Popular Science, 39, 40.
[3] Klee, Le Deluge, 143. [4] Braun, op. cit. [5] Klee, op. cit. 143.

ground, 111.5 feet in circumference and $11\frac{4}{5}$ feet above the surface, which was presented to the King of Denmark by a proprietor at Hesselagergaard in Fionia. The king had it uncovered to a further depth of $10\frac{1}{3}$ feet, but its base was not reached, and it went on increasing in size.[6]

Visitors to St. Petersburgh have seen the granite base of the statue of Peter the Great there. This vast block was brought all the way from Carelia. "Being placed," says Dr. Clarke, "on balls of brass fifteen inches in circumference, which rolled on sledges over a causeway raised for the purpose, it was moved every day, by four hundred men, with the assistance of pulleys and a windlass, over a space of ground equal to about half a mile. From the coast it was brought on a raft of a peculiar construction, to the city. The original size of the rock was 36 feet in length, 20 in height, and as many in breadth, but it was much reduced in size when shaping it."[7]

De Luc describes another of these famous stones, which was found near Eutin. He tells us how he was shown at the gate of the palace gardens there, four pillars about ten feet high, of a fine grey granite, with their pedestals of the same piece, which had been hewn out of a single block. The walls of the kitchen garden were entirely built up out of fragments of the same block, and they formed a quadrangle, having a gate in the middle of each side, between two pillars seven or eight feet high. Close by was a temple, supported by six columns, which, with the steps round the pavement of the temple and its cupola, had all been taken from the same block, and notwithstanding this only one half of the boulder had been used.[8]

The largest of all boulders known to me, however, are in England.

They are of chalk and lie in the cliffs near Cromer in Norfolk, embedded in, and surrounded by, clay and sand. One of them, according to the measurements of Mr. Searles Wood, is more than 800 feet long and 60 feet high.

If we turn to America, we meet with similar phenomena. "The size of some of our boulders," says Professor Hitchcock, "is very great. Those from 10 to 15 feet in diameter are

[6] Id. [7] Clarke's Travels, part 3, sect. 2, 496. [8] Travels, i. 279.

very common. Those from 20 to 25 feet are less common. I have seen several, as at Gay Head, and Bradford, in Massachusetts, and in Winchester, in New Hampshire, fully 30 feet in diameter; and one of conglomerate existed a few years ago, but is now a good deal blasted away, at Fall River, in Massachusetts, which must have weighed not less than 5400 tons, or 10,800,000 pounds. I also recently measured one in Antrim, in New Hampshire, whose horizontal circumference is 150 feet."[9]

Let us now turn to the various explanations which have been forthcoming to account for these boulders and the phenomena associated with them.

Unsophisticated philosophers like shepherds and other contemplative students of nature who live much in the open country, have not unnaturally invoked uncanny causes to account for the strange stones. The Devil, or Robin Hood, or some famous giant, or some still more potent god like Thor or Odin, have been supposed to have brought these strangers either through freak or for some good reason. Robin Hood's bed, the Fairies' Stone, the Devil's Quoit or the Devil's Marble, are names which recur in the mountainous districts of England. To mention one famous case; does not, or ought not, everyone to know that the great stones on Salisbury plain were brought thither by the magician Merlin and ranged in that great double circle at Stonehenge to afford speculation to antiquaries for all time? while the very rocks which the gods threw down upon the Titans, and with which they crushed them, still remain in Greece to testify to the story.

Professor Geikie, speaking of Scotland, says, "Numerous are the myths and legends connected with the great boulders of our country. They are 'the giant's putting stones,' 'the deil's burdens,' 'the witches' hearth stones' of the fanciful peasantry. Zealous antiquaries have occasionally claimed them as monuments set up by man in some long-forgotten age."[10] M. Falsan tells us how the strange outlines of many of these *erratic blocks* (as De La Beche named them), have arrested popular attention. In the midst of the white lime-

[9] Phenomena of the Drift, 166. [10] Prehistoric Europe, 171.

stones of Bugey, their deeper colour has especially aroused notice. Hence they are called *pierres bises* (*bis* or *gris*); they are also called "blue stones," "salt stones," "stones which do not yield lime." In the plains of Northern Germany, as Credner tells us, they are called "Findlings blocke" (i.e. foundlings). In other places their names recall other notions, as Pierre du Bon Dieu; Pierre du Diable, Pierre de Samson; Galet de Gargantua; Pierre du Mariage, etc.; and M. Ed. Piette, who has studied them from the point of view of archæology on the mountain of Espiaup in the valley of Oo, in the Pyrenées, has shown that they have been made the objects of a kind of religious cult, combining a number of superstitions and customs, some bizarre and some licentious, which are gradually dying out as education spreads.[1]

The first scientific person, so far as I know, who referred to these moss-grown wanderers, was that very remarkable man Emanuel Swedenborg. The latter part of his life was so entirely devoted to the mystical and visionary schemes which culminated in the foundation of the Church of New Jerusalem, that it will be news to many people to learn that in his earlier days he was an acute and distinguished man of science, a member of the Academies of Upsala, Stockholm and Petersburg, and that he filled the post of assessor of the School of Mines in Sweden, and this at a time when Sweden was the nursing mother of modern Natural History.

It was this official position that doubtless brought him into contact with the geological aspects of the surface deposits of the North, which he was the first to publish and to treat seriously. *Inter alia*, he not only refers to the boulders which are so conspicuous in Sweden, but he connects them with other phenomena, such as the long whale-backed ridges of sand and stones known to the Swedes as *äsar*, and he assigns a cause for both.

He describes these ridges as being very conspicuous in the provinces of Westmannland, Dalekarlia and Upland. They are occasionally of great size and as much as 150 Swedish ells in height. A Swedish ell being one English foot and $11\frac{7}{10}$ inches. Some of them, he says, are formed of sand

[1] Falsan, La Periode Glaciaire, 71, 72.

intermingled with pebbles, the latter being rounded and polished as if turned on a lathe.

Others of them, he remarks, are made up of pebbles only, without sand, all most curiously turned. Many again seem formed of enormous stones like fragments of mountains heaped together. These last ridges, he says, consist entirely from top to bottom of large fragments of stone, each probably weighing from ten to forty Swedish maritime pounds (skeffund), and he tells us he had often noticed them several miles off, although they were not more than thirty to fifty paces wide. Their sides are generally precipitous. The open country is further strewn with vast boulders, like ruins of mountains, even where there is not a hill in the neighbourhood. They are found in the plains, at the bottom of lakes, and even on the very summits of mountains. The provinces of Helsingland and Orebro are full of them, spread like buildings over the level country. He further noticed that the ridges run north and south, that they have shelving sides whose slope depends on the materials of which they are formed, those made of sand not being so steep as those made of stones; that the stones in question have been rubbed and polished as if triturated by the sea, and that they exist chiefly on the highest ground, as in the province of Orebro.

Swedenborg argues that these ridges can be accounted for only by invoking some very deep or universal diluvian ocean, the former existence of which, he contends, is proved by the mixture of substances of different kinds, sand, clay, smooth pebbles, large stones and masses of rock. Their slope proves that they were thrown up by the sea into great accumulations, and so formed into lengthened ridges, with sloping sides. The fact that these ridges run north and south, he says, shows that the same winds prevailed in the diluvian age as now, and it follows that the ocean referred to must have been universal and stood at a great height above the land, and that it had no limits, shores, or straits to warp the winds; and he appeals to the principles of hydraulics to prove that the sea could have done this kind of work. Its pressure depends on its depth and is more considerable at greater than at lesser depths; and as the bottom of the sea was then covered with different kinds of stones, sand and clay, it is not wonderful that very deep waves should have had sufficient power to tear enormous stones from their foundations, to strew them about in heaps, pile them up in mountains, and thus form ridges or cover whole provinces with them. He explains the existence of so many stones on very high ground on the principle that if the sea had sufficient power to carry them to the highest parts of the district they would stop there and travel no further, since the force of the water would diminish with the depth. This is shown by its frequently, when very deep, tearing up masses of stone and overthrowing walls of masonry; dykes, though built of stone, being overthrown when the water rises three or four ells higher: and he shows that just as the wind, when it becomes a hurricane, can blow down trees and houses, so can an ocean, when much agitated and in gales of wind, alter the configuration of great sandbanks and shoals. He also urges that this would be assisted by the fact that although stone, when in the air, is heavier than water, as $2\frac{1}{2}$ to 1, it is only $1\frac{1}{2}$ to 1 when immersed in fresh water, and still less in salt. On these grounds he concludes that the phenomena of the boulders and the *äsar* is to be explained by the operation of the diluvial waters.

Swedenborg's views, however crude in some respects, were far in advance of his time. They were first published in the years 1719-22.[2]

Some years later, namely in 1740 and 1742, Daniel Tilas drew attention to the great granite blocks existing in Sweden, some distance from their parent mountains. These stones he describes as of various kinds, some as big as houses or even small mountains, and others quite small. He remarks that the stones which have travelled the shortest distance are less rubbed and have their angles sharper than the others. He attributes their distribution to a great flood, adding that the

[2] In 1719 he published a small memoir, entitled "Om vattrens höjd och förra verldens starka ebb och flod. Bevis utur Suerige." In 1721 he published a paper in the "Acta Literaria Suecia," i. 192-196, and in 1722 issued his "Miscellanea observata circa res Naturales," Leipzig. These two last papers were translated in 1847 by Mr. C. E. Strutt, London.

larger ones were possibly carried by floating ice which came from the Pole, while all would be assisted by the specific gravity of the stones being altered when immersed in water. He says that stones in a rapid current do not sink to the bottom but are forced to the sides. He mentions the rounded and smooth outlines of the hills as evidence that they have been planed down. He had remarked, he says, how much more numerous the stones are on the high ground than in the low valleys, diminishing in size as we get further away from their source; and that this is natural, since they would be arrested in their progress by the ground; and he also noticed that on the low ground they are frequently covered with clay, sand, etc. He tells us they are often found tumbled together in heaps, sometimes a large stone lying on smaller ones, and that they often occur on the rounded and polished surfaces of the mountain tops. He also mentions the absence of these stones from the Russian and Siberian steppes, and their being so much more common in some districts than others. Thus he had seen hardly any when travelling along the Ladoga canal and the Sweti river towards the Onega lake, while in the Carelian uplands between Kontzoffer and Petrofskoi they abound. Contrasting their abundance further north with their infrequency as we go south, he says the people of Scania have difficulty in finding enough stones for their houses, while the Dutch would freely exchange gold for the granite boulders of Central Sweden. To the great flood which he invokes he also, like Swedenborg, attributes the long sand ridges so conspicuous in the north, which often traverse, he says, whole parishes and even provinces. He also reports how these deposits sometimes occur as collections of mounds of sand and clay, which he noticed consist of heterogeneous materials tumbled together and unstratified.

We next, namely in 1744, find Linnæus, whose quick eye and keen judgment did so much for natural science, writing: "There is, about forty miles from Hoburg, in Gothland, an immense micaceous stone, heavier than any which human power could move. It is not in its natal place, for no such stone is found there, whence it may be concluded that the sea brought it hither from Sweden or

Muscovy, when as yet Gothland was submerged under the sea."[3]

Erik Pontoppidan, whose knowledge of Scandinavia was so profound, refers to the erratic blocks, and tells us the largest known to him in Denmark was one near Herselagen in Fuhnen, fifty-four ells in circumference, twenty-two ells long, and fifteen wide. He also refers to the so-called Swan stone in Moen, which is twenty-two ells in circumference.[4]

T. Bergmann, writing in 1769, refers to the loose stones scattered over Sweden, and says they cannot well be attributed to earthquakes, since earthquakes do not occur there; nor again, he says, can we attribute them to the action of ice, since while ice will no doubt lift up stones and deposit them on higher ground, it could not have raised them to the tops of the mountains where they are now found. He accordingly argues that they were moved by water, and gives instances to show that Scandinavia is rising from the sea at this moment, a conclusion which has since been amply verified.

Elsewhere he mentions the so-called giant's cauldrons, and so far as I know he was the first to try and explain them.

He refers them to the working of stones arrested by some impediment and then made to turn round and round by water until the holes have been bored. He mentions six of them as occurring at Kokare Fiäiden in Finland: one of them, still under water and a foot deep, was only thirty years old, and its actual growth had been watched by Nordenskiold, and he urges that similar ones elsewhere, out of the reach of present streams, had the same origin. He mentions others as existing in Jemteland, in West Gothland, in Bohuslan and in America under Cohoes Fall on the Hudson, and elsewhere he remarks that they are always situated on the slopes of mountains, where water may have flowed, and never on their summits. They are sometimes two or three fathoms deep, round or oval, always smooth and occasionally the stones which made them still remain in them.[5]

It will thus be seen that although these researches have been almost entirely overlooked, the main features of the drift

[3] De Telluris habitabilis incremento, sect. 39.
[4] Kurzgefasste Nachrichten, etc., 1765. p. 52.
[5] Phys. Besch. der Erdkugel, 389-390, 403-404.

phenomena of the North, namely the scattering and distribution of the erratics, their occurring on strata of a different character, and their connection with the mounds and *äsar*, the possible intervention of floating ice to assist the diluvial waters in moving them, and the polishing of the rock surfaces, had been recognized in the earlier part of the last century.

Meanwhile, similar observations had been made elsewhere, and we will first turn to the Alpine and neighbouring lands.

In 1729 L. Bourguet, a native of Neuchatel, in his "Memoire sur la Theorie de la Terre," speaking of the district of the Jura, says that on mountains, hills and plains, there occur irregular masses of granite, and of another kind of rock, very hard, and composed of talcous scales, which are not attached to the solid ground, nor have they tumbled down from a neighbouring height, there being none such at hand; and besides, they are of an entirely different kind of stone to that found close by. Some of these blocks are of an immense size, and their number is so enormous in some places that they look as if they had tumbled from the clouds.[6]

In a letter, dated Meiningen, 10th May, 1735, and addressed to Mr. C. Mortimer, the Secretary of the Royal Society, which has been much neglected, Dr. Ehrhart calls attention to certain observations he had made in the Tyrol. In this memoir he tells us how at Meiningen there is on the mountains a vast congeries of rounded stones which have evidently been rolled by water, but in asmuch as no river reaches or could ever reach them, it must have been a powerful torrent. He goes on to say that these stones increase in size from Meiningen as we near the Alps till they become three and four feet in diameter; in the opposite direction they gradually decrease to the size of coarse sand. He further says that he had noticed among the Tyrolese Alps stones *in situ* in a continuous layer which were just like the scattered and broken stones occurring between the Alps and the Danube. These scattered stones, he says, are found of as many varieties as the rocks of the Alps. The cause which broke the Alpine rocks, and rolled the fragments into these rounded shapes, and distributed them over all that part of Alemannia where he dwelt, he did not doubt was the Noachian deluge. The waves of this flood, which he

[6] L. Bourguet, Lettres Philosophiques, Amsterdam, 1729, pp. 198-9.

says must have been twenty leagues in length and breadth, must have flowed from the south, northwards. The fragments which had been broken from the Alps (which were higher before the Flood) were triturated and rubbed until they acquired their present shape and size, whence those which are nearest to the Alps are the largest and most intact, and they diminish as we get away. The identity of the scattered stones and the Alpine masses is unmistakable. He speaks further of the Alpine erratics being scattered over Swabia as well as Alemannia.[7]

Our next authority is Arduino.

In 1759 we find that Italian geologist writing that among the Tuscan Hills were strewn a large number of stones, some of them foreign to the district, and which had come a long distance. Those, he says, which were derived from neighbouring mountains, had still sharp edges, while the others were rounded.[8]

It will be seen from these extracts that the phenomena of the drift had now been observed in the south as well as the north of Europe. So far as I know, the only theory which had, up to this time, been forthcoming to account for them, was an appeal to the Noachian deluge.

Meanwhile, a revolt against this Diluvial Theory had commenced, and was largely fostered, if not created, by Voltaire, in whose attacks upon the Bible it was found necessary to discredit in every way the evidence of a general flood; nor can his efforts, although supported by some strange and fantastic arguments, be considered as altogether fruitless. They largely led to the discriminating of different strata, representing different epochs and different causes, and to the proof that the great mass of fossil shells which had hitherto been traced to the Deluge, must have had a very different origin, and men's minds began to turn to more inductive methods of research, and to the study of causes actually in operation rather than to transcendental ones. It would seem that among some of these students an idea prevailed that the

[7] Philosophical Transactions, vol. xli. 549-551.

[8] Giov. Arduino, Lettera seconda al Sig. Cav. Ant. Vallisneri, written from Vicenza, March 30th, 1759, and published in Nuova Racolta d' Opusculi, etc., del P. Abb. Calogera, vi., clxv. and clxvi.

great blocks of stone to which we have referred were nothing more than the ordinary débris of flooded rivers. This view soon met with acute critics.

Ferber, in his "Letters on Italy," published at Prague in 1773, says of the hill country about Venice: "The fields near *Gallio, Asiago, Campo di Rovere*, and other places belonging to the *Sette Communi*, are situated in the mountains much above the level of the sea, and are covered with a great many detached pieces of granite, quartz and other vein rocks, which appear to be forced from, and belong to, the Tyrolian primitive mountains. These ruins appear at the same horizontal elevation in several other places of the Alps, as in *Feltrino (Stato Veneto)*, separated by the Brenta from the former; westward on the same Alps from *Astico* to the Adige River, and so on. Their quantity and variety of size is remarkable at *Tonezza* and near *Folgaria* in the hills under the Austrian jurisdiction; which, being entirely calcareous and entirely destitute of primitive strata, prove the more that these forced and blunted rocks, pebbles and sands are adventitious.

"This phenomena is more curious because the débris are precisely like those washed and carried down by the Adige and the Brenta from the Tyrolese Mountains, and the level of these rivers being at present one thousand feet lower, it is impossible to conceive how, in their course so deeply cut into the Alps, they could have rolled and deposited these stones at such a high elevation. But this and similar observations force us to conclude that their channels and beds have been, in former times, at the same elevations which are marked at present by the position of their deposits. No water can have carried, rolled and forced them up into their actual position, whatever allowance be made for its impulse, and whatever violent inundations we should please to suppose. This is fair evidence of the great ruptures in the Alps and the prodigious effects of the unwearied and never-ending course of rivers, and points also to still greater ones which may have been occasioned by violent ruptures and inundations in such places where earthquakes first stopped, and afterwards entirely changed, the course of rivers."[1]

Let us now turn to another part of Europe. The boulders of

[1] Op. cit. Engl. Trans. 43, 44.

North Germany must have early attracted attention, and it must have been well-known to travellers that they existed over a wide area. I have not, however, met with a reference to them by a scientific man before Guettard.

In two papers on the geology of Poland, published by him in the Memoirs of the French Academy for 1762, the immense number of boulders of granite, etc., found in the level plains between the Baltic and the Carpathians, is continually referred to. He mentions how these boulders are not distributed evenly over the whole country, but abound in some districts and are rare in others. In some places granite prevails, in others, quartz and other stones, and the boulders are used for paving most of the villages. M. Guettard says the boulders especially abound near Grodno, between Otrock and Osieck, between Soroez and Lukartof, and in the cantons of Niesrietz and Pinczovia in Lithuania, and he describes in detail the character of the various stones of which they are formed. He attributes their origin to the destruction *in situ* of high mountains chiefly formed of granite. "The sand in which the boulders occur," he adds, "is more like that which is formed by the decomposition of granite than of sandstone rocks.[2]

In 1768 Guldenstadt set out on his travels, the account of which was afterwards written by Pallas. In this he calls attention to the fact that in the neighbourhood of the city of Twerza, in Russia, where the native rock is limestone, there are found scattered in the bed of the Twerza and in the fields, great stones, which are always rounded. Some of these are of brown, others of green, and others of grey granite. There also occurs red felspath.[3] Guldenstadt offers no explanation of these stones.

In the same year Pallas himself commenced his famous journey. In his account of it he tells us how a hill which he saw at Bronitzoi Jam, not far from Old Novgorod, was covered with boulders formed of a red quartz and black blende. Many of them had been transported to Petersburg for building purposes. The largest of these stones was three toises in length. "Was it water," he says, "which rolled hither these masses of stone? We do not think it possible for

[2] Op. cit. 234, 293, etc. [3] Op. cit. ii. 460.

water to move such immense masses, and we think there must have been some other 'cause.'"

About this time another and more famous geologist devoted himself to a critical study of the erratic blocks in many countries, namely, J. A. De Luc. His long residence in England makes him especially interesting to us, and he may be looked upon as largely the founder of an English school of geology. He also had an advantage over previons inquirers in that his observations were made over a wide area, and he consequently escaped some of those errors of induction which are inevitable when a theory of the universe is based upon the examination of a gravel pit or upon the traces of geological revolution which are found in some small hamlet. If his theories are sometimes fantastic they are easily separable from his observations, which are shrewd enough. It must be remembered also that many of his observations were made in countries far removed from high mountains and where the theories which might avail to explain mountain phenomena are not tenable. As early as 1779 he called attention to the granite blocks of the Hartz, which he compared with the more numerous blocks found on the Alps.

Having revisited the Hartz in 1786, his attention was directed again to the blocks there, some of granite, others of quartzose and other primordial rocks, but all foreign to the beds on which they lie. He goes on to say that this phenomenon is a very general one and found on most continents.[4]

His observations extended over the greater part of Germany, and included the district of Luneburg, so fertile in these stones. In one paper he refers to the fact that the granitic part of the Hartz Mountains is the Blocksberg, against which rests the Bruchberg, a schistose mountain separated from the latter by a small hollow. On another side is the Rehberg, which forms a lower branch of the Blocksberg. The Rehberg is covered entirely with granite blocks, and it is, he says, of such small extent and so little below the Blocksberg, that it is impossible to conceive that these blocks could have descended from thence. "Whoever shall observe this scene of disorder, will acknowledge that there is not any cause actually operating

[4] Voyages, Fr. ed. i. 5-7.

[5] Lettres Physiques, etc., 337-340.

which could, in any length of time, have produced such an accumulation of these blocks." Whereas the Rehberg is covered with granite blocks, the Bruchberg, which is schistose in its structure, is covered with blocks of a quartzose rock, no stratum of which is to be found in any part of these mountains; and these blocks, as well as those of granite, are scattered also on many mountains of calcareous stone, as well as in the plains, and proceed from a class of strata whole mountains of which are found in other countries, as in Veteravia and in England. The blocks of granite are traced upon the summits, on the declivities, and at the bottom of the valleys of the schistose ridges, as well as of the calcareous mountains which rest against these last, and they abound in all the heaths of Lower Saxony and Westphalia, on the hills, on the plains, and even to the very coast of the sea.[6]

He goes on to say that on the declivities of the calcareous mountains of Hildesheim he had observed immense loose and insulated masses of a calcareous stone resting on strata totally different from the strata of the loftier parts of those mountains; and at the same time had found fragments of granite, and other stones equally foreign to the nature of these mountains, on their declivities and at the bottom of their valleys, and that nothing is more common than, on mountains, hills and plains, and within the loose strata on the surface of the soil, to find blocks, or smaller fragments and gravel, belonging to stony strata known to exist in some mountains elsewhere, but none of which is found in that country, either in eminences or under the soil of the plains at any attainable depth.

I have quoted these few paragraphs from De Luc's earlier works to show how methodical and clear his observations were. I will now describe the theory which he formulated to explain the erratic phenomenon.

His view was that the contour of the earth's surface was due to subsidences rather than elevations, and that this necessitates our postulating the existence of subterranean hollows or caverns, which, he says, must have been filled with some expansible fluid, perhaps with atmospheric air. The subsidence of masses of great magnitude must have caused

6 Id. 30-32.

great pressure in these caverns and as the compressed fluids could only escape through the intervals of the disunited masses, they must have rushed out with great impetuosity. "Here then," he says, "is a force of which we may form a sufficient idea without comparing it with that which projected the planets in their orbits. It is the same in its nature, though far superior in degree, with the force that raises lavas to the summits of the volcanoes of the Andes and by which rocks of an enormous size are projected some miles into the air, an effect which was observed during the formation of *Monte Nuovo* near *Naples*. . . . By this force, therefore, which nothing that was detached and loose could possibly resist, the fragments broken off by the collision of the lower parts of the strata were thrown up to the surface, in the manner of bombs, notwithstanding the resistance of the water, under which these explosions took place." He goes on to argue why so many of the blocks are of granite and other similar substances constituting the inferior strata. He further urges that, as the revolution in question took place at the bottom of the sea, there would be a tremendous conflict between the expanding fluids and the water which would account for the fragments being ground against each other. The combined effect would be to triturate the fragments and also to afford a cause adequate to the dispersion of gravel or rounded stones over the surface of our present land.[7] A somewhat similar view of the subterranean origin of the blocks occurred, it seems, independently to another observer.

In 1780 J. E. Silberschlag published a work entitled "Geogenie oder Erklarung der Mosaischen Erderschaffung." In this he discusses at some length the great deposits of sand and the erratics of North Germany, especially in the valley of the Elbe. He calls attention to their variety and wide distribution, and urges that they are the products of subterranean volcanic eruptions, and he professes to have found traces and remains of these volcanoes in North Germany, a conclusion which De Luc criticizes.

De Luc maintained his own position in many polemics, and it was supported with considerable vigour by his

7 Lettres Physiques, iv. 152, 153; Elements of Geology, 338-341.

nephew, the younger De Luc, in a paper published in the eighth volume of the "Annales des Mines," but in his later writings the latter considerably modified his theory and supplemented it by the then more generally accepted views of De Saussure and Von Buch.[8] In 1829, Engelspach-Lariviere published a paper on erratic blocks in which he attributed their dispersion partially to the same cause, and as late as 1835 a similar theory was advanced by Dassen, in a paper on the gravels of the province of Drenthe, in which he contended that the Scandinavian peninsula was elevated when the strata was still soft, except only its exterior envelope, which was broken, and its fragments were scattered through the air.[9]

De Luc's views were at once sharply challenged. Thus De Saussure, writing in 1779, says, "Who has seen an example of such explosions which would shoot through the air blocks several cubic toises in extent for a distance of 12 or 15 leagues? To cause such explosions the heat must have been very great, and would have vitrified or calcined the stones, and such volcanoes must have poured out lavas or other distinctly volcanic products, none of which results are present."[1] Elsewhere he urged that rocks of such a size shot through the air over such a distance as that from the Alps to the Jura, would have broken the rocks on which they fell or buried themselves in the ground, while they are found on the surface and often touching it only at a few points.[2] Von Buch also criticized the theory, and argued that if explosions had scattered the blocks they would have been found scattered promiscuously over the whole range, and would not be absent, as they are, from those places where other heights intervene.[3] De Saussure and Von Buch further showed that diligent search can always find the beds *in situ* whence the blocks have been detached, and that we need not seek for them in the bowels of the earth, but on the surface. That the problem therefore is limited to finding some means by which they could have been transported.

De Luc's explosion theory, I need not say, commands no

[8] Considerations sur les Blocs erratiques, 27.
[9] D'Archiac Histoire des Progres, etc., 59, 60.
[1] Voyages dans les Alpes, i. 150. [2] Id. i. 168.
[3] Abhandlungen., Berlin Acad., 1804.

support now. We will next turn to his great contemporary, the father of Alpine exploration, and, like himself, a careful observer, De Saussure. In the first volume of his "Voyages dans les Alpes," published in 1779, in discussing the gravels and superficial blocks, he begins by protesting against those who treated the pebbles in gravel, as they treated the bones found in the ground, as *lusus naturæ*, and then goes on to show that the rounded character of the pebbles is not their original form, as in the case of concretions, etc., but that they were clearly rounded by the action of water, and he appeals to what may be seen in most mountain torrents, where the process of making such pebbles is continually going on, and more remarkably so in the torrents underlying the glaciers, and on the shores of the ocean and of lakes.

He then goes on to contrast this rounded gravel with the blocks found on the glacier backs and on the flanks of the mountains bounding the glacier, where not a stone, he says, is to be found which has not its angles sharp and its edges unrounded.

He next urges that the greater part of the boulders and blocks found near Geneva are of granite, gneiss, etc., like the Alpine masses, while the beds on which they lie are formed of limestone, sandstone, etc., and therefore of an entirely different character.[4]

On the other hand, he points out that it is possible, by comparing these blocks with primitive rocks still *in situ* in the Alps, to trace them to their parent beds. He adds further that they are only found on those slopes of the Jura which face the Alps (a view in which he was mistaken), and he concludes that they had been transported from the Alps by some agency which had rolled them and mixed them confusedly.[5]

He next points out that the parts of the Jura which bear the largest number of these blocks, and at the highest level, are opposite the great Alpine valleys. Thus the most important deposits are opposite the Valley of the Rhone, and he remarked how he had discovered prodigious masses of them above Bonvillars, Grandson, La Sarra, to the N.W. and N.N.W. of the opening of that valley; while further south, above Nion,

[4] Voyages dans les Alpes, i. 145-150. [5] Id. 148, 149.

Bonmont, Thoiry, and Collonge, there were only very few such blocks, since the exterior chain of the Alps, above Gingouph, Meillerie, and Evian, forms an elevated and continuous barrier which intercepted the passage of the fragments from the great chain.

Similarly, Mount Saléve, opposite the Valley of the Arve, is covered with blocks to a great height, and has itself formed a barrier to their passage further into the Jura range. The same applies to the Valley of St. Joire, opposite whose opening at Montouse, on the southern flank of the Voirons, the blocks abound; while the northern flanks of the Voirons have none of these erratics on them at any height, the continuous outer buttresses of the Alps having prevented their passage.[6] Our author urges generally that wherever the country is separated from the Alps by a continuously high buttress, as the Valleys of Neuchatel and Franche-Comté, there the blocks are not found, nor where the Jura intervenes, but wherever gaps or valleys penetrate this barrier, opening up routes to the Alps, in such valleys the case is different, and he thus concludes that the blocks must have travelled to where they now occur along these valleys.[7] This was clearly an advance upon previous theories.

Having brought together these facts, and other similar ones, De Saussure concluded that water alone could have taken up and redeposited the boulders, and water alone could have deposited the gravels, since the large and small stones are often deposited in horizontal layers with mud intercalated.

But while water was the agent, he claimed that it was water acting, not in its diurnal methods, either pluvial or fluvial, but a great mass of water in violent motion. He nowhere quite distinctly formulates a theory on the subject, and we can only gather a general notion of what his view was.

Inasmuch as the foreign stones have been found scattered over Mount Saléve, and on the flanks of the Jura, to a height of 300 or 400 toises above the level of the lake, he affirms that the water which carried them must have risen at least to this height.[8]

[6] Id. 152, 153. [7] Id. 153, 154. [8] Op. cit. 150-152.

Writing in 1796, he speaks of the rolled pebbles of various kinds of rock, granite, limestone, etc., and some of them of a considerable size, which he found mixed with sand and gravel on the very summit of the "Abbaye"; all were rounded, and he says "it cannot be doubted that these pebbles have been deposited by a current of water on the flanks of the mountain, and such a current must have been very large to reach so high, and to have filled a valley more than half a league in width. It is unnecessary to add that it could not have been rain which produced a torrent sufficient to carry along with it such large stones. The nature of the stones, whose matrix is only found a long way off, also proves the strength of the current."[9]

He did not limit himself to explaining the rounded and travelled stones by the agency of water. He also explained by it, as Tilas had done long before in Sweden, the softened contour of the country, and especially noticed the succession of round-backed rolling rock surfaces, to which he gave the name *roches moutonnées*, from their fancied resemblance to flocks of sheep lying side by side.

In addition to the blocks and *roches moutonnées* he also mentions how, on Mount Salève, there are remains of horizontal fissures, four or five feet wide, twice this length, and one or two feet deep, with their borders having rounded curves, such as water makes, and he contrasts these fissures with those made by rain; the latter form excavations either perpendicular to the horizon, or following the greatest inclination of the rocks, while the others run along horizontally, and he attributes them to the current which bore the Alpine erratics. He also refers to rounded cavities in the rocks, two or three feet deep, whose opening is towards the N.N.E., and which he thinks were caused by currents of water beating directly and impetuously against the most salient and exposed parts of the rocks, namely, on their vertical faces. These have their bottom and edges also rounded. He very frankly admits that certain of the fissures are distinctly in a direction inconsistent with that of the current which he postulates, but he explains this by the fact that, both in the case of the sea and of rivers, the water near

[9] Op. cit. iii. 95.

the border of the currents forms backwaters, etc., which often move with considerable rapidity, and small whirlpools, whose excavating power is considerable. The waves also have an excavating force, and when driven by wind move in various directions with the wind, and he appeals to the experience of what occurs in ordinary rivers as a support to his theory. He attributes to the same agency the deep pits, thirty feet deep and eight or ten times longer, known as the Grottes de l'Hermitage, as well as the carving out of the so-called gorge of Monetier and the large gap separating the great and little Saléve, since the beds of it on either side correspond, and he could appeal to no other cause which could remove the intervening rocks now wanting,[1] nor explain the portage of the steep beds of white sand occurring on some parts of Saléve.[2]

Having appealed to water moving on a considerable scale, De Saussure then goes on to discuss whence this water came, and what gave it sufficient impulse to move such large masses, He urges that long before the historic period the sea covered the Swiss mountains to a considerable depth, when a violent cataclysm occurred which disclosed great cavities in the ground, previously empty, and caused the breakage of the strata. The waters thereupon rushed into these gulfs, carrying with them vast quantities of earth, sand, and fragments of different kinds of rocks, choking the valleys with them. Presently the violence of the waters somewhat abated, whereupon the looser materials were washed away, leaving the larger blocks behind.[3]

He concludes quaintly that his explanation would suffice as a demonstration for naturalists. "They know," he says, "that granite does not grow in the ground like truffles, and does not grow like willows on calcareous rocks; and if they have, as is possible, different ideas to mine on the cause of the aqueous movement which bore the débris so far, there will be few who will not believe in a great *débâcle*, or a current of considerable violence and extent which transported and deposited them where they are now found."[4]

While De Saussure's view, that the blocks were dispersed by water, gained many adherents, his theoretical explanation

[1] Id. 164-166. [2] Id. 169. [3] Op. cit. 150-152.
[4] Op. cit. 161-162.

seemed very inadequate. For granting that the contour of Switzerland was the same at the time of the dispersal of the blocks as now, which is apparently De Saussure's view, and that the general level of the sea was then much higher, it is very difficult to understand how the water could have been lowered.

As Charpentier says, we have no evidence that there exist in the interior of the earth empty spaces capable of absorbing such a quantity of water as would lower the general level of the ocean by 11,000 feet, up to which point fossil marine remains occur in the Alps; but, on the contrary, the evidence points to the mountains in which these remains occur having been raised violently above the present level of the sea.

Dolomieu, De Saussure's great contemporary, while accepting his friend's appeal to a mass of moving water as alone explaining the phenomena of the travelled stones, made the position more tenable by claiming, as Ferber had done before him, that since the stones were deposited the Alps had suffered great denudation, so that what were once inclined planes down which the water could roll the stones, were now uneven valleys, necessitating their travelling up and down hills. To quote his own words, writing in 1793, he says "those who attribute the foreign stones often found in valleys to the portage of the rivers flowing through them, forget that many of these stones are often ten thousand times larger than those which such rivers roll along even when most impetuous. They also forget that similar stones occur on the flanks of the valleys as those occurring on some of the hills of Dauphiné, a long way from the Alps; also the rolled stones found in the plains of Verona, and among them beautiful pieces of granite, and the various kinds of porphyry found in the Tyrol. These have been assigned to the work of the Adige, because, in order to reach the place where they are found from the Tyrol, they must have come through the gorge through which the Adige flows. But they far exceed in size any stones carried by the Adige even when most flooded."

Turning especially to the plain of Lombardy, he calls attention to the fact that it is covered with sands and transported stones which are foreign to it, and if they were removed the Adriatic would overflow it right up to the Alps and the Appenines. *Inter alia,* he mentions how the hill occupied by the hermitage of the Camaldoli at Turin is formed of an immense mass of erratic blocks of granite and other stone of great size, and much rounded, and imbedded in shelly marls, the fragments of primitive rock being agglutinated together with marine shells. This mountain, three leagues long and one league wide, rises up isolated between the valley of Aniers and the plain of Turin, and is at a great distance from the mountains whence the stones were derived. In between flows the Po, in a channel not more than a hundred feet wide. There is no proportion, he says, between the size of this river and the work it is supposed to have done by those who attribute to it the excavation of the valley.[5]

The dispersal of the boulders, many of them thirty cubic feet in size, and the mixture of the shells with the boulders, point to the sea having been there ; it was no ordinary sea operating through ordinary waves, which could not have reached up to the height of 200 *toisés* where the stones occur, nor carried and mixed together débris of such size and of such different origins. Like De Saussure, he was forced, he says, to appeal to the sea agitated by a quite abnormal force, which swept up over the Alpine summits, whence it broke off the masses of rock, then rolled and deposited them where they are now found. But, he adds, he does not demand that these waves should have acted contrary to gravitation, and raised the blocks more than 600 feet from a lower to a higher level, as they must have been raised if the contour of the country had been what it is now. He urges that the intervening valleys have been since excavated, and that once a more or less continuous plane joined the points whence the stones came and those where they now occur. The intervening valleys were previously occupied with débris, which was itself swept away by the water, and it was over this that the big stones rolled.[6]

Dolomieu's views were virtually the same as those which Hutton propounded in his famous "Theory of the Earth,"

Observations sur la Physique, xlii. 54-57.

[6] Observations sur la Physique, vol. xlii. (1793), 54-57, note.

which caused a great revolution in geological reasoning. The first edition of that work, published in 1788, did not deal with the problem which at present interests us; but, in the second edition of it, published in 1795, he had to face the facts which had been published by De Saussure, and to reconcile them with his Uniformitarian views. After quoting De Saussure's observations about the erratics, he proceeds:—

"These are facts which can only be explained in supposing that the valleys have been hollowed out of the solid mass, by the gradual operation of the rivers. In that case, stones, travelled from afar, will be found at considerable heights, upon the sides of the valleys at their under end, or where, as our author says, they terminate in plains."[7] This explanation was of course quite consistent with Hutton's general premises that the operations of Nature have always in times past been such as they are now.

The view propounded by Dolomieu and Hutton was advocated by a considerable school of geologists. Thus Playfair, in his "Illustrations of the Huttonian theory," published in 1802, urges that when the blocks, etc., were transported, the mountains were vastly higher than they are now, and that the valleys between them have been since cut out by the rivers and torrents from an immense rampart of solid rock.[8] Inasmuch as the blocks found on the Jura have had to traverse the space now occupied by the Rhone valley, he says, "Stones could not by any force that we know of, be made to ascend over this height. We must therefore suppose that when they travelled from Mont Blanc to Jura, this deep valley did not exist, but that such a uniform declivity, as water can run on with rapidity, extended from the one summit to the other. . . . The horizontal distance from Mont Jura to the granite mountains at the head of the Arve, may be accounted fifty geographic miles, though we suppose Mont Blanc and the rest of those mountains, to have been originally much higher than they are at present, the ridge of Jura must have been so likewise; and though probably not by an equal quantity, yet it is the fairest way to

[7] Theory of the Earth, ii. 249. [8] Playfair's Illustrations, 383.

suppose the difference of their height to have been nearly the same in former ages that it is at present, and it therefore may be taken at 10,000 feet. The declivity of a plane, from the top of Mont Jura to the top of Mont Blanc, would therefore be about one mile and three-quarters in fifty, or one foot in thirty; an inclination much greater than is necessary for water to run on even with extreme rapidity, and more than sufficient to enable a river or a torrent to carry with it stones or fragments of rock, to almost any distance."

In answer to the obvious reply that such a notion would not account for the fact pointed out by De Saussure, that the blocks on the Jura are found in groups opposite the points whence the central Chain of the Alps can be seen, he says that this coincidence is quite accidental, because from whatever cause the transportation of these blocks has proceeded, the form of the mountains, especially of Mont Jura, must be too much changed to admit of the supposition that the places on it from which Mont Blanc is now visible, are the same from which that mountain was visible when these stones were transported hither. It may be, however, that the passes which now exist in Mont Jura are the remains of valleys or beds of torrents which once flowed westwards from the Alps, and it is natural that the fragments from the latter mountains should be found in the neighbourhood of those ancient water tracks; and, in order to explain how it comes about that the granite of the Ornex has found its way across the intervening valley of Orsiere and over the watershed into the outlet of that of the Drance, he postulates that the valley of Orsiere was not formed when the stones were transported.[9] The same view was urged in a work on geology, published by Ebel in 1808. He adopts Dolomieu's view of an inclined plane, and he further urges that the great *débâcle* flowed from SS.E. to NN.W., and argues that this is proved by the direction taken by the erratics not only in the district round Switzerland but in Darmstadt and Central Germany. It is only, he urges, in Holstein, and the borders of the Baltic, that the erratics seem to have come from another direction. He urges that this southern origin of the flood is also

[9] Id. 384-388.

proved by the pebbles found in the Lower Rhine valley from mountains near its sources. The gravel which is found fifty French miles north of the Pyrenees points the same lesson, and while it is true that drift occurs both to the south of the Alps and Pyrenees, pointing to a smaller and secondary flood in an opposite direction, this drift is much less in amount and much less widely distributed than that to the north of the mountains.[1]

These same views were defended by Conybeare and Phillips in their well-known "Outlines of Geology," where we read: "Another circumstance connected with the distribution of these travelled fragments is, that we often find them in masses of considerable size, accumulated in situations now separated by the intervention of deep valleys from the parent hills (if we may so speak) whence we know them to have been torn. This appears to be a demonstrative proof that these intervening valleys must have been excavated subsequently to the transportation of these blocks; for although we can readily conceive how the agency of violent currents may have driven these blocks down an inclined plane, or if the *vis a tergo* were sufficient, along a level surface, or even up a very slight and gradual acclivity, it is impossible to ascribe to them the Sisyphean labour of rolling rocky masses, sometimes of many tons in weight, up the face of abrupt and high escarpments."[2]

It was not difficult to answer this theory, which not only invoked a quite transcendental denudation in recent times, but failed to meet the actual facts. The first person to challenge it was the elder De Luc, one of whose principal works is devoted almost entirely to a sharp criticism of it.

In the first of his letters to Blumenbach, dated from Windsor, Sept. 1, 1792, he says, *inter alia:* "We are not astonished to find great blocks of stone on the declivities of mountains, and in their valleys when their summits are precipitous, because the first and most natural idea is, that they have been detached from the upper parts. When afterwards similar blocks are found in the channels of torrents and then of rivers, still enclosed within the valleys, it may be thought that they have descended along their declivities, covered with

[1] Ebel, Bau der Erde, ii. 294, etc. [2] Op. cit. ed. 1822, xxix.

other fractured stones, which have yielded to their weight; when again we see those blocks even in the plains, carried away by the idea that they must all proceed from the same eminences, we are apt to forget the lakes, and so many deep and tortuous channels, through which it was impossible for the waters to propel them. But time is called in to supply the want of means, and from the habit of thinking that no other cause than running waters can have produced these phenomena, impossibilities are overlooked. Neither the enormous size of many of these blocks, nor the astonishing dispersal of smaller blocks, not only on the plains but on the hills; nor the variety of their species in certain grounds, has been noticed; above all it has not been perceived that gravel of the same kinds of stone is contained in many strata, mixed with marine bodies. . . . When we are within the mountains of granite, and find on their declivities great blocks of that stone, down to the beds of their torrents, and of their rivers, it seems very natural to think that these blocks belonged originally to the broken rocks observed in the upper parts of the mountains; and thus gradually to ascribe the same origin to the blocks and pebbles of granite discoverable in the beds of some rivers, even in the plains. But let us quit the granite mountains and proceed to some remote chain whose strata are of calcareous stone, as the chain of the Jura. In these mountains, where the strata are in the same disorder, and the heights present only broken rocks, we find also great blocks, not only on the declivities, but in the channels of the torrents, and of the rivers in the valleys: but do all these masses proceed from the rocks above? By no means: for, in the first place, these rocks are of calcareous stone, and the greater part of the blocks are of granite; and besides, these blocks are found on the summit of rocks of an entirely different nature, and even sometimes in large heaps, which reduces the hypothesis to an absurdity." [3]

In his famous travels De Luc devotes many pages to the description and discussion of the boulders of the Baltic countries, and to the polemics in which he was involved about them with Hutton and Playfair. He remarks on the great

[3] Lettres, etc. 27-29.

variety of *species of granite* heaped up as it were in the same places, and mixed with a variety equally great of species of porphyry, gneiss and other primordial stones scattered over the hills and plains embraced by the courses of the Weser, the Elbe and the Oder, and adds that "many observers as well acquainted as himself with the mountains whence those rivers proceed, knowing that they contain but few *species* of these different *genera* of stones, have been convinced that the scattered *blocks* and *gravels* could never have come from these *mountains*, and that their source must be sought elsewhere."[4] And he repeatedly refers to their being found aggregated in masses, forming islands sometimes several acres in extent, in the midst of a district otherwise more or less free from them; that they are found on plains as well as on hills. In some countries, he says, the blocks are more generally disseminated; but we always find their quantity differs considerably in different places, without reference to any particular point of the horizon.

In his travels De Luc states his objection to the running water theory. He says: "As the force of *running waters* diminishes in proportion as the declivity decreases, the largest masses should have remained nearest to the supposed mountains whence the blocks came, and should get gradually smaller and smaller. Now in reality nothing like this exists in the dissemination of these masses; besides, the blocks are found in great quantities in many places surrounded by large tracts of land containing scarcely any, and as for the size of the masses it does not follow any regular order connected with their situation."[5]

He further urged, to use his own words, that "ridges of *mountains* and *hills*, altogether foreign from the nature of these fragments, intervene between the places where they are dispersed, and every other mountain of their class."[6]

De Luc says that Playfair, by appealing only to normal causes, even granting that it is reasonable to suppose the valleys have been hollowed out since the blocks were deposited, fails in the first place to account for the blocks themselves;

[4] Op. cit. i. 56. [5] Travels (1810) i. 120-22.
[6] De Luc, Elements of Geology, 109.

and, secondly, fails to account for the necessary torrents of water, since torrents cannot originate on the highest parts of a country.[7] Nor again, he says, can any theory of inclined planes account for the accumulations of blocks found on level grounds, very remote from mountains.[8]

The views of Ferber, Dolomieu, Hutton and Playfair were also met by Sir James Hall, who says: "This hypothesis removes the difficulty of the intervening valleys only; for the transportation in this case would be scarcely less difficult than it would be to Geneva, as matters stand at this day." He corrects Playfair's calculation of the slope and shows it would be one in twenty-three, since the blocks are not found at the summit of the Jura, and then goes on to say that "the snowline is at present 7500 feet below the summit of Mont Blanc. If a sloping plane extended from that summit to the place where the blocks occur on the Jura, the limit of perpetual snow would extend along it horizontally for a distance of thirty-two and a half miles. But granite is not found *in situ* beyond a distance of ten miles from Mont Blanc at the Buet. Consequently the whole of the granite *in situ* would be buried deep in snow, and could not well have been found on the surface of the plane. Not only so, but inasmuch as the whole of the snow covering the granite would be much above the snowline, it is most improbable that streams could ever have run along its surface, much less torrents and rivers competent to carry huge blocks of stone." "To move a mass of granite of even fifty or sixty cubic feet and to carry it a few yards, would require the utmost efforts of the Rhone or the Arve, as they flow near Geneva, in their highest floods, but that such blocks could be conveyed by one of them along its whole course is contrary, I conceive, to all experience, and still more when we consider that these rivers are divided at their source from beneath the glaciers into forty or fifty small streams. Yet from the glaciers these blocks must have come; and when we take into account the magnitude of some of the granitic masses, it is clear that the task is beyond the power of any river that flows on the face of the earth. . . . These stones do not lie merely in the beds of rivers, but occur all over the country,

[7] Elements, 269, 270.

[8] Id. 272-3.

and on the summits of mountains, where rivers could least be conceived to have flowed; nor are they confined to that side of the country, or to the side of the lake of Geneva which lies next the Alps; for we find them in particular on the face of the Jura, which fronts the central ridge and the lake at an elevation of 2000 feet above the latter. A set of low hills also intervene, which occasionally hide that central ridge from the view; and it is principally where the snowy summits are visible from the face of the Jura by means of some depression in these intervening hills, that we find those travelled masses."[9]

This argument, which seems conclusive, was further pressed home by Charpentier. He urged that the mean distance of the Alps from the Jura is about twenty-five leagues, that the highest point where the erratics occur on the latter is about 4280 feet above the sea level, and, granting that the places on the Alps where the rocks occur *in situ* were 1100 feet higher, which is considerably above the mean, the inclined plane would have a slope of only 1° 8′ 50″, much too slight to allow of the sliding on it of the blocks as suggested. Inasmuch also as the upper limit of the blocks on the Jura is marked by a convex line, this also must have been the contour of the plane, which is very improbable. The only means again by which such a plane could be removed would be by the erosion of rivers, but the valleys of the Alps are apparently all older than the deposit of the blocks, nor are the greater part of them valleys of erosion at all, but valleys formed by ruptures and fractures of the strata.[1]

These arguments make it clear that an appeal to the diurnal action of rivers, whether flowing in their present channels or flowing much higher up, before, as it has been argued, they had worn down their present beds, is futile as an explanation of the distribution of the boulders, and that an explanation must be sought elsewhere.

Reverting to the order of our story, we will now turn to an inquirer whose memoirs on this subject have been somewhat over-looked. In 1790, J. H. L. Mierotto published a paper in the Transactions of the Berlin Academy on the origin of the

[9] Trans. Roy. Soc. Edinburgh, vii. 142, 143.
[1] Charpentier, Essai sur les Glacièrs, 173-175.

Baltic lands. In this he especially discusses the district watered by the Oder. He urges that the external features of the boulders found there, some of them rounded and some of them angular, have special lessons for us. He also says that the sand in which they lie is not composed of perfect crystals of quartz, etc., but has had its angles and edges worn off as it would have if one stone had been rubbed upon another as in a mill, just as the rounded stones point to a similar friction as exemplified in a sea strand. The sand dunes bordering the Oder and the great number of rolled stones point, he says, to the powerful action of water. If, as some argued, these effects had been produced on the spot and the boulders, etc., were the débris of mountains once existing here, then we ought, in digging into the ground, to find larger and larger blocks and eventually to come upon the unbroken strata; but this is not so, the largest blocks occur on the surface. If the sea had worn away some mountains of hard rock once here we should have had some débris of the old rocks still left *in situ*, and should also have had a rugged coast instead of a flat and sloping one; and he consequently urges that the débris referred to are not *in situ* but have been brought from elsewhere. He then goes on to argue that in order to learn whence they came we must examine the different kinds of rock of which the boulders consist, and he gives a list of such, and proceeds to argue that they may be traced to the mountains whence the river springs and that they are the débris of the weathering of these mountains subsequently rolled and distributed by the flooded rivers and the result of myriads of years of work. He also suggests that the blocks of stone may possibly have been assisted in their transport by being attached to masses of ice which bore them as far as the water. This view of Mierotto he afterwards modified. I have not been able to find any memoir where he published a revised theory, but both De Luc and Wrede refer to it, and he must have published it before 1800, when he died. Having apparently compared the boulders of North Germany with the rocks found *in situ* in the Silesian mountains and discovered that they were not derived thence, he suggested that they had come from Sweden across the Baltic, and had been carried by ice, at a time when the sea was sufficiently high to cover the surrounding hills. This sea, he urged, had been successively depressed, whence the large number of blocks on the coast. De Luc says that in confirmation of his view Mierotto had told him how blocks were sometimes seen engaged in the ice and that when driven on the coast by storms they remained there after the ice had melted.[2] I shall recur to this theory in a later chapter; here it will suffice to point out that Mierotto was the first to suggest that the erratics of Germany came from Scandinavia, a view which has been generally attributed to other inquirers.

De La Metherie, writing in 1795, devotes several pages of his well-known work, entitled "Théorie de la Terre," to the consideration of the travelled stones. He tells us that so far as he could learn from travellers, granitic blocks were not found on secondary mountains in Asia, nor in Egypt, Mauretania, or Africa, nor yet in America (! ! !). In Spain, France, outside the Jura range, England and Holland, granitic pebbles, he says, are found only in valleys, plains, and along the rivers, and he would limit the phenomenon to the calcareous chains about the Alps, the Carpathians and the Waldai mountains and generally over an area stretching from Turin to St. Petersburg and Moscow. He speaks of the difficulty of accounting for the dispersal of the blocks, chiefly because we have to postulate their having travelled across deep valleys, and because of the great distances to which they have been borne. He urges in the first place that the blocks must have been transported after the deposition of the secondary rocks. Secondly, that they were transported by water since they are all more or less rounded. Thirdly, that this water must have moved at the height where the blocks occur. It must also have included in its action the Alps, Carpathians, and Waldai mountains, and its chief force was exerted against the western faces of these mountains. He argues against the possibility of any river or other such agent having excavated the valleys, since the blocks were moved, as Dolomieu and others urged, and also against a mere local flood such as would be created by the bursting of an Alpine lake, and he attributes the distribution of the stones to the action of the sea, especially

[2] De Luc's Travels (1810) i. 57, & 219, 251.

quoting the power of the tides at St. Malo to move large stones and also to raise them up to a considerable height from the floor of the Channel to the points where they occur. And he says: "Granting that these waves were sufficiently great to include the whole district from St. Petersburg to the Alps in their action, their size and consequent force would be equal to the work suggested. This being supplemented by the action of earthquakes and other movements of the earth's crust which sometimes have transported matters a considerable distance."[3]

In 1797 M. P. Bertrand published his "Nouveaux principes de Geologie," in which he traversed some of the positions maintained by De La Metherie. In this work he urged that the erratics are not confined to the environs of the Alps, Carpathians, and Waldai chain. He says he had seen similar blocks from the Pyrenees, along the Garonne, and in the district of Pau; others from the Cevennes, along the Aveyron and in Dordogne; others from the Vosges at Bains. He further urges that no such cause of transport as the waves of the sea merely would account for the removal of such boulders as the famous monster out of which the pedestal of Peter the Great's statue at St. Petersburgh was carved. Its angles were still unrounded and its weight enormous. He argues that the only cause competent to move the blocks, whose large numbers he calls attention to, was a vast torrent of water consequent on a general *débâcle* or flood, and he argues, as Dolomieu had argued, that this was perhaps assisted by the valleys having then been filled up so as to form more or less inclined planes, and he quotes the fact of the sharp angles of many of the stones being against their having been worn by the sea, and to the further fact that the sea really only acts upon the rocks and stones over a depth of 200 feet about its margin and does not raise them from its own depths and lift them 3000 or 4000 feet upwards, as the facts necessitate it doing if it is to explain the position of many of the boulders.[4]

About this time, namely in 1798-9, M. De Lasterye made a journey to Sweden and Norway. He apparently did not publish the results of his investigations there till the year 1827, and it is not easy to separate what he discovered

[3] La Metherie, Theorie de la Terre, ii. 251-263.
[4] Op. cit. 160-170.

during his voyage from what he afterwards added, but it would appear the facts were mainly collected during his travels.

He urges that the waters of the sea were driven from the Pole towards the S. and S.W. of Europe, and that *en route* they broke the mountain tops and carried with them the débris and scattered them over the European continent and the sea bottom. Among the débris were boulders which it rolled, and others which were carried intact. The further one gets from the Pole the finer do the materials of the loose surface-deposits become. The mixed heterogeneous deposit of this flood exists in Spain, Italy, England, France, and South Germany. Further north the sands increase in quantity and in coarseness of texture. This is so in Westphalia and North Germany. In these sands rolled boulders often occur. In Holstein and in Denmark the boulders increase in number and size. In crossing the gulf of Flensborg the sea bottom seems as if it had been paved by human hands, and the rounded blocks of granite increase in numbers as we near Elsinore.

In regard to Sweden, he tells us that its plains consist for the most part of sterile sand and gravel, on which are scattered blocks of granite whose angles show that they have been less rubbed than the stones found further south. He remarks that the outlines of the mountains are rounded, which could only have been produced by the friction of the débris carried along by water. He noticed further, that while the northern faces of the mountains are rounded, the southern ones preserve their rugged aspect and have their angles intact, and he attributes this to the fact that the flood of water laden with hard materials came from the N.E. and flowed towards the S.W. He says he had noticed everywhere on the mountains and the plains that the granite surfaces were polished, wherever the superincumbent soft material was removed, as perfectly as if they had been of marble and polished by human agency. Everywhere over a space of 250 leagues, where the granite surface was not weathered it appeared as brilliant as if it had just issued from an artist's studio. In going north from Trolhätten the angular erratics are found mixed with rolled pebbles, but the latter diminish and the former preponderate as we go northwards until they occupy almost the entire surface. The two isolated mountains of Hanneberg and Halleberg stand on a low granite plateau showing the polish already named. On the summit of the former are many angular erratics and many rounded boulders, some of which are three metres in diameter, while the rock-surface on the summit is also polished. Boulders of the same kind of rock of which they are composed are scattered in various places to the south and south-west.

In going from Halleberg to Christiania he noticed the same polished surfaces and the same evidence of a smoothed contour to the mountains on their northern faces and a rough one on their southern ones, and that the rounded boulders are less frequent where the angular ones prevail.

He noticed in regard to these polished surfaces on the tops of the mountains that as far north as 62° N. lat., where the surfaces had been exposed they had been weathered, but when protected the polish had been preserved. After passing the 66th degree of north latitude he had not found any of these polished surfaces, although he noticed both rounded boulders and sharp-angled ones. While rounded boulders occur on the mountains, unrounded ones occur in the river beds, showing that the rounding had not been caused by river action, and that the rivers had not carried the unrounded ones far. The east of Sweden, he says, presents similar testimony to the west, and he argues that in going north to Lapland it is probable the same testimony to a flood of waters coming from the north would be available, and he argues from all the facts before him, that there must have been two catastrophes, and that the waters causing them flowed from the north towards the south-west.[5]

De Luc's repeated continental journeys, during which he occupied himself so much with drift phenomena, naturally opened his eyes to similar facts in England, where he lived so long, and where, so far as I know, he was the first to notice them. Had it not been that when here, he resided and travelled almost entirely in the district south of the Thames, he would no doubt have forestalled a good many later writers.

[5] Journal des connaissances usuelles et pratiques, 2nd ed. v. 6-14.

His only opportunities of seeing boulders was on his journeys to the continent on the eastern coast, where he mentions having noticed them, and also suggests that they possibly came from Norway. He was apparently also the first to refer to the Irish boulders. He tells us that his nephew, in writing to him from Ireland in 1796, noticed, in constructing an aqueduct at Kildare, that the materials, which were taken from one of two adjoining hills, about eight miles from Dublin, to make an embankment between them, consisted of loose débris, among which were rounded masses from two to six feet in diameter, of a calcareous stone very different from that of the strata of the hill. Similar rock appears at the surface at various places some miles off, *inter alia*, at Rahenny, four miles from Dublin, and Slane, twenty-four miles from the same city. He also mentions how, in the north of Clare, there is a plain thickly strewn with calcareous blocks, some of immense size.[6]

I only know of one other reference to the drift phenomena in these islands in the literature of the last century.

Mr. Arthur Aikin, in his "Journal of a Tour in North Wales," published in 1797, refers to the beds of rounded pebbles found on the highest parts of the slate mountains. Their present position, he says, could never have been that in which they were formed, for they consist almost universally of porphyry, quartz, serpentine, and other substances which lie in large masses composing the primitive mountains, their rounded shape too, like that of the pebbles on the sea shore, seems to intimate that they have been carried by the force of water to the places which they now occupy. Another circumstance which appears to point out the quarter whence they originally proceeded is, that, in proportion to their vicinity to the primitive mountains, is their size: this might be expected, since the further they were carried, the more they would be rounded and comminuted. "Still, however, there is a difficulty attending this hypothesis; namely, by what means could these rounded pebbles have been forced across the many deep valleys that intersect the mountains in all directions without first filling up the valleys themselves?

[6] El. Treatise on Geology, 319-321.

And, if this was the case, by what means were the valleys so entirely cleared of them afterwards, as they appear now for the most part to be?" He then goes on to suggest that when the slate mountains were formed under the water there were no valleys, but the whole mass was one uniform bank, the valleys being afterwards formed by the rivers as the waters subsided.[7]

In concluding this chapter two facts will be present to the reader; first, that the superficial and commonplace phenomena of the drift had been observed and recorded long before the time generally supposed, as judged by the current memoirs on geology; and secondly, that England and its science had very little part in these first excursions into the unknown land. The latter conclusion is a little unexpected when we consider what an amount of work has been devoted to elucidating the subject in later times by the geologists of our own country.

[7] Op. cit. 219, 220.

CHAPTER II.

THE PHENOMENA OF THE DRIFT.

The Champions of Water (Part II.), 1800–1825.

"The Cornish tinners hold a strong imagination, that on the withdrawing of Noah's flood to the sea, the same took his course from east to west, violently breaking up and forcibly carrying with it the earth, trees, and rocks, which lay anywhere loosely, near the upper surface of the ground. To confirm the likelihood of which supposed truths, they do, many times, dig up whole and huge timber trees, which they conceive at the deluge, to have been overturned and overwhelmed."—Carew's "Survey of Cornwall," 1602, p. 7.

Playfair, 1802—J. A. De Luc, 1809-1813—Von Buch, 1802-1818—Hausmann, 1807—Landt, 1810—Brochant de Villiers, 1819—Charpentier, 1841—Sir James Hall, 1812—Allen, 1812—Drake, 1815—Weaver, 1818—Imrie, 1818—Mitchell, 1818—Greenough, 1819—Escher Von der Linth, 1819—Fox Strangways, 1819-1821—Razoumofski, 1819—Henslow, 1820—Horace H. Hayden, 1820—Bigsby, 1821-1823—Hitchcock, 1822—Grainger, 1822—"N.," 1823—Dr. Buckland, 1823.

In the previous chapter I collected some of the opinions of the geologists of the last century who first scientifically examined the problem of the drift. I now propose to continue my story and to discuss the views of students in the earlier part of the present century who attributed the phenomena in question to a diluvial origin. Whereas the names of Englishmen are singularly absent from the former chapter, they now become conspicuous, and the progress of discovery in this difficult field is very largely traceable to our countrymen. The person who, as I have said, initiated a real interest in the question here was De Luc, who lived so long in England and who died at Windsor. It was the sharp difference which separated him from the Huttonian school which in a large measure led to his critical sifting of a problem which afforded a fair touchstone of that theory.

In 1802 Playfair published his famous Illustrations of the Huttonian theory. In this work he devotes a considerable space to the travelled stones. He argues that they diminish in size the further we get away from their source; and he further adds that many of the stones were already rounded in a former geological period and have been washed out of early conglomerates. For the most part, he quotes the facts already collected by De Saussure and others, but he also refers to a large granite stone on the sea shore of Arran, at least three miles from the nearest granite *in situ*, and with a bay of the sea intervening, and which must have come from Goat Fell.[1] As we have seen, he objects to invoking a great *débâcle*, or torrent, and accounts for the transportation of the stones by attributing a much higher altitude to the mountains formerly, by the argument that the valleys have been cut out of immense ramparts of solid rock by the rivers and torrents flowing through them, and by the fact that when the stones were transported, it was along inclined planes.[2]

This work of Playfair's, as I have said, largely incited the polemical spirit of De Luc, who, in his Treatise on Geology and the several volumes which he devoted to an account of his geological travels, adduced a large number of carefully observed facts in regard to the distribution of the travelled boulders in the country north of the Baltic. I have already referred to the general line of argument which these facts suggested to him, and will here content myself with mentioning the more important of his later conclusions.

His journeys round the Baltic enabled him to state very positively that all its coasts, as also the isthmus joining Denmark to the mainland, and the Baltic islands, including the islands of Funen and Zeland, as well as the small islands of Læsoe, Anholt, and Samsoe in the Kattegat, and the island of Rugen, are covered more or less with erratic blocks.[3]

"In the heaps of blocks," says De Luc, "found in Westphalia, Lower Saxony, and Pomerania, there occur an immense variety of species, not only of granite, but of porphyry, and other stones of the same class, entirely unknown in the mountains of Germany. This has induced some German mineralogists

[1] Op. cit 393-4. [2] Id. 383. [3] Travels, i. 384-386.

to argue that these blocks have proceeded from the mountains of Sweden; the ice in the Baltic having served them as a vehicle for traversing that sea of which, since then, the level has subsided."[4] He elsewhere quotes Professor Pfaff as an authority for the fact that in the mountains of the Black Forest, in passing over a horizontal tract a league or two in extent, he had observed as great a variety of species of granite as occurs among the blocks of Holstein and the countries bordering on the Baltic.[5]

Elsewhere he urges that the blocks are not geographically arranged according to their size. They are generally found associated with gravel formed of the same kind of rock, but in some places the blocks are infrequent while the gravels abound, and *vice versa*. The blocks again differ, in that some are angular and others are rounded. In regard to the sand in which many of them are buried in Westphalia, Lower Saxony, Pomerania and further eastwards, it is not, he says, *a detritus*, for it consists of grains of transparent quartz of irregular figures, polished even in their smallest cavities.[6] He further mentions how the blocks abound on the mounds occurring in Lower Saxony in the form of rings.[7] All this, he urges, is against the notion that the débris has been borne along by running water, and in favour of its having been extruded from below the ground by explosions.[8] Reverting to this sand, he says, "When examined through a microscope its grains are transparent and polished even in the hollow parts of their irregular figures, while granitic sand not inferior in fineness found on the sides of torrents among granite mountains always consists of angular fragments without any transparency, the grains having been rubbed against each other." This observation, he tells us, had also been made by De Saussure.[9]

In answer to the notion of Playfair, he remarked, that instead of the larger stones being always found nearest the mountains from which they sprang, as they ought to do by the diluvial theory, the district where he had observed the largest of about twenty feet in diameter, was that most remote from every granitic mountain, namely in Holstein.

[4] Elements, 301. [5] Id. 339. [6] Id. 62.
[7] Id. 64, 65. [8] Id. 65. [9] 124-5.

He also remarks, as contrary to that theory, that he had found blocks of granite, porphyry, and other stones of different kinds, accumulated in spots surrounded by tracts in which few or none occur, forming as it were islands in tracts of land free from them.[1]

While De Luc was engaged in his widespread researches, other inquirers were busy elsewhere. Leopold Von Buch, in his "Geological results of travels in Germany, etc.," refers to the boulders occurring in the country south of the Baltic, and points out that they grow in numbers as we approach that sea; and, like Mierotto, he suggests that this points to their having come from Scandinavia and not from the Silesian mountains.[2] In 1807 Haussman pressed the same view in his "Resultate geog. Versuch auf eine reise, etc." [3]

In 1810 Landt wrote an account of the Faroes, from which I take the following sentence:—

"Besides the large collections of stones which are occasionally found on the hills, there are seen sometimes in the valleys single stones six, eight, or ten feet in diameter, but in places where it is impossible they could have fallen down from the hills. Such stones are found also, here and there, at a considerable height on the hills when there is no other eminence in the neighbourhood from which they might have rolled down," and he goes on to discriminate between these stones and others which have obviously been detached from the local rocks *in situ*.[4]

On the 1st of October, 1811, Von Buch read a famous memoir before the Berlin Academy, entitled "Ueber die ursachen der Verbreitung grosser alpengeschiebe," in which he collected fresh materials for solving the riddle of the drifted stones, etc.

Following in the steps of De Saussure he not only showed that these blocks occur on the Jura, opposite the outlet of the great valleys and sorted according to their origin, but that they are limited to the upper parts of the Jura, while similar blocks are absent or very infrequent in the low ground intervening between that range and the Alps.

[1] Elements of Geology, 315 and 316. [2] Op. cit. i. 132.
[3] Gottingen gesell. anzeiger, 1807, 208 and 2074.
[4] Landt, Desc. of the Faroes, 1810, 8.

Not only do they occupy the upper zone of country, but they occupy it in a very curious way. They are not found at the same level everywhere, but, on the contrary, at very different levels. The culminating point, according to Von Buch, is the mountain of Chasseron above Yverdun, where they are first found at the village of les Bulets, at a height of 1900 feet above the sea level. At Chaumont, opposite Neuchatel, they are first found at 2400 feet above the sea level, on the hills bordering the Val de Ruz, they are first seen at the villages of les Hautes-Genevays, Dombesson and St. Martin, 1800 feet above the sea level, at Nods not far from the lake of Bienne, at 1360 feet. On the mountain of Boudry, opposite Neuchatel, at 1100 feet, at the upper end of the Val de Ruz, at 840, 850 feet, and so on. The culminating point at Chasseron is opposite the great Rhone valley.[5]

The fact of the occurrence of the blocks, only in the upper zone, and of their occurring in this way with a culminating point whence the levels at which they are found are lower in all directions, while they do not occur at all in the bottoms of the valleys, led Von Buch to the conclusion that only a violent projection or impulse by a rapid current of water would account for the facts; and also that their dispersal was due to *one* sudden effort, inasmuch as a continuous force acting for some time would have carried the blocks outside the two extreme limits of the embouchure of the Rhone. He urges that the Alps were then considerably higher, and that the force which impelled the stones was sufficient to overcome their gravity. He compares the effects of this force to that of powder on a cannon ball, and urges that in this way the blocks were projected across the valleys and deposited on the heights which formed barriers to them, and that if the Jura had not intervened they would probably have been scattered over Burgundy and Franche Comté.[6] As evidence of the violent transport of the blocks, he mentions how that frequently the big ones are broken into two or three pieces and sometimes surrounded with smaller ones whose angles correspond, and which are clearly fragments which have been broken from them. Among many others which he had seen he

[5] Abhand Acad. Berlin, 1804-1811, p. 166, 167. [6] Id. 168, 169.

especially cites a group on the way from Vanseyon to Valangin, another at the entrance to the valley of Corcelles, and others on the way to Sonceboz above Bienne.[7]

With great pains Von Buch traced the *provenance* of the blocks. He showed that they consist of various kinds of granite, among which the peculiar granite of Mont Blanc, of the St. Gothard, of the Grimsel, and of Grindelwald can be easily discriminated; and, following up de Saussure's researches, he showed how the blocks on the Jura and those of the Drance come from the Ornex, which forms the most northern promontory of the Mont Blanc range, which also supplied the granite for the blocks in the valleys of Ferret, Champeix, Entremont, Bagne, etc. At the other extremity of the Mont Blanc range are the sharp-pointed rocks above Montjoie, whence came the blocks in the valley of the Arve. Besides granite, other blocks on the Jura are formed of a conglomerate which he traced to Valorsine and Trient, on the line connecting Ornex with the Jura. The gneissic rocks on the same range he traced to the mountains of Saint Branchier and Martigny also on the same line; the black limestone and slate to the Dent du Midi and the Dent de Morcle, bordering the Valais near Saint Maurice; while the diallage and jade came from the valley of Bagne near the Drance. He thus showed that it was from the Rhone valley and its tributaries that the blocks of the Jura came;[8] those of the Aar, which came by way of Thun, Berne and Soleure, are traceable to the granites of Grindelwald and to gneiss like that of the Eiger. Those opposite the Limmat which traverses the greater part of the Canton of Zurich, are formed of the red conglomerate so common in the Glaris range, and the deposits opposite the mouth of the Arve below Geneva, of the granites from the southern extremity of the Mont Blanc range, etc., etc.[9]

Von Buch not only assigned a cause for the dispersal of the erratics, but also calculated the intensity of the force with which the torrent he postulated must have flowed. As the summit of Ornex is distant from Chasseron about 356,117 feet, and the difference of their heights is about 5100 feet, and

[7] Id. 170. [8] Id. 170-183. [9] Id. 184, 185.

as the blocks must have traversed the distance at one bound, it would take them the same time to go as it would for them to fall perpendicularly to the ground by gravity, that is to say eighteen seconds, so that they must have travelled at the rate of 19,460 feet a second. If a mass of water 5100 feet deep were suddenly to break loose, and if the lower blocks were conveyed by it in a different time to the upper ones, the pace at which they would travel would be 553 feet a second. To get a speed of 19,460 feet would require a depth of water of 6,311,526 feet, hence it is clear no appeal to a divergence of the Rhone over the Jura or the mountains of St. Maurice would suffice, and we must appeal to a wider catastrophe.[1]

Von Buch had now satisfied himself that the blocks found scattered over Brandenburgh, Pomerania, Mecklenburg and Holland came from Scandinavia, and he points out how a continuous stream of similar stones may be traced across the Danish islands into South Sweden where their number culminates, and he urges that they crossed the depression of the Baltic as those of the Jura crossed the intervening valleys. The blocks found over Northern Europe, he says, differ from the rocks of Saxony and Silesia. Besides which they become scarcer as we go south, and disappear long before we reach those districts. They occur infrequently near Leipzig, and not at all about Weimar and Erfurt, and wherever any low outliers of the Hartz intervene, there we find no blocks to the south. If we trace their distribution we must draw a half circle with Scandinavia for a centre. They are found in Eastern England, at Antwerp, and scarcely reach Brussels. In the woods of Breda there are many large blocks, as there are in Gröningen and Overyssel, Munster, Munden, Hildesheim, the Hartz, Leipzig and the frontiers of Upper and Lower Lusatia. In Poland they follow the Prussian boundary.

In Russia, Guldenstadt[2] had found masses of granite scattered as far as the Torschok, not far from Twerza near Tuer, but not farther south than Moscow, and Von Buch urges that while the phenomenon of the dispersal of the northern erratics is on a wider scale, it is of the same nature as that of Switzerland, and was probably due to the same

[1] Id. 183.

[2] Vide ante.

cause. He suggests that it may have been the same cause which overwhelmed the great elephants (i.e. the mammoths).[3]

In a letter to M. Brochant de Villiers (published in the Annales de Chimie, vol. x.) Von Buch supports his theory by the fact that other than granite blocks are found in the Pays de Vaud, whose original *provenance* was at a lower level than Ornex, and he argues that this origin at a lower level would cause them, with the same impulse, to touch the ground sooner. As to the extravagant calculation of the amount of water and its speed combined, to secure the required effects, he qualifies it greatly by showing that the immersion of the blocks in water would greatly diminish their speed, and he cites some experiments of Hook, on the rate in which objects rush in water to that effect. By this means he increases the time employed in the trajectory from 18 to 1020 seconds, and calculates the speed at 354 feet a second. This would be in water simply, but water charged with mud and débris would have the effect of still more reducing the speed to reasonable proportions, and he quotes the experiences of Escher Von der Linth and of Charpentier, that if we may take the famous *débâcle* of 1818, at Bagnes, for our guide, the amount of solid débris was probably seven-eighths of the whole flood. That famous *débâcle* (an account of which by Escher is contained in the Bibliotheque Universelle for August, 1818) is described by Charpentier as a torrent of mud preceded by a mountain of wood from the forests, of débris of houses, dykes, etc., etc. According to M. Escher this famous torrent flowed at the rate of thirty-three feet a second at the higest point, and gradually diminished in speed to 18·12, and lastly to six feet a second.[4] Von Buch nowhere, so far as I can find, specifically attributes this rush of water to a distinct cause, but he seems to infer that it was due to the sudden upheaval of the great granitic masses of the Alps.

The editor of these supplementary notes of Von Buch, M. Brochant de Villiers, adds his own criticisms. In these he admits with De Saussure, Dolomieu, Escher, etc., the Alpine origin of the blocks as indisputable. He also admits the probability of their transport by a violent *débâcle,* but he

[3] Id 186. [4] Annales de Chimie, x. 248-253.

cannot accept the rest of Von Buch's conclusions. He urges, with reason, that the difference of level between Ornex and the parts of the Jura where the blocks occur being 5100 feet, gives a slope of 0·014 only, while the torrent of Bagnes had a mean inclination of 0·017. This would indicate a speed of only 15 feet a second, instead of one of several hundreds; while, if it was a torrent loaded with débris as Von Buch suggests, it would have to be at a considerable greater level than the point where the blocks occur, and probably as high as the crest of the Jura, which would diminish the inclination unless the Ornex was itself much higher; but in such a case it would be difficult to account for the blocks existing on the *cirque* of Pierre at Voie, at Liddes, etc., and which seem to fix the limits of the height of the liquid mass near its origin. How, again, was it that the torrent did not drop any portion of its load of mud on lower heights and before it reached the Jura, or cover the blocks instead of leaving them bare. It would be difficult to believe, also, that such a torrent could retain the blocks in the middle of its current, and that a considerable number did not escape laterally and fall into the intervening low ground. For these various reasons he seems to incline to the notions of Dolomieu and others, namely that a continuously inclined plane once joined the summits of the Alps and the Jura, and he professes to find traces of subsequent denudations. This plane would do away with the necessity for Buch's immense watery trajectories, since it would account for the removal of the blocks with the beds on which they lay from the intervening valleys. He urges that the fact of the blocks being found at different levels is not an objection, as Von Buch supposed, to this view, since loaded currents have a tendency to deposit mud in their beds. A section of such a torrent would have an arched outline, and consequently when it met a barrier such a current would deposit its load in an arched curve also. This hypothetical torrent bed, our author apparently concludes, was removed by subsequent denudations on a large scale.[5]

Charpentier urges that if a huge flood was caused, as seems to be implied, by the sudden upheaval of the ground,

[5] Annales de Chimie, x. 254-264.

the blocks would not be projected horizontally only, so as to occur in well marked zones, but many of them would be shot into the air and be thus distributed indiscriminately on the mountains. Again, if they were shot out horizontally, it is difficult to understand how the uplifting of the granitic chain of Ferret could so project the débris as to reach the Jura when intervening ranges like those of Catogne, le Sex d'Armensier, le Mont de Charavex, le Salantin, la Dent du Midi, la Montagne de Foully and la Dent de Morcles stood in the way.

Again, the blocks found in the Rhone valley have come not only from the Ferret, but from all the valleys of the Valais, and it is impossible to understand how a shock such as is postulated could give these blocks a horizontal movement which should take them for a distance of sixty leagues, and not in straight but in crooked lines. The stones are found scattered over the low country, both south and north of the Alps, so that the shock must have had a propelling force in all directions at once. These blocks, again, have clearly followed the waved courses of the valleys, thus the great block of the Seeberg, which is 60,000 cubic feet in size, and which came from the valley of Binnen, changed its course during its progress of sixty leagues at least five times, two at least of which were in the reverse direction to the others.[6] So much for Von Buch's special theory.

We will now revert somewhat, and turn to Sir James Hall whose remarkable memoirs on Chemical Geology formed an epoch in the history of the science. In a paper on the revolutions of the earth's surface, read on March the 16th, 1812, and published in the seventh volume of the Transactions of the Royal Society of Edinburgh, he examines at some length the question of the origin and distribution of boulders. The paper is full of suggestive thought. He framed his conclusion on a wide induction, and although his principal data were collected in South Scotland, he applies his theory to the boulders of the Jura, and refers us for other examples to the granite boulders at Stainmore in Cumberland, which he traced to Wast-dale-crag and to a great block, four or five feet in diameter, in the streets of Darlington.

[6] Charpentier, op. cit. 197-200.

He attributes the results named to the action of a great diluvial wave. In the case of the largest blocks, he suggests their having had an increased buoyancy by being attached to ice. "The enormous masses already mentioned," he says, "which are found near Geneva and at the Coteau de Boisy, may now be accounted for; and the same system will apply also to the blocks on the Baltic, which may have been brought to their present place, not by a permanent and steady position of the ocean, varying by slow degrees, as has been alleged by M. Wrede, but by a sudden diluvian wave washing over some district, situated either at a sufficiently high level or near enough to the Pole to be the seat of glaciers."[7]

Speaking of the vast deposit of sand in Holland and that part of Europe, Sir James Hall says, "Its magnitude appears to me very far to surpass any deposition that could reasonably be ascribed to the present rivers. All this sand may be conceived to have been hurried along, by that mighty stream, (*id est* the diluvial wave) and deposited when the torrent began to spread, and lose its force by diffusion"[8] Hence why the blocks in Germany and the Baltic islands are found on the hills of diluvial sand. After describing in detail a number of boulders occurring near Edinburgh, he says, "I have thus pointed out fifteen specimens, within a circle two miles in diameter, surrounding Corstorphine Hill, each exhibiting both the large and the small features which indicate the action of water flowing with violence along the surface and carrying large blocks of stone along with it. The number of examples concurring in the same indications of direction might easily have been greatly augmented, as any person will see who examines the summit and western face of the hill."[9]

Sir James Hall was, I believe, the first to describe in detail the phenomena known as crag and tail. When an obstacle whose height is equal or greater than its breadth, stands in the way of a running stream, it generally causes a stagnation on the side towards which the stream is flowing, and a deposition is formed of the transported substances, constituting a tail or prolongation which extends in the direction

[7] Op. cit. 158 and 159. [8] Op. cit. 159. [9] Id. 191-2.

of the stream, by a gradual descent, to the distance frequently of eight or ten times the height of the obstacle, while a hollow or depression is frequently caused on the other side of the obstacle by the accelerated motion of the stream there. The isolated hill of North Berwick Law has a conspicuous tail running eastwards. A current of water acting upon sand,[1] or of wind upon snow, would equally produce these effects. He then applies this principle to the geological facts presented by the hills near Edinburgh, and shows how, in each case, such a tail can be traced on one side, and on the other frequently a hollow marked by a lake. In addition he refers to the numerous parallel ridges occurring in that district. Thus, he says, "From Corstorphine Hill to the eastward, the country embracing all the space between Edinburgh and the sea presents one continued series of ridges, upon one of which the New Town of Edinburgh stands. It is an important circumstance that these ridges maintain a very correct parallelism with each other, with the tail of the Castle rock, and of the Calton Hill, and with the alluvial prolongations that extend to the eastward from all the eminences of this neighbourhood; and a series of parallel ridges occur also on the south side of Edinburgh, extending from all the rocky eminences, as may be well seen on the road leading to Dalkeith, which passes over several of them; one of the most remarkable of which is that on which the village of Libberton stands." Hall urges that these phenomena could not have been caused by the action of diurnal waters flowing by the action of gravity, which would have produced depositions at right angles to those under consideration, and while diurnal waters are everywhere found in the act of corroding and altering these forms, they are nowhere seen to produce them, and he insists that they are consistent only with the action of a diluvian wave.

He further urges that small streams have a meandering shape, while the long ridges alluded to have a short curvature whose radius is of infinite length, whence we must infer that a cause very different from any now in activity, and far more powerful, has exerted its influence on the spot, and that a

[1] Id. 171.

current has once flowed there capable of overwhelming and disregarding objects by which the Nile and the Ganges would have been turned out of their course.[2]

Flowing water, he says, rounds the original angles of rocks over which it flows, not only the prominent but also the entering angles, the water excavating for itself a waving groove more or less longitudinal. It also excavates the surface into shallow depressions, which he calls scoopings, like those which would be produced on a soft body by the oblique blow of a spoon or scoop. These are probably caused by the action of eddies of inferior force to the main stream, but acting in connection with it in different and sometimes opposite directions. These various corrosions going on together, have each produced its particular effect, and most of them being concave, their meeting has given rise to the set of waving angular ridges which constitute the most unequivocal feature of a water-worn rock. These angular forms differ completely from those which occur in the broken surface of a rock. These last are acute, rectilinear, and abrupt; while the others are continuous, flowing, and have their angles very obtuse; so obtuse, in some cases, as not to be visible, unless the light strikes upon the rock in a particular direction.[3] These forms, he says, occur on the surface of water-worn rocks, and also on the surface of a mass of snow acted upon by wind. The very rocks over which, according to his theory, Hall says "torrents of water have flowed loaded with sand and gravel and large stones, and accompanied by streams of mud, are found to exhibit at their surface all the characters of abrasion above mentioned, the rotundity and flowing character, the excavation of hollows into the form of waving grooves, the concave scoopings, and the obtuse-angled and waving ridges."[4]

Again he says, "It is obvious that the power of mud, when flowing as a stream, in transporting heavy bodies, and in abrading assemblages previously formed, must bear some relation to the resistance which it would oppose to any object forced through it, and of course that its power in these respects must have been much superior to that of pure water. I conceive also,

[2] Id. 178, 179. [3] Id. 179, 180. [4] Id. 180.

that a deep stream must exert a greater power of transportation and corrosion than one which is shallow, flowing with the same velocity." [5]

He also attributes the grooves and scratches occurring on the surface of certain rocks to stones of considerable bulk carried along by torrents and acting as grinders, he quotes the case of a sudden torrent of water loaded with stones having assailed a house and gone through its lower story, and he observed that every stone, as it passed the house, left a rut or scratch behind it, upon the flags over which it passed.[6]

He then goes on to point out that when the mantle of clay, etc., which covers some surfaces is removed, we find the surface to resemble that of a wet road, along which a number of heavy and irregular bodies have been dragged; indicating that every block that passed, and every one of its corners, had left its trace behind it. These are rendered very distinctly visible when the surface is drenched with water. "In some cases," he says, "these grooves have resisted all the effects of the weather, and show themselves in rocks that have been always exposed, sometimes for many yards; occasionally, single scratches, and parallel sets of them, deviate by five or six degrees from the general direction, but such deviations are rare, the majority agreeing in parallelism with each other and with the general direction of the larger scoops and grooves, and also of the ridges and larger features of the district."

Lastly, he calls attention to the heterogeneous and unstratified character of the loose deposits as being consistent only with a diluvial origin. He points out that a torrent, such as he postulates, loaded with a multitude of blocks of every size, shape and quality, and with a quantity of clay (soon reduced to mud, through which the stones would be irregularly and confusedly scattered), would deposit its materials very much as we find them. The body of compact blue clay which we find in many places, some of which he specifies, is totally devoid of stratification, though frequently of great thickness, and bears every indication of having flowed as a mass into its present situation, together with

[5] Id. 174. [6] Id. 182.

the mixed assemblage of blocks of all sizes and shapes and qualities, some sharp and some round, confusedly scattered through it. He urges that these facts are inexplicable by any diurnal cause, and seem due to a torrent, some parts of which encountered less of the strata of shale and clay, deposited the comminuted sandstone in the form of sand and of gravel, and gave to the external contour of the deposits a smoothness and regularity forming a striking contrast with the abrupt and irregular dislocation which very commonly occurs in the solid mass within.[7]

Sir James Hall was very emphatically of opinion that the phenomena he thus describes were caused by one great catastrophe and not by several. "Thus," he says, "all the diluvian facts in this neighbourhood (i.e. Edinburgh) that have come under my observation concur in denoting *one* inundation overwhelming the solid mass of this district, which had been elevated into its present position by some still more ancient revolution of the same sort; this inundation being the last catastrophe to which it has been exposed . . . . my theoretical notions limit the action upon the hill under examination, to the passage of a single wave, embracing a period of time that could only be expressed in minutes; but during that short time, I conceive the water to have been urged forward with such force, and to have carried with it so many powerful agents, that it has produced effects equal to the work of ages under other circumstances."[8] In regard to the extent of this flood of waters, he concludes, from the way in which it seems to have passed over Corstorphine Hill without being deflected or affected by it, that it must have been incomparably superior in magnitude to the Nile or the Ganges. "This hill," he adds, "is stated to be four hundred and seventy feet above the sea, and postulating that the stream was double this depth, or 1000 feet, we should have a height sixteen times that of the wave at Cadiz, which was 60 feet." The phenomena of the Alps, he says, indicate a magnitude double of this.[9] In regard to the direction of the great wave in the neighbourhood of Edinburgh, as indicated by the medium result of a number of observations, he urges that it

[7] Id. 170-183. [8] Id. 195. [9] Id. 192 and 193.

appears to have been from 10° S. of W. to 10° N. of E. by its true bearings taken with a needle, and allowing 27½ degrees west of north as the variation. He had met with no case deviating more than 10° or 12° from that average on either side.[1] In other places this direction was not quite the same. Thus, he says, in the case of the diluvian ridges called Lowry's knolls, situated behind the promontory of Fast Castle in Berwickshire, the direction is west 35° north, thus forming an angle of 45 degrees with the general direction of the estuary.[2] Near Stirling Castle, again, Hall noticed a set of alluvial ridges of a true diluvial character which agree with the surface marks of the rocks in having a direction no less than 50 degrees north of west or south of east, and being thus very different to that observed near Edinburgh.[3]

The analogy of the events at Lisbon and Lima led him to suppose, he says, that several similar actions may have taken place in succession. At Lisbon five waves in succession followed the first. In the present case the influence of all but the first, would probably be confined to the western side; or, if any of the rest possessed elevation sufficient to surmount the ridge, still the portion of its water which passed over, and that which returned, would follow the same course with the corresponding portions of the first wave. And there is reason to believe that the action would be often repeated upon the side facing the inundation.[4]

In regard to the south-west of Scotland and Cumberland, Hall noticed the absence of the straight and parallel ridges and the forms of crag and tail, the diluvial débris being there arranged in a perpetual succession of knolls equally round at both extremities. To account for this contrast, he postulates that the diluvian wave flowed from a westerly or north-westerly direction and broke over the island, the great body of its water crossing the ridge of country separating the two coasts, overwhelmed the district near Edinburgh and discharged itself into the German ocean. Another portion of the wave being detained by that ridge of country, and being left to follow the impulse of gravitation, when the superior water flowed on, returned again upon the countries

[1] Id. 183 and 184. [2] Id. 199. [3] Id. 200. [4] Id. 202, 203.

which it had just crossed, like the back draught of a common wave. The eastern coast would thus be exposed to a single diluvial action, while the western would undergo two in succession, and nearly in opposite directions. "The first wave would create the phenomena of parallel ridges and crag and tail everywhere, the second and perhaps other subsidiary waves would when they operated, break up the ridges into successions of knolls, etc., and the crags would be filled up so as to equalize the declivity on all hands. Thus the phenomena in the west of Scotland would be accounted for, and this is consistent with what we know of the grooves and furrows there, which instead of keeping a uniform direction, follow directions so irregular as to baffle all arrangement, or if any such can be traced, it is found to be entirely local and to depend upon the course of the valleys in the immediate neighbourhood." [5]

In assigning a cause for this diluvian action, Hall appeals to the elevation of the land, which according, to him, "was performed by successive starts, similar to volcanic eruptions, though far more rare and more powerful. The percussions impressed by these starts upon the waters of the ocean were such as to form waves, sometimes of a moderate force, as those at Lisbon or Callao, sometimes of overwhelming magnitude, and capable of producing the effects described in the Alps, in Germany and in Prussia." [6]

Some of the phenomena observed by Hall in Scotland were also noticed in the Faroe islands in the same year, by Allen, who speaks of the smooth surface of the mountains; and notices that near Eide the rock was scooped and scratched in a very wonderful way not only on the horizontal surface, but also on a vertical one of 30 or 40 feet high, which had been opposed to the current.[7]

The phenomena we are discussing were now beginning to attract attention on the other side of the Atlantic. Thus we find Drake writing in 1815: "It is familiar to all persons in any degree versed in geological science that granite gneiss, mica slate and other rocks termed *primitive* are naturally inferior in situation to all the strata that have been described.

[5] Id. 203 and 204. [6] Id. 166-7. [7] Id. 265.

In the western part of Ohio, these stones are found on the surface of the ground, or partly embedded in the layers of soil and loam. They are sometimes solitary; at other times a great number of masses may be seen collected together and piled on each other, as in the township N.W. of the village of Eaton, Preble county. They are of irregular shapes, and of various sizes. The largest I have ever met with is in the town just mentioned. It is composed of quartz and mica, and was estimated by Mr. Jesse Embree, who ascertained the dimensions of that part which rises above the ground, to contain at least 300 cubic feet. The strata underneath are secondary siliceous limestone. These fragments of primitive rocks are said to be scattered extensively, over the State of Ohio, the Indiana territory and Kentucky." In regard to the force by which these blocks were scattered, Drake contests the notion of Kirwan that they were ejected by volcanoes, since the masses are too numerous and too large, their surfaces too free from vitrification, and their distribution too much in groups, nor does America contain any volcanoes or obsolete craters. He favours the notion that at some former period there were currents over this part of the continent from north to south. "By these currents the masses of primitive stone might perhaps have been brought down in cakes of ice, and deposited where they are now found."[8]

In the year 1818, Mr. Weaver read a paper before the Geological Society of London on the Geological relations of the East of Ireland. In this he urges that it is impossible to reflect upon the nature and disposition of the alluvial tracts on the surface of the earth, without recognizing the powerful agency of an agitated fluid in a state of retrocession; *inter alia* he mentions how Cronebane hill, which is made of slate, bears upon its summit a boulder of granite (called Motty Stone), 9½ feet high and 42 feet in circumference; while the sides of the hill are also strewed with boulders of granite, some of which are not much inferior in dimensions. "How then did these attain their present position? The nearest granite rock is that which extends from the eastern bank of the Avonmore towards West Aston hill; but this is very dis-

[8] Pictures of Cincinatti, by Daniel Drake, pp. 74 and 75.

similar in aspect to the granite boulders on Cronebane. The next granite, in point of distance, is that of Ballincarrig on the banks of the Avonbeg; yet in both these instances (with the exception of West Aston hill itself), the granite rock is found in a situation several hundred feet lower than the summit of Cronebane, which stands nearly upon a level with the bottom of Glenmalur; but the interval between these two places forms at present a curved line descending from that level until it obtains a depth of nearly seven hundred feet below, at the meeting of the two Avons."[9] Mr. Weaver goes on to urge that these rocks were moved to and left in their present position by the receding waters of a flood. Further on he asks, "Whence are the limestone, gravel and marl derived which we find distributed along the coast of the counties of Wicklow and Wexford? The nearest visible limestone rock on the northern quarter, is that which occurs at Williamstown and Booterstown, on the southern side of the Bay of Dublin, and to the southward the first rock of this description that appears, is on the coast south of the town of Wexford. . . . . It is worthy of remark that many of these deposits of limestone-pebble, gravel and marl are situated at distances from two to ten miles from the nearest part of the continuous calcareous tract; and at an elevation reaching to 200, 300 and 400 feet higher than the existing surface of the limestone rock itself; as, for instance, in the glen of Imale, and in the upper part of the Aghfarrel stream, the Liffey and King's rivers. The occurrence of débris of red sandstone conglomerate at an elevation considerably higher than the actual fixed position of this rock is an analogous case."[1]

"If we examine the borders of the mountain chains on the west and the south, we observe similar relations, and in the valleys which find their outlet in limestone plains calcareous débris may be remarked intermingled with those of the mountain tract, in some cases at the distance of two or three miles from the plain itself, and considerably elevated above it, while higher up they disappear altogether."[2] Mr. Weaver elsewhere points out that in descending the declining surface of the coal district of Lisnamrock in Leinster, the

[9] Trans. Geol. Soc. v. 294-297. [1] Id. 298-300. [2] Id. 300.

hollows are found partly occupied by a deposit of limestone gravel, mixed with débris of sandstone and shale, which gradually increases in thickness as we approach the lower grounds. In this manner, even large boulders of limestone are found in the hollows at the distance of two miles from the limestone flat;[3] and summing up the result of his observations, he says, "it is evident that the cause which produced these phenomena was posterior to the formation of all these rocks, since we find the débris of limestone not only resting high up, on the borders of the older tracts, but also intermingled with the débris of the newer, and covering, in fact, the more elevated surface of the latter. The limestone field itself abounds in rolled calcareous masses, pebbles, gravel, sand and marl, often raised into hillocks, or long extended ridges; which seem to owe their form to the action of eddies, and opposite currents. There is scarcely any part of the extensive limestone tract that is not more or less marked by them. Sometimes these ridges appear like regular mounds, the work of art forming a continuous line of several miles in extent. That which passes by Maryborough in the Queen's county, is a remarkable instance of the kind; and similar mounds, hillocks and ridges occur also in the counties of Meath, Westmeath, Kildare, Carlow and other portions of the limestone field, in which the calcareous gravel and sand frequently exhibit a stratified disposition, the alternate layers being very distinct from each other.[4]

Contemporaneously with the paper just cited, giving the first details about the Irish drift phenomena, Colonel Imrie published a paper on the Geology of the Campsie Hills in Scotland, in the Transactions of the Wernerian Society. In this, he says, *inter alia*, "In all situations of this district, where the trap has disappeared, the vegetable or surface soil rests upon a strongly tenacious clay, much mixed with water-worn stones; and this blue clay rests upon sandstone. Among the water-worn stones embedded in the clay, I seldom found specimens of the native rocks of the district; those which I examined, consisted mostly of rocks generally deemed of the oldest formations, such as quartz, porphyries,

[3] Id. 293. [4] Id. 300, 301.

granite, etc.; the native beds of which are far distant to the north and west of that part of the country."[5] He attributes the disappearance of the trap here mentioned to the effects of the attrition of heavy bodies set in motion by a great force of water in rapid movement, and says that in certain places in the glens and narrow vales where it had not entirely disappeared, he perceived upon its surface strong indications and marks of attrition. "In some places the surface of the trap was smooth and had evidently received a considerable degree of polish; and this polish was almost always seen marked by long lineal scratches. In other places, there appeared narrow grooves, apparently formed by the rapid movement of large masses of rock having been swept along its surface; and I remarked that these striæ or scratches, were very generally, in a direction from west to east, excepting when inequalities of the surface, and sudden turns in vales had partially influenced the course of the current. In the eastern part of the district, there occurs a small elevated plain, slightly undulated. Here the surface of the trap, in some places, had lost its covering of soil, and was left bare for inspection. Upon this plain I again detected some of these scratches which were in this situation, generally in a direction from west to east, sometimes deviating a little towards the south-west and south-east; but in taking the average or medium of these directions, their general and true course seemed to me to be from west to east. Upon the surface there were scattered immense masses of trap, which from their apparent weight, seemed perfectly capable of forming the scratches and grooves above described, had they been put in motion and impelled along the surface. Upon examining some of these huge masses I found their surfaces scratched and worn in such a way as to prove sufficiently indicative to me, that they had been long subjected to attrition in water, and I also observed that many of them presented their principal or most projecting angle towards the west, and sometimes towards the north-west, which, according to my opinion, strongly implies the direction of the current which left them in the position in which they

[5] Mems. Wern. Soc. ii. 35.

now are. . . That such currents as were capable of the effects which I have endeavoured to describe, have flowed over the surface of our globe, is to me clearly evident, and the scratches and grooves here mentioned are some of the minor, but clear proofs of its action."[6]

In Dr. Mitchell's appendix to the American translation of Cuvier's works, published in the same year, 1818, and entitled "Observations on the Geology of N. America," we find him arguing that the inland lakes of America are the remains of a once vast and continuous sheet of water. This water, he urges, burst its barriers at various points and thus distributed the North American drift. Speaking of one of these breaches he says, "The thousand islands and the whole of the scenery in their vicinity, bear witness to the mighty rush of waters which at some former period prostrated the opposing mounds and left them as scattered monuments of the ruin. . . Great masses of primitive rocks from the demolished mound or dam, and vast quantities of sand, mud and gravel, were carried down the stream to form the curious mixture of primitive with alluvial materials in regions below."[7] Speaking of another breach, namely, that made by the Hudson, he quotes Governor Chutons thus: "As you approach the falls you pass through immense rocks, and actually of granite interspersed with limestone. In various places you observe profound excavations in the rocks, made by the agitation of pebbles in the fissures. . . . The rocks are composed of granite and many of them are 30 or 40 feet thick. You see them piled on each other like Ossa on Pelion; and in other places huge fragments scattered about, indicating evidently a violent rupture of the waters through this place as if they had been formerly dammed up and forced a passage; and in all directions you behold great rocks exhibiting rotundities, points and cavities, as if worn by the violence of the waves or hurled from their ancient positions." Dr. Mitchell similarly refers to breaches which he argues were made through the barriers in question by the other rivers of the States. He urges that another vast dam confining the waters in former ages extended from the extremity of the Cumberland mountain to the Missouri hills,

[6] Id. 36, 37. [7] Op. cit. 334-5.

and that its demolition caused the vast tract behind it to be drained, leaving lakes Erie, Huron, and Michigan as relics, while the rivers of sand and soil were carried down the Mississippi valley.[8] Later in the same work, after describing the remains of shells and other marine débris found in the loose beds, he reverts again to the travelled stones and mentions the large number of loose rocks rolled and worn, consisting mostly of granite and gneiss, of actinolite and black schoerl, etc., etc., found abundantly on Long Island, New York. Their great weight and bulk must, he says, have required extraordinary power to detach them from their primitive beds and put them where they are. On the ridges of hills traversing the island he noticed that while the north side was occupied by great masses of these stones, not a pebble as large as an egg existed on the south, and he suggests that they were dispersed by the waters of the lake forcing its barriers, "pouring over Joe's Hills, and carrying along the loose materials! sand and stones being urged to the greatest distance, moderately weighty rocks not quite so far, and the heaviest ones loitering in the rear or concealing themselves under the water of the sound where they annoy navigators."[9] Another portion of the flood, loaded with rocks, stones, sand, and water was, he suggests, apparently driven along the valley between Bergen and Newark, and rolled up a huge pile of materials on the north side of Staten Island; while another mass of dislodged materials appears beyond the Raritan in the form of the Neversink Hills, on the summit of which are pieces of sandstone, some many feet in length and of considerable thickness, none of it *in situ*.[1]

Later on he says: "The ground on which New York stands is sand, gravel, rounded stones fit for paving, and loose rocks, some of them of enormous magnitude . . . . the rounded rocks are sometimes six feet in diameter . . . . they consist of rolled or rounded masses of schoerl rock, quartz and schoerl, stellated asbestos, granite and gneiss, and schist. These boulders, etc., are similarly arranged as those already described beyond the Highlands, and the boulders can only be viewed as members torn by violence from the body to which they were

[8] Op. cit. 332-345. [9] Id. 382. [1] Id. 383-4.

once attached." Dr. Mitchell, in accounting for the cause by which the barriers of the great sea were broken, assents to the view of a diluvial catastrophe, and speaking of America he says: "What agent so capable or so likely to wash up the sand and other materials into such ridges as our mountains present? The impulse of an ocean upturned from its bed, rolling impetuously over the land, and carrying everything before it, may be supposed competent to such a work." [2]

Mr. Greenough, then president of the Geological Society, writing in 1819, calls attention to the wide distribution of the drift phenomena. The boulders, he says, do not occur only in Switzerland and the Baltic lands. "At Glenmalur, in the county of Wicklow, huge masses of granite rest upon the mica slate. Along the valleys of the Garonne and Gave de Pau, you find granitic blocks derived from the Pyrenees; along the valleys of the Aveyron and Dordogne from the Cevennes; at Bains from the Vosges. In the department of Morbihan, the number of blocks is estimated at 4000, and some of them are not less than 20 feet high. Near Turin, the calcareous hills are covered by blocks of granite, some of them of the size of 30 cubic feet, although no mountain of that substance is found within many leagues. Gerenna in Grenada, is famous for its boulders, which suggest the idea of a shower of stones . . . I am informed that blocks of granite extend for more than one hundred miles on the south of lake Huron, in North America, and appear in islands twelve miles from its margin. Granite boulders, therefore, are not of partial occurrence, nor is the theory tenable which supposes those found in the north of Germany to have slid thither upon the ice." . . .

After quoting with approval Hall's answer to Hutton in regard to the incapacity of rivers to move the Alpine blocks, Greenough adds, "The phenomena of the blocks which have been traced to the northern parts of Europe, are still more adverse to the hypothesis of Hutton. This, not merely on account of the vastly greater distance to which they must have been carried, but also on account of their lateral extension. They are scattered," he says, "over the continent

[2] Id. 410.

from Holland to Petersburg and Moscow. It is perfectly incredible, that these blocks, extending over so immense an area, and found on the opposite sides of lakes and seas, should be the waifs and estrays, either of a single river, or of any number of rivers." [3]

Greenough, after discussing the efficacy of seas and rivers to produce the phenomena of the drift, and discarding them as inadequate, says the only remaining cause to which these effects can be ascribed is a *Débâcle* or Deluge.[4] He goes on to say, "The universal diffusion of alluvial sand, gravel, etc., proves that at some time or other, *an* inundation has taken place in all countries; and the presence of similar alluvial deposits, both organic and inorganic, in neighbouring or distant islands, though consisting often of substances foreign to the rocks of which the islands are respectively composed, makes it highly probable, at least, that these deposits are products of *the same* inundation."[5] As to the direction of this flood, he says, *inter alia*, "blocks from the Cambrian mountains have travelled eastwards as far as Pierce Bridge, and northwards as far as Staffordshire. Chalk flints occur in alluvion on the north coast of Cornwall, and even at the Land's End."[6] He elsewhere refers to the parasitic gravel and soil of Malta and Gozo, the blocks of primitive Norwegian rocks scattered over north Germany, Russia, Holland, and occasionally met with on the east coast of England, and pieces of granite on Staffa as evidences of the non-existence at the time of their dispersal of the neighbouring seas or lakes.[7]

In 1819 M. Escher Von der Linth published a memoir on the Natural history of the Erratics, which was printed in the first volume of Leonhard's Taschenbuch, 631, etc. In this memoir he developed the notion that the cause of the dispersal of the blocks was the bursting of huge Alpine lakes; one of these, he considered, filled the Valais or upper Rhone valley to a depth of 5000 feet, and was dammed in by a vast barrier above St. Maurice. This, he says, would constitute a lake containing 50,000,000,000 cubic fathoms, each 1000 feet in extent. If such a lake were to empty itself at the rate of

[3] Examination, etc., 137, 138. [4] Id. 149. [5] Id. 155. [6] Id. 159. [7] Id. 177-8.

500 feet a second, it would discharge 25,000,000 cubic fathoms per second, and take half an hour to empty. He compares the results of such an outburst with that which followed the breaking loose of the lake in the Bagnes valley in the Valais in 1818 by the giving way of a glacier, and urges that the effects would be such as we see in the distribution of the blocks and débris of the Alps.[8]

In order to explain the fact that the débris from different valleys is not mixed, but sorted separately, he urged that the barriers which formed dykes to the several mountain lakes burst simultaneously, and thus the current in each valley prevented that in the adjoining valley from interfering with it. Charpentier, in answering this view, argues that the outer chains of the Alps which would form such barriers, are calcareous, and as the lakes must have been very deep, they would necessitate correspondingly deep and wide barriers, which would leave immense masses of débris when disintegrated, a result which we nowhere find. The phenomena, again, which we are discussing is not limited to the Swiss Alps, but the Alps of the Tyrol, the Bavarian and Austrian Alps present similar effects as do the Italian Alps. All of these must have had barriers, which is a very improbable fact, and one not supported by other evidence. How, again, can we account for all these barriers having been ruptured at the same moment? Even if they were the result of the bursting of a barrier, the result is very different to that which we find in the Alpine valleys. When a breach occurs in such a barrier, the effects of the rush of water are only traceable immediately in front of the breach and for a short space. Within certain limits it sweeps everything away, and carves out deep ravines, and when its force diminishes it deposits débris along its course, but this is outside the barrier. Inside the barrier, and within the lake basin itself, there are no such results. There the only result is the lowering of the level of the lake. All this was amply shown in the case of the rupture of the glacier restraining the lake du Gietroz in the valley of Bagnes on the 16th of June, 1818; the current of water which poured out was three quarters of a league wide and 200 feet deep. The devasta-

[8] Op. cit. 665, etc., etc.

tion caused by the progress of this column of water outside the barrier was terrible, but inside, the basin of the lake itself suffered no denudation. Except close to the breach, the water did not displace any of its bed, which had once been a wide pasture, and the Châlets of Torrembey, built of rude stones without mortar, and which the lake had submerged, situated at a distance of two gun-shots above the breach in a line with the direction of the current, were not destroyed. It is clear, therefore, that a rapid fall in the level of a lake would have but a small effect on its floor or on the basin in which it is situated. We cannot, therefore, account for the distribution of the great erratics in this way. And if we could do so, as M. Escher supposes in the Valais, by postulating a barrier between the Dent de Morcle and the Dent du Midi near Saint Maurice, this would not account for the blocks in the valleys of Avançon, of La Grand Eau, of Illiez and de la Tiniere, which open into the Rhone valley below St. Maurice. The famous block of Devens, 161,000 cubic feet in extent, came out of the valley of Avançon.[9]

On the 16th of April, 1819, Mr. Fox Strangways read a memoir before the London Geological Society on the Geology of the district round St. Petersburg. In this he speaks at some length of the so-called Diluvium of that district, a term which, he says, expresses that superficial deposit which covers everything and is composed of almost everything. He remarks that it is usually thickest in the valleys or on the flat summits of some of the hills, and is composed of two sets of materials, one local and another of "a set of rocks totally foreign to the country, the analogues of some of which have been recognized *in situ* at a vast distance, while others remain yet to be identified. . . . All the primitive boulders, the analogues of which have been recognized *in situ*, appear to have been transported from the north. They are scattered universally over hill and dale; on the highest summits of the hills, and on the shores, and under the waters of the gulf; and are found not only on both sides of the Neva, but far to the south, to the east and to the west. Their water-worn surfaces, their enormous size, and the remoteness of their

[9] Charpentier, op. cit. 202-207.

ancient seat, seem to leave no other supposition to account for their appearance in their present situation, than that of their being gigantic pebbles, rolled and accumulated by the greatest aqueous revolution to which this globe has ever been exposed, the Mosaic deluge, of which they furnish the most striking of the material evidences. . . . The environs of Petersburg, as well as every other country, show those traces of the superficial action of vast currents of water which can only be attributed to the same cause; for instance, the Ropsha and Crasnoe Celo denudations and the numerous combes and valleys now destitute of water, along the edges of all the escarpments. . . . One of the most remarkable circumstances respecting the boulders of this stratum, is, that rolled pieces of the smaller-grained chamite beds are found on the summit of a steep ridge forming the extreme point eastward of the heights of Shulcowa. . . . The great elevation of this ridge, the geographical distance from the nearest point where the chamite beds appear *in situ*, and the great thickness of the limestone in that part of the district, make it a question of no easy solution, how rolled fragments of a rock three hundred feet or more below that on which they are found lying, could ever have arrived at their present situation." The limestone boulders are never found north of the outcrop of that stratum whence they are derived and rarely on the table-land itself. They increase in numbers towards the south. On the surface of the limestone plateau, particularly south and west of the upper lake of Doudorof, are several low ridges, which appear to be entirely composed of heaps of rolled pebbles of various descriptions. They are on a large scale, exactly like what are deposited by counter currents and eddies in running water.

"At the quarries near Mishkina . . . . the surface of the limestone, when scraped bare of superficial accumulations which are ten feet thick, presents hollows and steps from one bed to another, the edges of which are all rounded as if water-worn. The diluvium in this instance is entirely free from stones.

"From the situation of the parent rock of such of the primitive boulders as yet have been identified, and from the increase in quantity of the secondary pebbles on approaching the south, it seems that the great *débâcle*, or stream of transportation, came from the north; while from the great height at which the rolled blocks are found, it is evident not only that portions of rock, whence they have been brought, once existed at a higher level, but also that they were deposited in their present situation before the excavation of the Gulf of Finland and the vale of the Neva. These great denudations, together with those smaller ones that open into them, were not hollowed out till the diluvian waters began to draw off and drag the slopes of what had been the bottom of the ocean. The heavy blocks had already settled and the streams carried back with them only the softer materials. This accounts for the seeming incongruity in the direction of the diluvial currents; just as a great wave of the sea will transport pebbles large and small in a great body in one direction; then retreating, drop the heavier, and scoop channels in the lighter materials which it has itself just deposited, and in whatever is below them, into some of which channels a few large pebbles fall back again."[1]

In giving a list of the various kinds of rocks composing the boulders found in the neighbourhood of St. Petersburg, Strangways points out where the parent rock has been found *in situ*, thus granite of a large grained variety, containing much black hornblende and red felspar, has been found in quarries belonging to Baron Nicolai, near Wyborg; a coarse-grained gneiss, grey and red with iron, is found at Imatra and the interior and western parts of Finland; yellowish green felspar rock with large garnets, is said to be found between Kexholm and Cerdopol; a red felspar thickly spotted with hornblende is found at Borgo in Finland; black mica slate is said to be found near Cuopio in Finland; and basalt is said to occur in the north of Finland, Lapland and Norway; and he adds: "By the specification of the principal varieties which I have attempted to give above, it will be seen that all the primitive boulders the analogues of which have been found *in situ*, appear to have been transported from the north."[2]

In the same year Count Razoumofski published a memoir entitled *Coup d'oeil géognostique* sur le Nord de l'Europe en général et particulièrement de la Russie." In this he presented very much the same views as those urged by

[1] Trans. Geol. Soc. vol. v. 425-436. [2] Id. 430-433.

Strangways, founding his induction, however, on a wider basis. He mentions how a careful examination of the various blocks, etc., proves incontestably their transport from Scandinavia and Finland and the country of the north. He states how widely dispersed they are between St. Petersburg and Moscow, on the Waldai mountains as far as Bronitza, i.e., sixty-three French leagues from Wyborg, where similar granite is found, and in other parts of Europe as between Breslau and Berlin, and near Grossen, their furthest point, which is 140 leagues from the south point of Scandinavia. He mentions how the groups of stones are found in a kind of parallel allignments, and that straight lines passed along the axes of such groups would include a number of them, like beads on a string. He says the blocks of syenite found between Breslau and Berlin have their parent rock in Sweden, and those of primitive rock found in Hanover also in Scandinavia, and he assigns the blocks found in the neighbourhood of St. Petersburg to an original home in Finland. He found between St. Petersburg and Riga, and between Teiletz and Riga blocks of a red granite like that existing *in situ* in Ingria and Finland, and similar blocks near Memel in Prussia. He noticed that in Esthonia the boulders occur chiefly where the country has a broken contour, and rarely when it is of a continuous horizontal character; commonly on hills and rarely in the plains, as if the stones had been in some way arrested. He remarks that although the mountains and valleys of Scandinavia are on so much smaller a scale than the Alps, yet that the rocks which have travelled thence have travelled a great deal further. Thus he himself had found pieces of northern granite between Straszof and Ivaniska, thirty-two leagues from Cracow. He had also found at Memel enormous blocks and pebbles of the quartzose sandstone, hard and red, which the Germans call "rothe todtliegende," and which occurs *in situ* on the banks of Lake Onega, 245 French leagues off. He also remarks on the absence of secondary rocks among the boulders found near St. Petersburg, and that these secondary rocks only begin to occur at Dorpat between Little and Rana Poungern.

He lastly states that in one district, which he does not specify but which he says is 130 French leagues in extent, there is an immense lacuna in the general distribution of the blocks and that those occurring near there, instead of showing a huge current flowing from N.E. to S.W., show one from N.W. to S.E., and he accordingly postulates the existence of two currents, unequal in mass and size and starting from different points.[3]

Mr. Fox Strangways, writing again in 1821, says: "Throughout the whole of Finland, the evident traces of diluvial action are on the most astonishing scale. Without dwelling on the stupendous size and universal distribution of primitive boulders, it is impossible not to perceive, that the top of every rock *in situ*, every tor, every hill and knoll of granite, or primitive rock, from its first appearance in Carelia, till it sinks beneath the Gulf of Bothnia, presents a surface as much rounded, and as visibly water-worn, as the boulders or colossal pebbles that lie round their bases. Where the rock *in situ* does not rise high above the soil, and where the boulders are at the same time thickly scattered and of vast size, it is scarcely possible to distinguish one from the other. This is particularly the case between Wyborg and Fredericshaven, where they totally prevent the culture of the earth and barely allow the passage of a carriage over a most tortuous and narrow road. The islands off Abo possess the same rounded character, and another fact connected with this subject, and well worthy of remark, is, that in the south of Finland, where we recognize *in situ* the parent rocks of the commonest boulders of the neighbourhood of Petersburg, we find new varieties occurring in rolled masses, probably brought from rocks existing still further north."[4]

One of the most remarkable facts with regard to the central parts of Russia, is the quantity of siliceous boulders which are scattered over the government of Moscow, Vladimir, Tver and the neighbouring countries. They are found on the Waldai Hills, but rarely north of them, although they occur there *in situ*; a strong proof of the direction of the diluvian current from north to south. . . . They are rarely found north of the town of Waldai or east of Kostroma. Their boundary on the south and west is unknown.[5]

[3] Annales des Sciences Naturelles, xviii. 133-147.

[4] Trans. Geol. Soc. ser. ii. 1, 9 and 10.

[5] Id. 18, note.

In Mr. Henslow's Observations supplementary to Dr. Berger's account of the Isle of Man, read before the Geological Society April 7th, 1820, we read: "Blocks of granite similar to that of Slieu-ny-Clough lie dispersed for two miles over the low plains between St. Mark's chapel and South Barrule, nor have they the least appearance of resting on their birthplace, so that no cause appears so likely to have placed them there as that which scattered those of Dun. This disposition of the granite boulders appears to indicate a current having set over the country (at least on the eastern side of the central chain) in a direction from N.E. to S.W."[6]

In 1820 Mr. Horace H. Hayden published his "Geological Essays," a remarkable work for its time, which is nowhere referred to, so far as I know, in English geological literature. In this work he tells us the surface deposits of the United States offer stronger evidence than can be found elsewhere of a general current flowing from N.E. to S.W. having prevailed over the whole of that continent, and perhaps over every other.[7] He contends that neither the sea nor rivers working in their normal methods are competent to have distributed the deposits in question, which must have been deposited by a current flowing impetuously across the continent. *Inter alia* he quotes the rolled or water-worn pebbles and the wave-like undulating appearance of almost every section of alluvial formation, whether perpendicular to the surface or inclined.[8] "These pebbles are piled up into hills 100 or 200 feet above the rivers near which they lie, and sometimes spread over many square leagues of country over which the current of no one river upon earth has ever flowed. It is both morally and physically impossible that such results could, by any means, be produced by any river flowing through the district of country where they lie. Nothing short of a universal current could have produced such effects; and it must have been of such extent and rapidity as to have hurled them into motion with almost as much facility as the leaves of trees are raised into the air by a whirlwind. The direction of the current is shown by the fact that the pebbles are found in the greatest quantities on the W. or

[6] Trans. Geol. Soc. v. 503. [7] Id. 3. [8] Id. 45.

S.W. banks of these rivers and streams and valleys." Again, he remarks that most of the pebbles lie from three to five miles, and sometimes more, in a S.W. direction from the original gangue or matrix. Of these facts he gives several instances, and notably mentions Baltimore, where he says he had counted among the pebbles in a gravel ridge upwards of thirty masses of granite, micaceous schist and greenstone. "Well defined granite does not occur," he says, "within a mile to the north of where these masses now lie, and the greenstone range does not occur on the north within three miles, while to the south of the ridge neither granite nor greenstone were ever known to exist in place within 500 or 600 leagues, whence it follows they must have been deposited by a current from the north." Similar phenomena occur elsewhere from New York to Georgia. At a little distance north of the U. S. Bank in Washington is a circumscribed spot of about an acre, covered with masses of rock and rolled stones of various sizes. Among these Mr. Hayden says he discovered in February, 1820, rolled masses of amygdaloid, and of hornblende-porphyry, containing epidote, both peculiar to the Blue Ridge or south mountains in Maryland and Pennsylvania, and which cannot be found in place, within sixty miles of Washington city. There were also masses of a kind of granular quartz, weighing probably from two to five hundredweight, which is not found nearer than Herkomer county in New York, far beyond the north mountains in Pennsylvania.[9]

To the current which moved the gravel he attributes the clearing of the northern parts of America of its coating of soil, leaving the surface, as Hearne describes it in his travels, "a jumble of barren hills and rocks, destitute of every kind of herbage except moss." The alluvial covering of a large part of this area, he contends, was swept along by the great current which deposited it in more southern latitudes.

Mr. Hayden also refers to the giant's cauldrons of America. Thus Mr. Mackenzie, in speaking of the portage of Chaudière des Francois, says, "It must have acquired the name of Kettle, from the great number of holes in the solid rock of a cylindrical form not unlike that culinary utensil; at the bottom of

[9] Id. 49 and 50.

them are generally found a number of stones and pebbles. These holes are upwards of ten feet above the present level of the water at its greatest height, and are to be seen along every great river throughout this extended country."[1] Henry also remarks that the phenomenon is not restricted to La Chaudière Francois, but is observable at almost every carrying place in the Otaonais, evidencing the retirement of the waters from a higher level.[2] On the Mohawk river these cylindrical excavations in the rocks are numerous and deep, and afford, says Hayden, the most unequivocal proofs of the violent agitations of currents of water, at an elevation much above the present bed of the river.[3]

The flood of waters which distributed the American débris, he urges, started in the Arctic regions. Following somewhat in the steps of St. Pierre, he suggests that if the earth's axis were changed from north to south, so that the sun should pass immediately over thè two poles, there would be "a certain and inconceivably rapid dissolution of the immense hemispheres of ice there," and the water must accordingly have rushed in wild confusion into the adjacent seas. "From the north pole there are only two outlets, the one through the narrow channel of Berings Straits, the other through a very wide one into the Atlantic ocean, and by far the greater portion of the rushing water must have flowed into the Atlantic, its tide would consequently rise, accompanied by a current increasing in power and force, which would flow over the adjacent continents in riotous disorder. At first the current would be divided by the craggy heights of Spitzbergen, a part of it would be thrown into the White Sea, while the other, directing its force against Lapland and the rocky cliffs of Sweden and Norway, would be thrown back on the eastern and southern coast of Greenland; from thence in a S.W. direction until it struck the S.E. coast of Labrador, along which it swept through the Straits of Bell Isle across Newfoundland, Nova Scotia and along the Atlantic coast into the Gulf of Mexico." "In a short time," says our author, "the S. and E. coast of Labrador, over which the current was urged with increasing force, was desolated. Its soil was stripped off and carried

[1] Mackenzie's Travels, 37. [2] Henry's Travels, 31.
[3] Op. cit. 180.

southwards, depositing the blue clay containing shells of existing species and rolled stones of granite and gneiss with barnacles adhering to them which in the State of Maine exists at a depth of twenty feet below the present soil, perfectly resembling that which is taken from the borders of creeks and bays of salt water in odour and other properties." [4]

In reviewing Hayden's work, Professor Silliman, writing in 1821, says: "The almost universal existence of rolled pebbles, and boulders of rock, not only on the margin of the oceans, seas, lakes, and rivers; but their existence, often in enormous quantities, in situations quite removed from large waters; inland, in high banks, imbedded in strata, or scattered, occasionally, in profusion, on the face of almost every region, and sometimes on the tops and declivities of mountains as well as in the valleys between them; their entire difference, in many cases, from the rocks in the country where they lie, rounded masses and pebbles of primitive rocks being deposited in secondary and alluvial regions, and *vice versa*; these and a multitude of similar facts have ever struck us as being among the most interesting of geological occurrences, and as being very inadequately accounted for by existing theories. Pebbles may, in given instances, be formed (possibly) by decomposition of the angular portions of a stone, by various chemical agencies, aiding those of a mechanical nature, but an immense number, and in our view the immensely *greater* number of pebbles, present unquestionable evidence of having been brought to their rounded form by attrition. The attrition of the common waters of the earth, and even that exerted during the comparatively short period of the prevalence of the deluge of Noah, would do very little towards producing so mighty a result, and we must assign this operation to the more recent periods of the prevalence of the great chaotic deluge whose existence is distinctly recorded in the first chapter of Genesis, and equally admitted by all geologists." [5]

In a letter dated March 21st, 1821, by Dr. Bigsby, published in the third volume of Silliman's Journal, he says the bed of Lake Huron is covered with the débris of distant countries, its rocks are furrowed and abraded. And he explains these

[4] Op. cit. 63-67. [5] Silliman's Journal, iii. 49, 50.

and other facts by an enormous body of water having rushed over these countries and swept from distant lands the colossal fragments of rock so frequent in the lake.[6] "These fragments," he says, "are incredibly numerous in Lake Huron, and may be divided into two geological classes, the foreign and the native. The former are the more plentiful and are round and smooth. They are seen everywhere, but are collected principally in the interior of the coasts and islands, either in confused heaps, or in parallel ridges and crowning the highest acclivities in great numbers, and the fragments are of various dimensions. They belong almost exclusively to the older orders of rocks, and are therefore of a northerly origin, granites, gneiss, mica slate and porphyries prevail, of kinds which I never saw *in situ*, although I have skirted the north shore for 200 miles and have traversed the wilderness to the E.N.E. for 600 miles. Mica slate I never met with in a fixed state, excepting a few strata of the black variety at the Falls des Chats, on the Ottawa."[7]

In the spring of 1821 Dr. Bigsby read another paper before the Geological Society of London, in which he says *inter alia:* "The shores and bed of Lake Huron appear to have been subjected to the violent action of a flood of waters and floating substances rushing from the north. That such a flood did happen is proved not only by the abraded state of the surface of the northern mainland and scattered isles of the Mainton line range, but by the immense deposits of sand and rolled masses of rock which are found in heaps at every level, both upon the continent and islands, and since these fragments are almost exclusively primitive, and can in some instances be identified with the primitive rocks *in situ* upon the northern shore, and since, moreover, the country to the south and west is secondary to a great extent, the direction of this flood from the north seems to be well established.

"The boulders of granite, gneiss, mica slate (rare), greenstone, porphyry, syenite, and various amygdaloids, are principally," he says, "of such rocks as I have not met with *in situ*, either in the neighbourhood of Lake Huron, or in a journey of 600 miles which I made to the E. and N.E. of the lake,

[6] Op. cit. 255. [7] Id. 256.

through the forests of the river Ottawa. . . . It can scarcely be doubted that these rocks will be found *in situ* somewhere on the northern shore of Lake Huron,. between the Missassaga and Pelletau's channel. It is there and on the isle of St. Joseph that these boulders most occur. . . . Passing on to the southern division of the lake, sixty-four miles south of Cabot's Head, the limestone cliffs of the Manitouline range are succeeded by cliffs of clay. From this point beds of clay, covered towards the upper part of the river St. Clair by thick beds of sand, extend for 150 miles to Lake Erie, and thence along the northern shore, which presents a series of clay cliffs and sand-hills, to the N.E. extremity of the lake. The whole of the intervening shores and woods are strewn with rolled blocks of gneiss, porphyry, conglomerate, and greenstone, such as prevail on the northern shore of Lake Huron. In a S.W. direction, the clay beds prevail over the Michigan territory, and the States of Indiana and Illinois, to an unknown distance. In the two last named States (which I have not visited) rolled blocks abound." [8]

Hitchcock, in a paper on the Geology of Connecticut, read before the American Geological Society in 1822, tells us how, along the Connecticut river, large boulders are frequently found far removed from their original sites; stragglers, he says, may be found everywhere, and among all the rocks none seem more scattered than the granite ones. Generally the boulders, he says, are like the subjacent rock, but there are many exceptions; thus, a few miles west of New Haven, in Woodbridge and Milford, the surface is covered with rolled masses, sometimes quite large, of primitive and secondary greenstone, mica slate, gneiss, granite, and almost every other rock except that which is in place, viz. chlorite-slate or argillet. The diameter of the fragments varies from an inch to 20 and 30 feet, and they are usually rounded, showing attrition. Some of the highest of these boulders are found insulated on the pinnacles of our mountains. Masses of greenstone are found at a greater distance on the W. of the ridges than on the E., and the boulders of Woodbridge and Milford have evidently been brought from the N. He agrees with the view

[8] Trans. Geol. Soc. 2nd ser. i. 204-206.

of Hayden in his essays, that the alluvion of the middle and S. states was formed by a current or currents that formerly flowed across the continent from the N.E. to the S.W.[9]

In the same year Mr. Grainger communicated to Silliman's Journal an account of a grooved or fluted rock at Sandusky bay, Ohio. The surface of the rock, which was of limestone, was covered by several feet of earth. When this was bared it was found to be fluted in a direction from east to west with grooves differing in width and depth, perfectly straight and parallel with each other. "It appears to have been polished as if by friction; and this polish is still retained in a considerable degree. The grooves were traced over a distance of a mile." The author, speaking of the fluting, says: "It has to me the appearance of having been formed by the powerful and continued attrition of some hard body. It resembles, to some slight degree, the sides of a saw-gate which has been for a long time rubbing against the posts which confine and direct it. . . . the flutings in width, depth and direction are as regular as if they had been cut by a grooving plane. This running water could not effect, nor could its operation have produced that glassy smoothness, which in many parts it still retains."[1]

In the year 1823 there was addressed, under the signature "N," a letter to Professor Silliman on Boulders and rolled stones. In this the author tells us how, in his younger days, when rambling over the mountains and valleys of Connecticut, he had been led to inquire why so many rocks of such various sizes should have become so perfectly rounded; why they should be found on the highest mountains, as well as in every valley; why piled in such immense ridges, as well where no stream of water was ever known to flow, or, according to every appearance, ever had or could have flowed, as well as in those places where larger and smaller streams still existed? "Why, again, the uncovered faces and angles of the granite and other rock should bear the same marks of having been worn and ground as the rounded rocks." An answer was suggested, he tells us, when he went down to the sea at Newport, R.I., after a storm, and found the mighty process still going on; and he

[9] Silliman's Journal, vi. 83-85. [1] Id. 179, 180.

concluded that the whole country must have been covered at one time to a great depth with this now retired ocean. Proofs of this fact he claims to exist everywhere. "Thus," he says, "if we follow the Grafton Turnpike road, granite in place is first found about three miles from the meeting-house in Lyme. After that, it is everywhere seen in blocks, in rounded rocks, as well as very often in place. . . . Presently the road crosses a little brook; following this up, you observe large hills of gravel and of sand, apparently the residuum of what once filled this valley to the depth of seventy or eighty feet. When you arrive at the source of this stream, you have to ascend a steep hill of perfectly rounded gravel to the height of from sixty to eighty feet, and then pass on a level about thirty feet to a smooth coarse-grained granite in place." This gravel bore all the traces of having been water-worn. The rocks also had their corners rounded, while their hollows were filled with gravel, and he concluded that only the ocean itself when in movement in the valley could have produced these effects. He also remarked that the rock was marked with circular cavities, three feet across, such as are worn by water, and which were filled with gravel. The cavities were in a line, and not more than twenty or thirty feet apart. He noticed that the diameter of these cavities diminishes as we go down, and in one instance that a piece of rock which was harder than the rest still projected into the hole. He urges that these cavities were worn by the action of gravel and water.

Speaking of the rounded blocks and the smoothed rocks, he says: "The rivulet that descends into the valley is but a small mill stream, and never could have had any agency in rounding, carrying down, and depositing such an immense mass of rocks, gravel, and sand. . . . How could these blocks become detached from their natural bed, how become so perfectly rounded, and that in immense quantities, in the course of three or four miles, affording at the same time by their attrition, in that short distance, hills of coarse and finer sand? How could they become deposited in alluvial hills, gradually diminishing in quantity as they recede from their natural bed, at the same time becoming evidently more and more perfectly rounded?—how could all this have been effected, and much more, but by a slowly-retiring ocean?" Our author then goes on to suggest that there may possibly be cavernous spaces underground into which the water has escaped, and he continues: "The fact being once established that the earth was long covered with water, it would seem to account at once for those rounded and other rocks found everywhere in ascending streams and hills, although no similar rocks should be found in the higher situations immediately contiguous. Those blocks, perhaps, were detached by ice from their native beds, and being rounded and transported by the currents, might be deposited anywhere at random." He tells us how he had noticed that valleys shooting up between mountains have invariably at their outlet a deposit of gravel. This he attributes to the flow of the water, which also smoothed the sides of the valley. The gravel in one instance he names was spread out so as to be almost 1000 feet wide on the western side of the ridge, and nearly 200 feet high on the eastern side, reaching out 400 or 500 feet from the ledge. The mass is nearly of a semicircular form, and the eastern side of it was deposited in that upright position which banks of sand or sand and gravel always assume when deposited in still water . . . the materials of the gravel in the instance quoted must have originally come from some distance, for they are composed almost entirely of grey granite or gneiss. There is no rock like this close by nor anywhere to the west, while it is found *in situ* to the east; and he concludes finally that only a retiring sea aided by ice could explain these phenomena, which, as he tells us, occur at heights of a thousand feet and over, above the present sea-level.[2]

In 1823 Dr. Buckland published his famous "Reliquiæ Diluvianæ," a work written by a distinguished geologist with the avowed purpose of proving an Universal Deluge by means of evidence strictly scientific. Among the other proofs adduced by him are the erratics. Thus he says of the stones found in the so-called drift-beds of eastern England: "The pebbles are of two classes; (1) the wreckage of the adjacent inland districts of England; (2) large blocks and pebbles of many varieties of primitive and transition rocks which do not occur in England, and which can

[2] Silliman's Journal, ix. 28-39.

only be accounted for by supposing them to have been drifted from the nearest continental strata of Norway, by a force of water analogous to and contemporaneous with that which drifted the blocks of Finland granite over the plains of Russia, and the North of Germany. A diluvial current from the north is the only adequate cause that can be proposed, and it is one which seems to satisfy all the conditions of our problem. The pebbles of iridescent felspar, like that of Labrador, which are found on the coast near Bridlington, and resemble similar fragments near Petersburg, can only be referred to the primitive districts of the most northern parts of Europe. Many of the other pebbles of the English coast can be identified with rocks that are known to exist in Norway, and must have been drifted hither at the time of the deposition of the masses of clay and gravel through which they are disseminated; it is impossible to refer them to any action of the present sea, because they occur on high table-lands of the interior as well as on the coasts, and because the cliffs themselves, being composed of clay mixed with the pebbles in question, are undergoing daily destruction, and receive no addition from the action of the present waves. These foreign and probably Norwegian pebbles on the coast of England are mixed up with the wreck of the hills composing the interior of each district respectively; and the component fragments of the latter are less rolled and more angular than those which have come from the continent. . . . In the diluvium of the numerous valleys of Yorkshire, that unite to fall into the Humber, there is a similar admixture of the débris of strata composing the adjacent country with rounded fragments of distant rocks; and in the county of Durham I collected within a few miles on the north of Darlington, pebbles of more than twenty varieties of slate and greenstone rocks, that occur nowhere nearer than the Lake district of Cumberland. In the street at Darlington, at the north end of the town, is a large block of granite of the same variety with those at Shap, near Penrith. Blocks of the same granite lie in the bed of the Tees at Barnard Castle, and near the highest points of the pass of Stainmoor Forest. Similar blocks are found also on the elevated plain of Sedgefield, on the south-east of Durham: in all these places they are mixed with blocks of greenstone. The nearest point from which these blocks and pebbles could possibly have been derived is the Lake district of Cumberland, and the only place in which this peculiar granite occurs *in situ* is the neighbourhood of Shap, just mentioned, from which they are at present separated by the lofty ridge and escarpment of Cross Fell and Stainmoor Forest. If the difficulty of transporting them over this barrier be thought too great, the only remaining solution will be that they have come from Norway, like the other pebbles before mentioned as abounding in the diluvium of the whole east coast of England. I am disposed myself to adopt the opinion of Sir James Hall that they have come from Cumberland."[3]

Buckland then goes on to describe the foreign rocks occurring among the pebbles in the gravel-beds of Derbyshire and of Central England, attested by several observers, such as Aikin, Conybeare and Sedgwick, the last of whom discovered a joint of a basaltic pillar between one and two feet long in the gravels of Cambridgeshire and Bedfordshire. The same facts have been observed in the gravels of the Thames valley, attesting, as Buckland says, a violent rush of waters from the north which has drifted pebbles of quartz rock from the plains of Warwickshire and other central counties over the whole country intermediate between them and London . . . . so that we have pebbles of the porphyry and greenstone of Charnwood Forest, at Abingdon and Oxford; and pebbles of the rocks near Birmingham, at Maidenhead and in Hyde Park. It appears then, he continues, "that we have evidence, that a current from the north has drifted to their present place, along the whole east coast of England, that portion of the pebbles there occurring, which cannot have been derived from this country; a certain number of them may possibly have come from the coast of Scotland, but the greater part have apparently been drifted from the other side of the German Ocean. It appears also that there are proofs of a similar current having passed over the central and south-eastern parts of England; and if we examine its western side, we find similar evidence of a violent rush of waters from the north; in the pebbles and blocks of granite and syenite of

[3] Op. cit. 192-194.

a very peculiar character, that have been drifted from the Criffle mountain in Galloway, across Solway Firth to the north base of the mountains of Cumberland, where I have seen them at a place called Shalk, between Ireby and Carlisle; whilst pebbles and large blocks of another kind of granite have been drifted in still greater numbers from Ravenglass, on the west of Cumberland, over the plains of Lancashire, Cheshire, and Staffordshire; their course is marked on Mr. Greenough's map of England, and they lie in masses of some tons weight on the west of the towns of Macclesfield and Stafford, and between Dudley and Bridgenorth." [4]

Turning to Wales, Buckland quotes Mr. Underwood as having noticed diluvial scratches and scorings on the surface of slate rock where it is immediately covered with a very thick bed of gravel, in a section of the new Irish road at Dynas Hill, about a mile east of Betws-y-coed. He says further that the whole low country of South Wales from the shore of the Severn is strewn over irregularly with large boulders and beds of gravel derived by diluvial action from the mountains that flank this district on the north.[5] In regard to Scotland he merely quotes the evidence already cited from Colonel Imrie and Sir James Hall.

Buckland, in referring to Mr. Weaver's paper already cited, mentions how the distinct and lofty ridges of gravel mentioned by him also exist in Holderness in Yorkshire. There they are locally known as barfs, and are composed chiefly of rolled chalk flints, and a few primitive pebbles (apparently Norwegian), the most remarkable being near Brandsburton, north-east of Beverley: it stretches across the plain like a vast Chesil bank on the sea-shore, being about fifty feet high and 100 broad at the base, and nearly a furlong in length, and has at first sight the appearance of an artificial military mound of enormous magnitude: it bears marks of having been applied to military purposes, but is clearly of diluvial origin.[6]

In another chapter he refers to the American evidence collected by Dr. Bigsby in the paper already cited, and which had not then been published, and goes on to tell us how he had

[4] Id. 198, 199. [5] Rel. Dil. 206.
[6] Rel. Dil. 209, 210.

been informed by Sir Alexander Croke that the summits of the highest hills in Nova Scotia, which are composed of slate, are strewn over with large blocks of granite, the present position of which can only be accounted for by supposing them to have been drifted from the nearest granite districts by the same rush of waters which transported those described by Dr. Bigsby in the districts of Lake Huron and Lake Erie. Buckland also quotes a passage from Dr. Meade of Philadelphia, in his account of the mineral waters of Ballaston and Saratoga in the State of New York, where he says that the surface of the ground, which is there composed of shale and limestone, is covered with large insulated masses of stone, consisting of large blocks of quartz and rolled masses of other primitive rocks. "These," he adds, "must have been transmitted from the neighbouring mountains, as they are not attached to the rocks *in situ*, and have no connection with them: they are found in every country, and only prove the action of an extensive flood of waters. . . . In this dispersion of blocks of granite and beds of gravel in North America we have evidence of a *débâcle* by the diluvian waters in the Western hemisphere, analagous to that we have been examining in Europe.[7]

With Dr. Buckland, the originator of the term diluvium and the most famous champion of diluvial causes, we may fitly bring this chapter to a close.

[7] Rel. Dil. 217, 218.

CHAPTER III.

THE PHENOMENA OF THE DRIFT.

The Champions of Water (Part III.), 1825-1840.

"It is in vain that a vast duration is ascribed to the influence of an agent, unless it can be shown that its action has a tendency to produce the alleged result. If it has a tendency to produce a different result, that difference would be augmented in proportion to the duration of the action. Now diurnal operations are everywhere found in the act of corroding and altering the forms alluded to, but they are nowhere seen to produce them. This class of facts, on the other hand, all conspire in giving probability to the hypothesis of a diluvian wave, which affords an easy explanation of all the large features of this country."—Sir J. Hall, *Trans. Roy. Soc. Ed.* vii. 178.

Chabrier, 1823—Hibbert, 1825—De Lasterye, 1826—Phillips, 1826-29—Sedgwick, 1826-31—Taylor, 1827—Jameson, 1826—"A," 1827—Hildreth, 1827—De Kay, 1828—Omalius D'Halloy, 1828—Brongniart, 1828—Hausmann, 1829—Elie de Beaumont, 1829—Pusch, 1830—De la Beche, 1831—Trimmer, 1831—Ware, 1831—Thompson, 1830—Lapham, 1832—Woodward, 1833—Kloden, 1834—Strickland, 1835-37—Murchison, 1836—Home, 1835—Milne, 1837—Lartet, 1838—Sefstrom, 1838—Forchhammer, 1838—Jackson, 1838—Shepard, 1838—Hayes, 1839—Durocher, 1840.

Before proceeding with my regular story, I must here intervene a short notice of a work, the author of which had views which were at least original.

In 1823 M. F. Chabrier wrote a memoir entitled "Dissertation sur le deluge universelle," in which he denied that the blocks found in North Germany came either from the Hartz or from Scandinavia, or that they were distributed by water. He declared them to be aerolites, and he treated them as the débris of a small planet broken to pieces by a comet. This rain of celestial stones was accompanied by a real diluvial downpour from without the earth which constituted the deluge.

In 1825, Dr. Samuel Hibbert read a paper before the Royal Society of Edinburgh, on the Dispersion of stony fragments remote from their native beds, as displayed in a stratum of loam near Manchester. The bed referred to was situated near Strangeways Hall, and traversed by the new red sandstone cliffs at the confluence of the Irk and Irwell. This clayey bed contained innumerable fragments of rocks, some of them several tons in weight, many of which belonged to primitive or transitory beds foreign altogether to the neighbourhood (which is underlaid by new red sandstone), which had evidently been transported from a great distance. Granite abounded, most of it containing hornblende. Also several kinds of trap and greenstone equally unknown *in situ;* also grauwacke slate, "loosened relics of the hills of far remoter districts," and quartz. Dr. Hibbert affirmed his opinion that these rocks had come from Westmoreland. He states that he had never known any pieces of the porphyritic granite found at Shap, and inferred that the fragments had therefore been transported from a different place, perhaps from the neighbourhood of Dufton, near Appleby. Other specimens of newer rocks also occurred, as, fragments of dark-coloured limestone different to that of Derbyshire, and which he supposes were detached from limestone hills in the north. Also pieces of sandstone shale and coal, which had probably come from the extensive coal district of Lancashire, intermediate between the Manchester red marl and the Westmoreland hills. These facts pointed, he urged, to the various foreign rocks having been transported from the north. Most of the transported fragments appeared to be water-worn; some fragments, however, showed few or no marks of attrition. Dr. Hibbert concluded that, as no rocks are found *in situ* like the older specimens named nearer than eighty miles from Manchester, it gives no small degree of support to the conclusion, that an overwhelming force, most probably from the north, far greater than any which can be attributed to existing causes, transported these boulders to a situation so very remote from the place whence they were originally detached.[1]

De Lasterye, writing in 1826, tells us he had noticed in various parts of America, that when the soil was removed the

[1] Ed. Journ. of Science ii. 208-212.

surface below was found to be polished, the harder rocks naturally preserving the polish better than the softer ones. This effect he noticed on secondary as well as primary rocks, and even in the conglomerates found near New York the surfaces were found thus polished, which, he says, can only be explained by water having flowed over it; and he adds that several writers had remarked that currents of water had traversed America from north to south, while in Asia they flowed the other way.[2]

On November 7th, 1826, Professor Phillips read an important paper before the Yorkshire Philosophical Society, on the direction of the diluvial currents in Yorkshire. In this he cites some interesting facts, *inter alia,* that Shap granite can be at once distinguished, and is different to all other kinds of granite found *in situ* in Cumberland or in Scotland, its nearest relative being that found at Ben Nevis. In the summer of 1826, he tells us, he noticed large boulders of this granite a mile south of Carlisle, and on an elevated red sandstone fell north-west of Kirk Oswald. Abundance of boulders of it lie some miles north of their native site, near the villages of Shap and Great Strickland. In a southerly direction such large boulders lie on the hills of Kendal and Sedbergh common in the Lune valley, and were found of a ton weight in the vast heaps of diluvial matter cut through in the canal south of Lancaster.

"Eastward of Shap fells this granite has been rolled by the towns of Orton and Brough towards the hollow in the great summit ridge at Stainmoor, on the top of which pass the blocks remain to attest the direction and force of the transporting waters. From this point the granitic boulders seem to have been dispersed in different directions. We find them in Teesdale, at Cotherstone, below Barnard Castle, near Greta Bridge, and Darlington, at Scotton, south of Richmond, at Catterick and Leeming, at Stokesley and Thirsk, and several places below the range of oolitic rocks between Thirsk and Pocklington, and commonly in gravel pits and on the surface of the country round York. I have observed them of great size at several points on the Yorkshire coast between Redcar and Scarborough, where they are very plentiful on the

[2] Journal des Connaissances usuelles et pratiques, vol. iv. 242.

shore, while they lie on the cliff at Flamborough Head above 100 miles from their native situation."

Professor Phillips urges that it is easy to understand how the bursting of a lake, as Dr. Fleming had suggested, should have forced the blocks into the vale of Eden or that of the Lune, but he goes on to say that he cannot admit this as a cause when we have to explain the transport of great blocks across the deep valley of Eden, over the summit of Stainmoor, and down the valleys and over the hills of Yorkshire to Scarborough and Flamborough Head. "This case," he says, "has always appeared to me decisive in favour of the opinion, that the dispersion of gravel over large tracts is attributable to the deluge." From the facts cited he concluded that the flood which moved the Shap boulders went in an eastward or rather south-eastward direction, and that the dispersion of fragments in other directions may have been caused by minor currents.

Another rock easily recognized, he says, is a white compact felspar and large-grained hornblende, with here and there some magnetic iron found *in situ* on the sides of Carrock fell. To the north boulders of great size of this rock are scattered between Hesket, Newmarket, and Bolton. Near Kirk Oswald, on high ground west of the Eden, they abound about Lowther Castle and in Durham.

Thirdly, a conglomerate formed of pebbles of light-coloured mountain limestone united by red sandstone, and locally called *brockram*, occurs *in situ* about Kirkby Stephen and at Stenkrith Bridge. Phillips says he had "seen pebbles of it at Scotton, south-east of Richmond, and at several places on the Yorkshire coast, especially at Scarborough, Bridlington, and south of Owthorne, in Holderness. These examples all point to the current having flowed in an east or a south-east direction, but there is also evidence of a current from the north. The large blocks of limestone trap, millstone grit and other sandstones, coal and magnesian limestone which appear on the coast of Holderness may well be admitted to have been derived from north-west Yorkshire, where such rocks abound. But mica slate with garnets, which nowhere occurs in Cumberland, is probably derived from Scotland; the radiated fetid limestone clearly claims origin from Building Hill, near

Sunderland, and the lias fossils have been rolled from Whitby. These effects seem to require the admission of a powerful southward flow of water, and in the vale of York, lias fossils from the north, chalk and oolite fragments from the north-east, are mixed with the débris of rocks washed from the north-west part of the county."[3]

I shall now devote a short space to the views of Professor Sedgwick, whose grasp of phenomena was so firm and so catholic.

Writing in 1826, he says, "Transported boulders of considerable magnitude occur at the very top of several parts of the Derbyshire chain, which overhangs the great plain of Cheshire. For example, many large smooth boulders of primitive or transition rocks lie scattered over the surface of the ground on both sides of the high pass leading from Buxton to Macclesfield. These facts speak the same language with those which I have already quoted. They show the generality of the causes which have produced the superficial detritus."[4]

Again, "Every part of the Yorkshire coast and the whole neighbouring region, bears witness to the operation of similar causes. The numberless valleys of denudation in the eastern moorlands, the immense accumulation of transported materials on the hills as well as in the valleys—the whole contour of the vale of Pickering—the enormous cap of diluvium containing rounded masses of primitive rocks many tons in weight, and resting on the chalk near Flamborough Head—the external form of the Wolds—and the continuous mass of *diluvium* extending from Bridlington to Spurn Head, and from the chalk downs to the sea, are so many monuments of the gigantic powers which were let loose upon the world during the epoch of the diluvial gravel." Again, speaking of Holderness, he says, "The lower part of the cliffs, to the height of about twenty feet, generally consists of a stiff bluish clay, which in many places passes into a dark brown-coloured loam. Through the whole of this mass are embedded an incredible number of smooth round blocks of granite, gneiss, greenstone, mica slate, etc., etc., resembling none of

[3] Phil. Mag. ii. 138-141.
[4] Annals of Philosophy. New Series, x. 23 and 24.

the rocks of England, but resembling specimens derived from various parts of the great Scandinavian chain. Irregularly mixed with the preceding are found, in perhaps still greater abundance, fragments of carboniferous limestone, of millstone grit, of lias, of oolite, and of chalk, torn up from the regular strata of the country, and driven into their present situation by a great eastern current, which has left its traces on every part of the neighbouring district. In regard to the embedded fragments above mentioned, two things appear to deserve notice. 1. They exist in equal abundance in the upper as well as in the lower portions of the diluvial loam. This fact, though difficult of explanation, has been remarked in other similar deposits, and seems to prove the gigantic nature of the forces by which the materials have been drifted into their present position. 2. The boulders derived from distant countries are rounded by attrition; but those which are derived from neighbouring rocks are little altered in form. The hard Norwegian rocks are smooth and spheroidal, but the fragments of oolite and lias, and still more the fragments of chalk, are often sharp and angular. . . . The following conclusions may, I think, be fairly deduced from the facts above stated. 1. The diluvium of Holderness and of the whole coast of Yorkshire, is due to a set of causes which have acted over the western moors, and over all the great central plain of the county. 2. The diluvial currents which produced the gravel of Holderness were probably contemporaneous with other more powerful currents which drove large masses of primitive rocks from Scandinavia to the plains of Yorkshire. And it seems probable that the same currents were contemporaneous with those mighty propelling forces which have driven innumerable fragments of the Scandinavian rocks over the great plains of Russia, Poland, and Germany." [5]

Again, "From the foot of Stainmoor to Solway Firth, through the whole plain of the new red sandstone, the incoherent materials under the vegetable soil are spread over the greater part of the surface, and are often of such an enormous thickness as entirely to conceal all the subjacent strata. These accumulations are not partial or irregular; but seem to have been rolled out over the surface of the country by an inundation

[5] Id. 24-26.

which acted at one moment over the whole district; and like all similar deposits, they contain an incredible number of large boulders, principally derived from the neighbouring mountains. . . . From St. Bees Head to the southern extremity of Cumberland, the region bordering on the coast is formed of one almost continuous mass of diluvium, interrupted here and there by low hills of blown sand, and by other recent formations. In this part of the country the cliffs are of a deep red colour, caused by the presence of innumerable embedded fragments of the subjacent new red sandstone. With these fragments, boulders of granite, porphyry and greenstone, are scattered throughout the whole diluvial covering; sometimes in such abundance as to give it the appearance of a true conglomerate. . . . Some of the granite blocks embedded in the cliffs are of great magnitude. In the diluvial cliffs near Bootle, I found one of a rude prismatic form, which was twelve feet long, six feet wide, and five feet and a half high. All specimens of this kind of rock have been drifted to the coast from the granitic region which extends from Wastdale foot through Muncaster fen to the neighbourhood of Bootle, and occupies a part of Wastdale head and all the lower parts of the valleys of the Mite and the Esk. If we cross the estuary of the Duddon to the shores of Low Furness, we find an exact repetition of the phenomena we have left behind. All the country bordering on the western shore is covered by an enormously thick deposit of red-coloured diluvial gravel, containing innumerable rolled fragments of rocks derived from every part of the lake mountains, and all the neighbouring islands are composed exclusively of the same materials. A rolled mass of Eskdale granite, which had been embedded in the highest portion of the diluvial cliff near Rampside, fell down upon the strand in 1822. It rose nine or ten feet above the rubbish in which it was at that time partially buried. At the base of the cliffs of the isles of Barrow and Foudrey, among innumerable boulders of granite, and of other Cumberland rocks, were some specimens of a beautiful variety of compact felspar, which I afterwards found *in situ* near the top of Scafell and Bowfell. . . . The phenomena above described have obviously been caused by a violent rush of descending waters. Whatever forces may have put these waters in motion, it is, I think, obvious, from the facts already stated, that they have not acted partially, but have swept over the whole cluster of the neighbouring mountains."[6] He describes the syenite blocks which have travelled from Carrock fell, and the granite ones from Shap fell, one of the former, the Golden Rock on High Pike, being twenty-one feet long, ten high, and nine wide; the green slate from Kentmeer and Long Sleddale; masses of St. John's vale porphyry found near Penruddock, and descending the valleys into the Eamont; boulders of Shap granite on the calcareous hill of Appleby, some twelve feet in diameter. Shap granite, as he points out, is peculiar and cannot be confounded with other granites elsewhere, and blocks of it are found drifted over the plain of new red sandstone right over the Pennine chain into the plain of Yorkshire, embedded in the *detritus* of the Tees, and even carried to the eastern coast.

Sedgwick goes on to say, "I think it has already been proved that diluvial torrents have swept over every part of the Cumberland chain, because we find water-worn masses derived from the highest elevations of the country, embedded in the diluvial loam which covers almost all the neighbouring plains; and because we find large boulders of the same rocks scattered over many parts of the mid regions of the mountains, in situations to which they could never have drifted by any less powerful agent than that to which they have been ascribed."[7] As a general conclusion Sedgwick says that "the floods which produced the diluvial detritus swept over every part of England, that they were put in motion by no powers of nature with which we are acquainted, and that they took place during an epoch which was posterior to the deposition of all the regular strata of the earth, and prior to all known accumulations of alluvial matter. We have evidence enough," he adds, "to justify us in extending the same conclusions to every part of the European basin, and there is some evidence which makes it probable that they may be extended to the remotest parts of the earth's surface. Indeed, the mighty disturbing forces which produced the accumulations of diluvial *detritus* between the western extremities of Europe and the central plains of Asia, must probably have

[6] Id. 26-29. [7] Op. cit. 30.

acted with sufficient energy to leave some traces of their power over every quarter of the globe. On the continent of America the succession of formations seems to be very nearly the same with that of our own country, and over all the regular strata, there occur in many places alluvial and diluvial formations in every respect like those of Europe. It is, therefore, to say the least of it, probable that the diluvial phenomena of Europe and America belong to the same epoch."[8]

In some papers on the Geology of East Norfolk, published in the Philosophical Magazine for 1827, Mr. R. C. Taylor was, I believe, the first to notice some of the very remarkable features of the drift deposits of East Anglia. He says, *inter alia*, "The crag has, at the last of the geological epochs, been subjected to abrasion by diluvial currents. Portions, probably from its western edges, have been swept away. Their fragments, mingled with those of the chalk and preceding formations, piled in enormous heaps, form the cliffs of Cromer and Trimingham, 250 or 300 feet in thickness, upon the original crag, which rests *in situ* at their base. The proof of the disruption and transportation of more ancient strata may be observed in the enormous detached masses of chalk, in these diluvial cliffs, at various elevations *above* the crag. Near the lighthouse hill at Cromer, one of these insulated patches is 150 feet high, and has a kiln upon it, in which lime of an excellent quality is burnt. Further on at Runton is a large mass eighty feet thick; another rises to the height of 100 feet; and at Sherringham is another still higher. In all these cases they rest *upon* the crag, proving alike the breaking up of the older strata, and the continuity of the later."[9] In the same paper he refers to the variety of travelled rocks occurring in the Cromer cliffs, such as blocks of several varieties of granite, basalt, porphyry, trap and micaceous schist; sandstone of various kinds, chert breccia, beside limestone, claystone, etc. The diluvial fragments of the later series are from the chalk, plastic clay, London clay, green sand, Kelloway's rock, the oolites, lias clay and marlstones. They are of all sizes up to four tons in weight, one block on the shore was observed to be six feet high, another standing six or eight feet high was known to the fishermen as Black Meg.[1]

[8] Id. 32. [9] Phil. Mag. i. 288. [1] Id. 346.

Turning to America, we find in the appendix to the narrative of Parry's third voyage, a memoir by Professor Jameson, dated April 24th, 1826. In this he mentions how numerous boulders were strewn over the surface of some of the islands in the Arctic ocean. Thus the surface of Igloolik, a limestone island, is strewn with blocks of primitive rocks. The island of Neerlo Nakto, principally composed of limestone, is also strewn over with primitive blocks; and in Amherst Island, in which greywacke and greywacke slate are almost the only rocks, rolled masses or boulders of granite, gneiss and quartz rock, are not uncommon.[2] According to Dr. Neill's report, the whole limestone country, extending as far north as Cape York, and south as far as Cape Fitzgerald, was more or less covered with boulders of primitive rocks, which were either rounded or angular. Similar boulders were observed strewn over the limestone on the west side of Prince Regent's Inlet in North Somerset. They included granite, syenite, gneiss, indurated talc, hornblende-rock, actynolite rock, a beautiful quartzy iron-glauce and brown hematite. The first three were the most numerous and largest. Some of the boulders near Port Bowen were upwards of fifty tons in weight, and lay fully 400 feet above the level of the sea, and Dr. Neill saw a boulder of granite, two or three tons in weight, on the summit of the highest limestone hill near Port Bowen. They were numerous on the slopes of the hills on both sides of Prince Regent's Inlet; everywhere resting on secondary limestone, but on leaving the coast they gradually diminished in number and size, and fourteen to sixteen miles from the sea were comparatively rare, and in general not larger than a closed fist. These masses were most like the primitive rocks of the Whale Fish Islands, and the nearest known fixed primitive rocks were those at Cape Warrender, more than 100 miles distant.[3] Jameson argues that they afford evidence of the passage of water across the country since the latest tertiary times.

In a letter signed "A," dated April, 1826, and published in Silliman's Journal, reference is made to the fact that the surface of the rocks in America, where it has not been exposed to the atmosphere, is worn quite smooth, whether the

[2] Op. cit. 144. [3] Id. 148. [4] Id. 150.

covering be shallow or deep, of whatever kind of rock the mass may consist, or however unequal and irregular may be the form it has assumed. The appearance is as if worn down by some powerful but not very delicate agent. The harder parts have in some instances, especially when forming veins in a softer stratum, the *feeling* of being polished, but the general character of the surfaces, although smooth to the eye, is somewhat rough to the touch, with slight grooves or channels, running in a uniform direction, very nearly north and south, but from a little west of north to a little east of south. The author connects these scratches with the masses of rolled gravel often found at a great distance from any river and often on mountain tops. He suggests that some sudden change has sometime taken place in the velocity of the earth's motion about its axis, when the whole mass of water on the earth would rush forward with inconceivable violence. The Pacific Ocean would thus sweep over the Andes and Alleghanies into the Atlantic, which in the meantime would be sweeping over Europe, Asia and Africa: a few hours would cover the earth with one rushing torrent, in which the fragments of disintegrated rock, earth and sand would be carried along with the wreck of animal and vegetable life in one all but liquid mass. The appearances, he says, favour this conclusion, the outer more friable strata torn out from the solid mass, rounded into boulders and pebbles of every size or crushed into atoms, the surface of the harder strata ground, pounded and worn. The distribution of the materials favours the same view. The more elevated ridges would arrest the more ponderous detached rocks, together with the confused mass of pebbles and sand accompanying them, while the lighter portions of the latter would be carried off to a lower level. Thus we find the largest boulders, with immense hills of coarse gravel, piled up in mountainous districts of country, frequently nearly equal in height to the ridges of rock about them, except when the latter shoot up into sharp peaks. He remarks that he had noticed these phenomena in all parts of New England, and he adds that the polished surfaces are not confined to the primitive rocks but are nowhere more striking and beautiful than on the pudding stones near Boston, the embedded pebbles being worn down and polished as in the columns of the Halls of Congress in the Capitol.[5]

Dr. Hildreth, writing in 1827, describes certain parts of Ohio as consisting of a yellowish loam, twenty to fifty feet thick, based on tough blue clay. "But," he continues, "the most singular feature of all is the abundance of detached fragments of blocks of primitive rocks, with which this region is filled. All are rounded or worn by attrition, and lie in that confused state which they must be supposed to exhibit if brought there by an immense current of water. I picked up pieces of granite, hornblende, greenstone, gneiss quartz, limestone, etc. Some blocks of granite are large enough to make a pair of mill-stones, and are used for the purpose. . . . In reflecting on this formation, the impression is irresistible that it is the result of an immense current or body of water pouring down from the north, sweeping the south side of Lake Erie and all the Scioto country and Miami valley, as those regions are said to furnish specimens of the same primitive kind."[6]

In a paper by J. E. De Kay, published in vol. xiii. of Silliman's Journal (1828), he speaks of the southern part of the State of New York being composed of gneiss covered with sand. Promiscuously scattered over the surface and embedded in the sand are boulders of different sizes up to several tons in weight. They are totally unlike any rocks in place in the immediate vicinity, being chiefly greenstone and trap, occasionally slaty, sometimes granitic, and not unfrequently calcareous. He says, "Dr. Akerley has referred the schistose rocks to the region above the Highlands, the greenstone to the pallisade rocks on the Hudson, and others to rocks similar in composition and structure nearest to the place where the boulders occur." He himself believes that when the boulders consist of secondary rocks and are scattered over primary ones, that they represent the remains of strata disintegrated and destroyed *in situ;* when they are of primitive rocks, he says, we need not look for their parent strata several hundreds of miles away. We know that primitive rocks frequently thrust themselves through all the superincumbent strata, and if their peaks have been destroyed by some convulsion or by

[5] Silliman's Journal, vol. xi. 100-103.
[6] Silliman's Journal, xiii. 39.

time, and their origin concealed by the detritus, we can account for their débris occurring in a region to which they are apparently strangers.

Reverting to Europe; Omalius d'Halloy, writing in 1828, refers to the granite blocks scattered over the district of Campine in Belgium, which, he says, are either found on the surface or buried in the sand, and in immense numbers. He argues that their shape is due to weathering, like that of the well-known tors found in granitic districts, that they were detached from beds of granite which lie below them, and that the reason they are found at the surface is that, when a mixture of unequally divided matter is tossed about, the big pieces naturally come to the surface, and that earthquakes have probably shaken the deposits in which they occur; and he says he is confirmed in this view, because the blocks only occur in the sand and not in the clay which covers part of Frisia, Groningen and Overyssel.[7]

In 1828, M. Brongniart published a paper on the Swedish erratics, in the Annales des Sciences Naturelles, vol. xiv. p. 1. In this he mentions that although the blocks occur isolated from each other, they are generally found in groups. That sometimes they are found, as on the Alps and Jura, planted on the base rock, at others buried in sand, as on the plains of Westphalia; sometimes their angles are quite sharp, at other times their edges have been somewhat rubbed and weathered. He remarks that the sandy plains of Westphalia, Hanover, Holstein, and Zealand, Mecklenburg, Brandenburg, the plains and shores of Pomerania and Prussian Poland as far as the district between Warsaw and Grodno, and as far east as the Neva, are covered with these blocks, in some places more abundantly than in others. Thus between Lestadt and Hamburg there are a large number; then follows a sterile zone, then another group, and so on, and the same is the case in Sweden, where more or less sterile zones alternate with zones where the blocks are abundantly distributed. In East Friesland and Groningen they are buried in the sands underneath the turbaries, and are a fertile source of stone in those sandy districts where stone is otherwise scarce. Here also they occur

[7] Memoir pour servir a la description Géologique des Pays Bas et de la France, etc., 205-6.

in zones. At the other end of the Baltic there are found blocks of limestone containing trilobites, orthocerites, etc., which are dug out and burnt into lime. These blocks are apparently derived from beds with similar fossils in Sweden and Norway.[8] He goes on to show that as the calcareous blocks are shown by their contents to have come from Scandinavia, so M. Hausmann and others have shown that the granitoid blocks found in Pomerania and Westphalia are mineralogically identical with Scandinavian granites, and contain, *inter alia*, Wernerite. Whence it is pretty clear that the blocks of Northern Europe have been derived from Scandinavia, and not from the Hartz or elsewhere, and the Baltic depression which intervenes presents the same difficulty in explaining their transport that the valley of the Aar does in Switzerland. That they traversed the Baltic seems clear, for we can trace them step by step. The sandy soil of Holstein is covered with them, so is that of Zealand. They abound near Copenhagen, as far as Elsinore, and are used for building purposes in these districts where sand abounds and stone is so scarce. On crossing the narrow but deep Sound, we meet with them again at Elsingborg on the other side. The soil of Scania is as much strewn with them as that of Zealand, but there they are freer from sand and more exposed. Further north in Sweden they abound still more, and are occasionally found piled upon one another and intermixed with sand in peculiarly shaped mounds called äsar and sandäsar by the Swedes, according as the blocks or sand prevail. Brongniart says he found these ridges more especially in Scania, Smaland, Sudermannia and Upland. They are seldom a hundred metres high, are long and narrow, and a little wider and higher at one end than the other. These hillocks are widely distributed in Sweden, and differ somewhat from each other. Generally, in the south they are formed of gravel or sand, either quartzose or granitic, and of granite blocks of a considerable size. Further north, especially north of Upsala, they are more abundant and more sandy.

A remarkable feature about them is that they all range from N.N.E. to S.S.W., and are thus parallel over a large area. They also remain continually of one height and width, so much so that the roads in some places, as from Upsala to Wendel,

[8] Op. cit. 1-10.

from Enkoping to Nora, from Hubbo to Moklinta, etc., run along their crests. Brongniart compares them to the trains of sand, etc., which are often found in streams behind some obstacle, such as a large stone, etc., and he urges that the basaltic hill of Kinnekulle on the south-east borders of Lake Wenern, which has a tail of blocks detached from itself, that have not travelled far, of sand, etc., ranging north and south, forms an instructive example of the general law he would invoke, and he looks upon the äsar as the trains formed behind obstacles when the great watery *débâcle* sped over the land; a view which, as we have seen, had already been urged by Hall.

He attributes to the transport of the moving materials over the rocks *in situ* their polish and mammillated contour, and the fissures and grooves, etc., which mark them, and which, as he says, are most frequently found in the province of Gothenburg and on the borders of Norway and Sweden, near Stromstadt, Hogdal, etc., and he mentions the remarkably polished surfaces which are still to be found wherever the covering is removed. He also remarks that the grooves, like the äsar, have a constant direction from N.N.E. to S.S.W. He compares the polished surfaces to similar ones found in Upper Egypt, in the United States, in Cumberland and Westmoreland, etc.[9]

He further claims to have traced the blocks to the granitic and gneissic mountains of Scandinavia, and says that, as might be expected, the point from which the force emanated which dispersed them is the one where they are most infrequent, while they are found most abundantly on high pieces of ground which would naturally arrest their passage. M. Brongniart records that he had found it universally true that the higher one mounts the larger the blocks become. This, he says, he tested on the Supergue, a mountain near Turin, in the Eastern Jura, about Lausanne, and in the Apennines of Castel Arquato, not far from Piacenza, and adds that M. Schultze, in his History of the erratics of Pomerania, had also remarked that the blocks are more abundant on the heights than in the valleys.[1]

In 1829, Phillips published his "Illustrations of the

[9] Id. 10-17. [1] Id. 17-19.

Geology of Yorkshire," of which the second edition appeared in 1835. In this work, Phillips was the first to describe in detail the well-known boulder deposit of the Yorkshire coast. He describes it as extending in a connected mass over nearly all Holderness. "In the highest cliffs on this coast," he says, "its thickness is not less than 130 feet. Its composition is remarkably uniform. We everywhere observe it to be a solid body of clay, containing fragments of many kinds of rocks which vary in magnitude and in the degree of roundness to which they have been reduced. The fragments are in general not so numerous as to touch each other, but are scattered through the clay as plums in a pudding. . . . The rocks from which the fragments appear to have been transported are found, some in Norway, in the highlands of Scotland, and in the mountains of Cumberland; others in the north-western and western parts of Yorkshire, and no inconsiderable portion appear to have come from the sea-coast of Durham and the neighbourhood of Whitby. In proportion to the distance they have travelled is the degree of roundness which they have acquired. All the fragments of granite, porphyry, mica slate, and clay slate, which can be compared with no fixed rocks nearer than those of Cumberland and Westmoreland, are rolled to pebbles; the angles are worn away from every mass of limestone which has been drifted from the north-west hills of Yorkshire, but those which have been brought from the nearer points of the chalk range have yielded much less to attrition. Some attention is required to the original hardness of the stones; we find solid masses of ironstone and quartz, much less worn than granite; limestone is less rounded than millstone grit; and flint retains uninjured angles, whilst chalk and magnesian limestone have lost their original surfaces.

"Few substances, originally soft, are carried by water to a great distance in a solid form. The sandstones of the western and north-western parts of Yorkshire are plentiful in the gravel of the vale of York; but only the hard 'galliard' of Leeds and Bradford, and the solid millstone grit of the moors, can be recognized in the clay of Holderness. . . ." Speaking of the clay, he continues, "We might have expected to find the finer particles at the top, and the solid fragments of rocks lying beneath, according to their individual magnitude and weight. As nothing of this kind is observable, we must suppose the flood to have moved with such extraordinary violence that its spoils, when heaped together, were little influenced in their arrangement by the direct form of gravitation."[2] In vol. ii. p. 193 of the same work, published in 1836, Phillips says: "The extraordinary phenomena of the dispersion of the Cumbrian rocks to the east, will require the operation of great and violent oceanic currents excited by sudden subterranean movements."

Hausmann, in his dissertation entitled "De origine Saxorum per Germ. septentrionalis regiones arenosas despersorum commentatio," after passing in review the various theories accounting for their origin, goes on to discriminate certain blocks which have come from local strata, including also flints, and which may generally be distinguished by their sharper corners from the great mass of stones which have clearly come from Scandinavia, and he very elaborately describes the various localities in Sweden where the parent rocks are found *in situ*, and concludes, *inter alia*, that the blocks have come from the north, some from Scandinavia, others from the borders of the Baltic; that they were translated at one time, and by one effort, and by the agency of water; that the phenomenon was a general one, extending over the greater part of the north, and that the sand in which many of the blocks lie is very largely the result of the disintegration of tertiary beds.[3]

M. Elie de Beaumont, writing in 1829, argues that it was by the sudden melting of great masses of Alpine snows, caused by the outburst of gases, such as originated the dolomites and gypseous rocks of the Alps, that great diluvial currents were constituted whose effects are seen in the dispersal of the boulders, etc.[4]

Charpentier urges against this view, that even if it explained the phenomena of the Alps, it would not explain those of the Pyrenees, for it depends on the fact that the Alps were a con-

[2] Op. cit. id. 1835, vol. i. 20 and 21. [3] Op. cit. 3-34.
[4] Recherches sur quelques unes des revolutions de la Surface du globe, 285, note.

siderable range, and capable of nursing reservoirs of snow before their present contour was finally settled, a fact which is probable, but in the case of the Pyrenees there is no evidence that there was any range at all, but, according to M. Elie de Beaumont himself, they were raised by one effort from the surrounding horizontal strata. Again, the elevation of the Pyrenees took place between the deposition of the chalk and tertiary times, which is too early for an explanation of the diluvian phenomena.[5] I would add that no invasion of hot gases would melt the glaciers sufficiently quickly to cause the transcendent floods necessary to explain the transport of the boulders, etc.

In a paper on the Geology of the South of Poland, published by Pusch in 1830, he adduces fresh facts to show that the majority of the blocks found in that country have not come from the Carpathians, but from Scandinavia. Their size and number diminishes as we leave the Baltic and go southwards. Their southern limit extends along the environs of Czenstochau, by Pizedborz by the foot of the mountains of Sandomirz, and descending the Kamiona by the Vistula upon Lublin and Lubartof across Southern Lithuania and into Central Russia, that is a long way from the Carpathians, nor are they formed of materials found in the Carpathians, but they are in fact identical with those found in Sweden, Finland, and Southern Russia. Between St. Petersburg, the Duna and Niemen they consist of a red granite found at Wiborg, a felspathic granite found in Ingria, a red quartzose sandstone found on the borders of Lake Onega and the limestone of Esthonia and Ingria. In Eastern Prussia and in Poland, between the Vistula and the Niemen, of three kinds of granite which are found in Finland, at Abo and at Helsingfors, of an amphibolite either from Central Finland or from Sweden, of a quartzite exactly like the Fjall sandstone occurring between Sweden and Norway, and of porphyry from Elfdalen in Sweden. From Warsaw westwards towards Kalisch and Posen the red granite of Finland becomes scarcer, amphibolite, gneiss, and porphyry increase in numbers, there are few specimens of Finland and many of Swedish rocks. These facts agree with those forthcoming from North Germany, the Low Countries and the East of England, whose blocks have also come

[5] Charpentier, Essai sur les Glaciers, 209-213.

from Scandinavia, and he concludes that the evidence forthcoming from Prussia, Livonia, Courland and the North of Russia, concurs in proving a *débâcle* coming from the N.N.E. to the S.S.W., which also carried the blocks into the country N. of the Carpathians.[6]

De la Beche, writing in 1831, calls attention to the fact, that on Great Haldon Hill, about 800 feet above the sea, there occur in the superficial gravel pieces of rock which must have been derived from lower levels, and he says he "found pieces of red quartziferous porphyry, compact red sandstone, and a siliceous rock, not uncommon in the graywacke of the vicinity, where all these rocks occur at lower levels than the summit of Haldon, and where certainly they could not have been carried by rains or rivers, unless the latter be supposed to delight in running up hill. . . . Not only have gravels been brought from various distances, but even huge blocks, the transport of which by actual causes into their present situations seems physically impossible." After referring to various facts quoted by different authors, he continues: "The probability, therefore, as far as the above facts seem to warrant, is, that masses of water have proceeded from north to south over the British Isles, moving with sufficient velocity to transport fragments of rock from Norway to the Shetland Isles and the eastern coast of England, the course of such masses of water having been modified and obstructed among the valleys, hills and mountains which they encountered; so that various minor and low currents were caused. De la Beche says, referring to Elie de Beaumont, "The appearances are precisely what would have been expected from the passage of a rock-charged mass of waters down the respective channels, the largest fragments having been transported the shortest distances, being most angular, while the smaller and most rounded have been carried the furthest. Thus in the valley of the Durance, the transported substances become more angular and of greater volume as we proceed from the great mass of pebbles called the Crau to the mountains beyond Gap, whence the débris, judging from its mineralogical characters, have very clearly been derived. Similar phenomena will be observed up the valley of the Drac, which proceeds by another course to

[6] Journal de Geologie I. 251, etc.

the neighbourhood of the same mountains, the two streams of débris not mingling till they join in the Crau. De la Beche then goes on to refer to his own observations near lakes Como and Lecco, and notably on the north face of Monte San Primo, which, he says, presents a bold front to any shock coming from the north, leaving open passages to the right and left of it, one down the south part of Lake Como and the other down that of Lecco. Not only in front, facing the high Alps, but also round the flanks and shoulders of the mountain, and even behind it, where the eddy current would have transported them, blocks of granite, gneiss, mica slate, and others, from the central chain, of various sizes, and often accompanied by smaller fragments and gravel, are seen in hundreds, nay thousands, scattered over the dolomite limestone and slate of the mountain, and nearly filling up a previously existing valley which faces the north, the direction whence the rock-charged fluid descended. Proceeding down the side valleys, partly occupied by the lower lake of Como, and the lake of Lecco, we find the evidences of such a current in the presence of blocks occurring, as they should do, where obstacles were opposed to its course, or in situations where eddies would be produced behind the shoulders of the mountains. One very remarkable instance of such occurrence is behind, or on the southern side of, the Monte San Maurizio, above the town of Como, where numerous blocks are accumulated on the steep flank of the mountain, precisely where a body of water, rushing down the great valley, would produce an eddy at its discharge into the open plains of Italy. The blocks, though no doubt many have descended from their first positions in consequence of the long-continued action of atmospheric agents, occupy an elevated line, and occur also on other but lower heights in the vicinity, which opposed more direct obstacles to the *débâcle;* seeming to show that the blocks floated near the surface of the fluid mass, and were whirled by the eddy, at nearly the same level, against the steep sides of the calcareous mountain, as well as thrown against the more direct obstacle of a range of conglomerate hills." [7]

Having referred to these facts, he proceeds: "Therefore, proceeding south, the course of the waters seems to have con-

[7] Id. 194-6.

tinued in that direction, over the low districts of Germany, to the Netherlands, depositing huge blocks in their passage; these blocks prove by their mineralogical composition to have been derived from rocks known to exist in the northern regions. Such a movement as this over part of Europe would, if the supposition of a mass of waters were correct, be observed in other northern regions, for the waters thrown into agitation would cause waves around the centre of disturbance, and he accordingly refers to the great masses of such erratics in America, and says that, as they also point one way, we can scarcely refuse to admit that the course of the disturbance or disturbances was towards the north, the undulations of the waters having been caused by some violent agitation, perhaps beneath the sea in those regions, for it is by no means necessary that it should have been above its level.

"A convulsion or convulsions of this magnitude, reasoning from the analogy of those minor agitations which we term earthquakes, would be felt over a considerable portion of the globe, and the waters over a large surface would be thrown into agitation. A part of the earth would be greatly disturbed, and we should expect fractures and faults produced in strata where the convulsion was most felt, as similar minor effects are produced at present from the exertion of a less intense force."

In June, 1831, in the first volume of the Geological Proceedings, Mr. J. Trimmer describes the mantle of gravel covering large parts of Carnarvonshire. "A large proportion of this gravel," he says, "is composed of pebbles and blocks of various sizes derived from rocks that occur in Carnarvonshire, many of which are less rolled than pebbles of another class, that are mixed with them and which have come from a greater distance, and must have been drifted upwards by some violent inundation, in a direction contrary to that of the rivers which descend from Snowdonia into the Menai. Among these pebbles are several which can be identified with the granite, syenite, greenstone, serpentine and jasper of Anglesea: other granite pebbles agree with no rock in Anglesea or Wales, and resemble the granite rocks of Cumberland; some may have come from Ireland or the south-west extremity of Scotland. There are also some flints which can have come from no other source than the chalk of Antrim. The diluvium occurs in great thickness in the lower region of the valley of the Ogwen, usually from sixty to 100 feet. . . . Beneath the diluvial deposit of this district, when the surface of slate rock is laid bare, it is found to be covered with scratches, furrows and dressings, like those observed by Sir James Hall on the summit of the Corstorphine and other hills near Edinburgh. These furrows and dressings were noticed some years ago by Mr. Underwood. They are referred to the action of the diluvial currents which overspread the country with gravel; some of the larger blocks amid the gravel have also deep scratches upon their surface." [8]

In a paper by Dr. S. Ware on the Direction of the Diluvial Wave in the Shetland Islands, published in 1831, he remarks on the absence there of the large diluvial deposits found in England and Scotland, and which are represented there merely by detached boulders on the sides of declivities or strewed on the plains. He found on Papa Stour, at a place where the rock *in situ* was sandstone and secondary porphyry, numerous fragments of hornblende schist and actinolite schist which is nowhere met with in the Shetland Islands but at Hillswick Ness, at a distance when measured across the Bay of Saint Magnus of at least twelve miles, that is in a direction about north 47° east, a line pointing to the north-west boundary of Norway. He also noticed some large blocks, the largest twenty-three feet high and ninety-six in circumference, situated near the mansion of Lunna, on the east of Shetland, named the stones of Stefis, which seemed to have been moved a mile or two by diluvial currents in a south-west direction. On the east coast of the Island of Yell, which is composed of gneiss, were many large fragments of serpentine and euphotine, which have evidently been drifted from Unst and Fellar. On Roeness Hill, which consists of red granite and is 1447 feet high, were an immense number of boulders of a primary greenstone or trap which seem to have come from a site two or three miles off, and to have been rolled in a south or south-west direction up a gradient ascent of three or four miles on the summit of Hillswick Ness. He also noticed a surprising granite block, mantled with grey moss, removed from a

[8] Proc. Geol. Soc. I. 331-2.

rock the nearest site of which is about two miles north. It is many feet in dimension. "This Thor stone, as it was anciently called, evidently owed its original displacement to diluvial torrents." He also noticed in Shetland "fragments of stones which are strangers to the country, and which have probably been transported from foreign shores, as in the case of some detached boulders in Soulam Voe, an unsheltered harbour open to the Northern ocean; one of them, three or four feet high, is of a kind of granite quite unknown in the country, and popular imagination affirms that it was thrown here by the devil as he stood on some high hill in a neighbouring parish; and he concludes that the great currents which deluged the British Islands as well as some parts of the continent had in Shetland a north-east origin and a south-west direction, and that the evidence of the fragments of rocks from Cumberland found in Lancashire and Cheshire point to the current, after encountering the headlands of Scotland, England and Ireland, having had its direction changed from a south-west to a south or south-east course." [9]

Reverting again to America, Mr. Thompson writes, Dec. 22nd, 1830, in regard to the Alleghany range in Sullivan County, N.Y., "For twenty years past I have been forcibly struck with the fact that wherever the earth has been removed, leaving the horizontal graywacke bare, scratches and deep grooves are observed running a few degrees north of a due east course. I have observed this fact in more than fifty places where the earth has been removed in the construction of turnpikes, common roads, mill works, etc. But they scarcely ever appear on rocks which have been exposed to air, rain, etc. When these scratches first attracted my attention . . . I could not resist the inference, that they were made by heavy rocks or boulders, driven over the surface of the upper layers of graywacke by the waters of the deluge. I was even inclined to infer that the scratches indicated the direction of the oceanic waters. I supposed that the *general* direction was everywhere the same, but that the particular configuration of mountains, valleys, etc., might give local variations to the course of the mighty movements of water many miles in depth." [10]

[9] Ed. Journ. of Science, New Series, IV. 85-91.
[10] Silliman's Journal, xx. 125.

In a paper on the primitive and other boulders of Ohio, by D. and E. A. Lapham, published in 1832, they describe the gravel of greenstone and granite found in the Ohio valley and not in the neighbouring hills. "It is found mixed with clay and sand, forming banks and hills of moderate elevation, entirely beyond the influence of the Ohio or any other existing current of water. As we travel north the gravel predominates over the clay, and the fragments grow larger. In general, when they exceed the size of one's head, they are called boulders by geologists, but there is no distinct mark by which they are separated, they gradually pass into each other. Proceeding further north the boulders become more numerous, are of larger dimensions, and less worn. From all these facts it is evident that we must look beyond the Great Lakes for the origin of our primitive fragments."

"The surface of the diluvion everywhere presents a wave-like undulated appearance. These swells are generally more abrupt about Circleville than north of it. One elevation near that place has been mistaken for an ancient artificial mound. . . . The boulders occur more frequently in the beds of ravines and of rivers and the tops of hills. They seldom occur on the surface of the diluvion where it remains entire; only when the earth has been washed away. . . . The largest boulder we have seen was on the summit of a hill near Lancaster, and was about six feet in length. We are convinced that a great flood has at some remote period swept down from the north and inundated the whole country."[1]

In Woodward's Geology of Norfolk, 1833, he assigns as the wreck of the Deluge the transported gravel, numerous boulders of many tons' weight, and disintegrated portions of all the superstrata of the antediluvian period. The beds of gravel he describes as ten to twenty feet thick, consisting of small water-worn pebbles, in some places cemented by oxide of iron. The blue clay constituting the cliffs from Hasboro' to Cromer, is evidently, he says, the wreck of the blue lias brought from the interior of the kingdom by the diluvial waters.[2]

In 1835, Strickland wrote a paper on the deposits of transported materials usually termed diluvium, which exist in the counties of Worcester and Warwick. In this he discriminated

[1] Silliman's Journal, xx. 300-303. [2] Op. cit. 14-16.

between what he called the *local* and *fluviatile* drifts and the general drift. The latter, he says, has been produced by an agency far more powerful and extended than that which has carried the other two varieties.

The general drift, again, he divides into a quartzose and a flinty variety. Of the former he says: "From the northerly position of the rocks which have supplied fragments to the quartzose drift we may infer that the current which has placed these gravel beds in their present situation came from the northward. This is further proved by the fact noticed by Sir Roderick Murchison, that as we proceed from Worcestershire towards Cheshire, the boulders are found gradually to increase in size." In regard to the flinty drift, he brings the current which brought this gravel into Warwickshire from some point between N.E. and S.E.[3]

On February 3rd, 1836, Sir R. Murchison read a paper before the Geological Society, in which he pointed out that in South Wales and Siluria the surface is completely free from the far-transported rocks constituting the diluvium of other districts, and this area is covered with local débris only, which, however, is spread over the country irrespective of the drainage. This, he supposes, was deposited when the land was generally submerged under the sea. Differing entirely from this local débris is the deposit spread over Lancashire, Cheshire, and North Shropshire, which consists of granites, porphyries, and other hard rocks which have been derived from the mountains of Cumberland, and a few perhaps from Scotland. This drift also contains much sand and clay. Its distinguishing feature is the recurrence at intervals of large blocks or boulders of northern origin, some of which have been propelled as far as the edge of the Silurian rocks south of Shrewsbury. They prevail in vast quantities in the high inland district between Wolverhampton and Bridgnorth, from which latitude they begin to diminish in size; but coarse gravel, composed of the same materials, is prolonged southwards like the tail of a delta, through Worcestershire until it dies away in the fine silt and gravel of the vale of Gloucester. *Not a fragment of any such detritus enters into the region of Welsh and Silurian drifts.*

[3] Memoirs, 91.

From the occurrence of remains of recent shells in so many places associated with this drift, he infers that the sea must have formed a strait separating Wales and Siluria from England, and that the drift in question was not driven over the land by diluvial waves but accumulated under the sea, and was brought to its present position by powerful marine currents. Against the notion that the larger boulders were in this district borne by ice rafts, he refers to the rounded and worn exterior of the boulders, and to the fact that they diminish in size and quantity in going from north to south.[4]

To these views Strickland gave his adhesion.[5] In a second paper, read before the Brit. Assoc. in 1837, Strickland says: "A considerable portion of the surface of this island is covered by patches of gravel and clay transported from distant localities by forces which are no longer in operation. On examining these masses of drifted material at various points, many remarkable distinctions of character are perceived. The presence or absence of certain varieties of rocks affords a clue to the direction of the forces which have brought them to their present sites. In some cases the pebbles and boulders are derived from the immediate vicinity, in others their origin must be sought for at a distance of many hundreds of miles. Sometimes these beds are found wholly unstratified; in other cases they are finely laminated, and a violent or tranquil state of the transporting current is inferred accordingly. Some varieties of drift occupy the summits of hills, and are independent of the present configuration of the surface; others occur on the sides or bottoms of valleys, with a constant relation to the present line of drainage. The former he distinguishes as marine drift, and he definitely concludes that the great mass of the erratic gravel occurring in England has been brought by a northerly current at a time when all or the chief part of England was under the sea.[6]

In the account of Berwickshire by Mr. A. Home, in the New Statistical Account of Scotland, published about 1835, we are told that across the moor in the upper part of the parish of Greenlow, there runs an irregular ridge called the Kaimes.

[4] Proc. Geol. Soc. II. 232, etc.; the pagination jumps from 232 to 333.

[5] Memoirs, 102.

[6] Id. 109.

It is upwards of two miles long, about fifty feet broad at the base, and between thirty and forty high. "In accounting for it we must remember," says Mr. Home, "that the stones scattered over the fields towards the Tweed consist principally of graywacke which must have been detached and carried thither from the rocks of the Lammermoor hills by the agency of a current of water setting in from the north towards the south. As the Kaimes are composed of similar materials reduced to a very moderate size, their formation is to be ascribed to the same cause. . . . The waters subsiding at the universal deluge or some more partial inundation, rushing with great impetuosity from the Dorrington Laws, which are about two miles northwest from the Kaimes, would carry all the wood and underwood with them till the collected materials would form a kind of dam or weir through which the waters could not penetrate, and would force up against the dam or weir the gravel and sand which form the ridge in the same manner as is frequently done by rivers in certain parts of their courses during great floods. The materials of which the ridge is composed, and its shape, which is somewhat like a horseshoe with the hollow towards the hills, favour this explanation.[7]

In 1837, Mr. D. Milne published an elaborate paper on the Geology of Berwickshire. In this he says: "Over almost every part of the surface of Berwickshire, and particularly in the eastern part of the county, extensive and deep accumulations of loose sand and gravel are found, consisting of waterworn fragments of rocks of every age and description. These gravel and sandy deposits are very generally in the form of high ridges running in an E.W. direction. Good examples of such may be seen at Hardacres, in the parish of Eccles; and at Kaimknow, betweeen Newton Don and Kelso. Sometimes they form well-rounded knolls, as between Lamberton Shiels and Marshall Meadows; these accumulations are also more frequent on the east side of the rocky hills in the district than on the western side. On the estate of Milne Graden, in Coldstream parish, there is a bank of gravel and sand, which runs for about a mile in an east and west direction, and is about half a mile in breadth. The sand lies over the gravel, and is between fifty and sixty feet thick. The

[7] New Stat. Acc. of Scotland, II. 41 and 42.

fragments embedded in the sand and gravel vary from small gravel to large blocks many tons in weight, the hardest are the largest, but all are more or less worn down to a globular shape. Large blocks of *mica slate* (a rock which does not occur in the south of Scotland at all, unless perhaps in Galloway) have been found near Dunse Castle, at Blanerne (east side), and at Coldstream. Blocks of granitic greenstone may be seen at Billie and at Blackburn, four miles off. These must have come from Cockburn Law, about eight miles W.N.W. from Billie. The characteristic yellow trap of the Dorringtons is found in Langton Burn, which is five miles to the E.S.E. The *Kyleshill* trap has been found at Pittenhill Quarry, six miles to the E.S.E. The felspar porphyry of Cowden knows-hill is strewed all over the plains which slope down from it to the eastward. Blocks of greenstone about two centuries ago covered all the country on the east side of Dunse.[8] They have long since been carried off. Similar blocks are spread over the country at Gordon, and can be seen sticking out of the moors to the N. and N.W. of Greenlaw. Blocks of cherty sandstone are found in the rivulet at Blackburn, while none of this rock exists *in situ* except at Cumledge, seven miles W.S.W. of Blackburn, and at Hordens, four miles further in the same direction.

"These facts," says Milne, "afford irrefragable evidence of the nature of the agent by which these alluvial débris have been formed; that agent could be no other than a great accumulation of waters, which moved chiefly from the westward. Other facts point to this not having been the only current. Thus fragments of coal a good deal water-worn are found at Coldstream and on the Tweed near Milne Graden House. The coal nearest this place is six miles to the south, the next nearest localities being Berwick, the Carter and East Lothian. The probability therefore is that these pieces of coal came from the south. Again, the northern face of a hill, two miles north of Spottiswood, is entirely free from soil, while on the other side it is covered with boulders and deep alluvial débris.

[8] In the border history an account is given of a *rencontre* between one of the Homes of Wedderburn, and a Frenchman of some celebrity who was slain by him near the spot. The Frenchman had fled; but the historian relates that his horse was impeded and obstructed by numerous blocks of stone, which lay scattered over this part of the country.

In a small valley near St. Abb's Head pieces of red sandstone occur in the débris, which have probably been swept from the northward, where they occur five miles off. These facts seem to warrant the inference that, whilst the general movement of the diluvial waters was from the westward, there were many minor or accidental currents which had different directions."[9] In a later page Milne, referring to the Kaimes already noticed, mentions there being more sand at their eastern extremity, where the ridge is lowest, than elsewhere. The highest parts consist of the coarsest gravel. "No one," he says, "can doubt the aqueous origin of these deposits, and that the water came from the west is shown by the gravel consisting almost entirely of graywacke."

In a letter addressed in 1838 by M. Lartet to M. Arago, and published in the Comptes Rendus, he describes the diluvian deposits of the Lower Pyrenees and the adjoining low country. This diluvium, which overlies the cretaceous beds, he derives from the Pyrenees themselves, and says it could only have been deposited by a great inundation which probably coincides with the upheaval of the ophitic system at the feet of the Pyrenees. He refers to this inundation as the Pyrenean flood. Its violence differed in different localities. It was most marked on the Gers (Auch) and the Baise (Mirande), where it tore up and denuded the calcareous beds, less so in the valley of the Gimone (Simorre), and still less so on the Save (Lombez). There is not a single granite pebble in the gravels which are found in these valleys. These gravels are like those on the plateaux, and have clearly been derived from the northern faces of the Pyrenees. In the valleys of the Adour, the Garonne, the Save, etc., which drain the central part of the chain, granite pebbles abound. The dispersion of the granite pebbles he attributes to a later flood, which was limited in extent by the mountains bounding the valleys last mentioned. In regard to the origin of the water of this flood, M. Lartet agrees with M. Elie de Beaumont in attributing it to the rapid melting of large masses of snow; and he accounts for the former larger deposits of snow by the increased elevation of the mountains in old days, urging that this Pyrenean flood

[9] Prize Essays of the Highland Society, v. 226-228.
[1] Id. 230-233.

completely destroyed all the animals living in the district which it reached.[2]

In 1838, Sefstrom published an important paper in the Transactions of the Swedish Academy on the furrows occurring on the Scandinavian mountains. This was translated into English by Mr. Thomas. He had noticed, he says, at Fahlun, where the mountain-slopes are largely uncovered and the smoke from the calcining furnaces prevents all vegetation of mosses, that the furrows have all the same principal direction from north to south. This agrees with the fact that fragments of the porphyry of Elfdalen, north of Fahlun, are to be met scattered about Fahlun, and that the blocks of granite and limestone met with in Pomerania and Brandenburg bear marks of having come from the Scandinavian granite mountains and the limestone beds of West Gothland. Sefstrom accordingly determined to see how far this was a general law, and he tabulates a large number of careful observations in various parts of Sweden, and the result was to show that there is a very considerable divergence in the direction taken by the furrows, and this divergence is not at haphazard but has a certain method in it.

He had noticed, like De Lasterye before him, that the northern sides of the Scandinavian hills are all rounded and destitute of peaks and angles, while on the southern side sharp corners and peaks still remain, and he had named the former, the opposing side, stötsidan, and the latter, the lee side (läsidan). He was now able to correlate the furrows with the other phenomena and to show that they run over the mountains between the two slopes, as if the steep northern part of the mountain had offered a certain resistance to the flood by which it was smoothed down, while the south side had been protected.

The flood to which he appeals he thus describes. "A mass of large and small stones, sand and gravel, by the action of water have been rolled and washed forward over the already exposed surface of the earth, by which these stones rolling against each other in their course, have produced the pebbles which lie collected in heaps on the extensive hills we call *äsar* (ridges) and of which the heaviest, less subject to roll

[2] Comptes Rendus, vi. 377, etc.

under the pressure of the weight of the masses of stone lying above them, have slidden over and around the surface of the mountain and furrowed it in the same manner as a polished surface of marble is furrowed by grains of sand under the pressure of the finger when drawn briskly over it. This flood he named a Boulder flood (Rullstensflod) from the boulders which show signs of rolling and are quite different from the erratic blocks with sharp angles seldom or never rolled, which he says appear to have been conveyed by an agency quite different, and he further suggests as a scientific name for this flood the Petridelaunian flood, from πετρίδιον a little stone, and ἐλαύνω I roll forward.

In regard to the date of this flood he urges that it must have been later than the rocks which are furrowed. On the other hand, the fact that Napoleon's expedition found the quarrymen's marks still fresh on the quarries whence the stone was got for the Egyptian pyramids, shows how long-lived such scratches may be. The fact, again, that at the great and little falls of Avesta, uncommonly beautiful furrows exist which make an angle of from 75° to 86° to the present direction of the Dalelfven, which itself passes over them carrying sand, stones and gravel without having succeeded in abrading them, justifies us in attributing them to no very recent period. That the contour of Scandinavia has not altered since the boulder flood, is shown by the subordinate or side furrows agreeing with the present slopes of the mountains, and if any change has taken place, several combined localities must have been elevated above their former level, for neither large nor small fissures or faults are to be found. Older fissures which are filled with trap often appear, but the trap is furrowed equally with the softer rock.

Sefstrom not only attributed the distribution of the boulders to his flood, but also the crushing to a fine sand of enormous masses of hard rocky substances, such as gneiss and granite, with which wide districts of Europe are covered, often to a considerable depth, and also the formation of the so-called giants' cauldrons. Hence he postulates that the flood must have been long continued.

That the boulder flood was of great violence he argues from certain facts, such as the removal *up hill* of large rocks at

Blekinga from the beds where they once rested and the shifting of such an enormous stone as that near the church of Pelarne in Smaland, which, it is calculated, weighs seven millions of pounds, and of the various large stones in Germany and Denmark which have come from Scandinavia. The fact that the upper layers of the solid rocks now covered by loose materials are so often found shivered and broken, while in other cases they have been removed and reduced to fragments, is further evidence of the same fact. "It is besides," says Sefstrom, "very commonly to be seen that the boulders are conveyed upwards over the mountain slopes, which have from 33° to 40° inclination, to effect which not a little force must certainly have been required."

Since the furrows have been found twenty-one feet below the level of the sea at Carlskrona and more than 1500 feet above it near Särna in Dalecarlia, the depth of the boulder flood must have been very considerable.

In regard to the direction of the flood he points out that the evidence is contradictory. The direction of the furrows round Fahlun, Uddevalla and Westervik, seem to show the flood came from the highest land and flowed down on all sides. Snähättan, N.W. of Christiania, is the highest land in Scandinavia, the furrows at Christiania, however, go to the S.W. instead of the S.E. At Gefle and on the northern coast of Uppland again, the furrows run to the S.W. and not down hill but ascending, and at many places over very steep slopes and hills. The same is the case east of Carlskrona. These facts point to the flood having had a generally N.E. direction. When we examine mountains such as Omberg, Taberg, Billingen, Kinnekulle, and Stora Tiveden (east of Finn rödja church), we find that while on the summits the flood, as measured by the furrows, has followed the general direction, near the foot of the mountains they follow the irregularities of the country.

In regard to the äsar, Sefstrom says the lesser ones are oftenest found on the lee side of some rock. They generally follow the line of the furrows, but are constant in direction, whereas the latter diverge and converge, and he illustrates some of their abnormal shapes by those of the sand and gravel banks continually forming under our eyes in streams and by great floods.

In regard to the cause of the flood, he states that until further observations have been made it is not possible to say whether the boulders have themselves been impelled against the solid land by a *vis a tergo*, or whether (by the stoppage or alteration of the earth's revolutionary movement about its axis) it was the earth which moved against the boulders, but he seems to favour the latter view.

Sefstrom did not content himself with an examination of his own country. In 1836 he travelled through Germany and Switzerland, making observations on the erratic phenomena. He mentions the abundance of drifted flints in Pomerania, while at Montzberg, near Dresden, he remarked that the solid rock seemed much battered by the boulder stones, but there were no furrows. So far as could be judged from the phenomena of the lee side and the stoss side in the mountains of Saxony, notably of that of Winterberg, south-east of Schandau, the boulder flood went from north to south there. So it was in the country higher up the Elbe, from Teschen to Toplitz, where the boulders of sandstone, basalt, gneiss and porphyry are found to the south of the same rocks *in situ*. Round Prague, and as far south as Budweis and Linz, are boulders evidently, he, says, of a northern origin, notably blocks of pure quartz which have travelled from the high mountain of Jeschen, south-west of Reichenberg, in the extreme north of Bohemia, but neither there nor in the Saxon Switzerland did he notice any furrows.

In the Austrian Alps he found both the granitic and calcareous hills with rounded tops, and he makes the shrewd remark that the reason why the Alps in landscape drawings have always a craggy appearance arises from the artists generally choosing a deep valley for their station, from which they cannot see the plateau; and he also remarks that the Archduke John held that these plateaux are distinctly furrowed on their rounded surfaces, often to the depth of the cart-ruts of a public road. In the Taunus mountains he also noticed smoothed and rounded surfaces, except on the higher peaks.

As a proof that the boulder flood had traversed the Hartz, Professor Hausmann quoted to him the fact of certain rents occurring in the shell limestone, which he attributed to it. At

Berlin it was mentioned to him by Professor Rors, how, at the limestone quarries of Rüdersdorf, east of the city, on removing the covering of earth, the upper surface of the limestone beds was furrowed and polished.

Lastly, in regard to Finland, he tabulates some observations made by M.M. Hiriakoff and Rochette in 1836, showing that the furrows there, instead of following the general course of those in Sweden, follow the lines of the Finnish water-courses, both those which fall into the Gulf of Finland and those which fall into the Gulf of Bothnia.

The Danish geologist, Forchhammer, divided the drift deposits of his country into those formed of clay and those of sand, both containing stones. In both cases, he tells us, the beds contain marine débris. In the former of deep-sea, and in the latter of littoral species, and he argues that this can only be accounted for by their having been deposited by a greatly agitated sea. Hence he also explains the irregularity and frequent interruptions of the beds, the longitudinal slope of the hills or kames, as that between Heslov and Lystrup in North Zealand, a similar one between Nestved and the hamlet of Mogenstrup, the hillocks of Refsnoev in the northern part of Samsoé, and of Helgenoes in Jutland. He thus explains why the boulders prevail in the Swedish ridges, which are nearer their original home, while sand prevails in those of Denmark.

Elsewhere he points out that in some countries the largest blocks are found near the sea, while in others they diminish in size as we leave the mountains. He points out how areas occur several leagues in extent in which they do not occur while all around they are abundant, and that while most of them are rounded, a larger number, especially those which are buried, have their angles sharp; and he is very emphatic that all the blocks found in Denmark and Germany came from Sweden and Norway, those of Prussia and Poland from Finland, while those of North Russia, as far as the Niemen, came from either Finland or the high land about Lake Onega. He says that there is no doubt some of the blocks in Yorkshire came from Norway. "Others," he adds, "came from Labrador, while others came across the English Apennines."

In Holland are sand hills fifty metres high above the sea-level. In the sands as far as the Scheldt are found numerous erratic blocks which came from the north. About Amersfoort and in the province of Utrecht are blocks of granite and porphyry like rock found *in situ* in Scandinavia. They increase in size in going towards East Friesland, as they do in going from Scania to Smaland. About Mooker, on the other hand, are blocks of pyromacic silex which could not have come from Scandinavia. These blocks are not dispersed, he says, along definite lines, but seem as if they had been distributed like the spray from a fountain in semi-circular lines.

In regard to the äsar or sandäsar, he says, that while in Sweden they generally range north and south, in Finland their direction is more variable and sometimes they cross each other. He attributes to the action of stones borne along by water the polishing and striating of the rocks in Sweden, the striations being generally parallel, and from north to south. In other countries they go from west to east, and sometimes affect different directions on the same mountain. All the heights between 40° and 43° of long. and 60° and 62° of lat. are rounded and deprived of salient angles, and in the case of granite are rounded like the successive waves of the sea, and when protected by gravel are polished so as to reflect the sunlight, and marked by scratches.

Leonhard and others had argued that these phenomena were caused by violent invasions of the sea, which after inundating Scandinavia and Finland, transported the débris far and wide. Forchhammer somewhat modified this view in supposing that the rounded outlines of the Scandinavian hills are due to the ordinary action of the waves when the country was submerged, but he appeals to violent upheavals and subsidences such as are evidenced by the faulted and broken strata of the isle of Moen, the inclined beds of Halgoland, etc., as setting in motion the necessary currents to transport the blocks as we find them.

Let us now turn for a short space to America. Dr. Jackson, in his remarks on the geology of Maine, writing in 1838, says, "How much I should be delighted could any of those geologists in Europe, who feel sceptical on the subject, visit with me the numerous and most decided proofs which are presented in Maine, demonstrating the general direction and power of the diluvial current. I think that no reasonable person would venture to stand in opposition to the facts which we can present in illustration of this doctrine. . . . I have found strong diluvial marks extending across the whole state of Maine, and the transported boulders are always traced to their origin in the direction of the furrows worn on the rocks. . . . . The direction of the diluvial current in Maine is proved to have been from N. 10° to 20° W., to S. 10° to 20° E., the mean being N. 15° W. to S. 15° E."[3]

In an elaborate geological report on Maine published in 1839 by the same geologist, he says, *inter alia*: "In various sections of this report may be seen recorded the proofs of diluvial transportation of rocks, far from their parent beds, and we have every reason to believe that this removal was effected by a tremendous current of water, that swept over the state from north 15° W. to south 15° E.; and we have adduced in testimony that such was the direction of that current, numerous grooves, furrows or scratches upon the surface of the solid rocks, in place, and have shown conclusively that the rocks that we find thus transported were portions of ledges situated to the north of the localities where their scattered fragments are found." Dr. Jackson then goes on to urge that the fact of rocks only weighing half as much when immersed in water makes the transporting of big blocks much easier to realize, and he continues: "From the observations made upon Mount Klaadu, it is proved, that the current rushed over the summit of that lofty mountain, and consequently the diluvial waters rose to the height of more than 5000 feet. Hence the ancient ocean which rushed over the surface of the State was at least a mile in depth . . . we find that the whole mass of loose materials on the surface has been removed southwardly . . . all the markings on the surface of the rocks, and the scattered boulders of granite and gneiss which abound in that soil, indicate its origin to have been in the north 15° or 20° W. . . . Diluvial grooves on the rocks are exceedingly common in Maine, but I know of few localities where they are so distinct as at Hope and Appleton. Here they may be observed running in a N.W. and S.E. direction, while they are very deep and perfectly defined.

[3] Silliman's Journal, xxxiv. 71-72.

Their direction, it will be remembered, does not coincide with that of the stratification of the rock, and could not have resulted from disintegration of the different strata. Three-quarters of a mile S.E. from a hill in Appleton, they may be seen forming deep channels in the rocks, to the depth of a foot, and six inches in width. Since the direction and appearance of these grooves correspond with those observed in other parts of our country, I feel no hesitation in attributing them to a similar origin. They are certainly the results of an aqueous current, which once prevailed over New England, and probably over the whole world. The current, from similar grooves seen in other places, appears to have proceeded from N. to S., or from N.W. to S.E." [4]

In regard to Illinois, Dr. Shepard, writing in 1838, says: "The occurrence of boulders on the rolling prairie had often been mentioned to me, under the significant and original appellation, bestowed upon them in this region, of 'lost rocks.' Their abundance, however, surpassed my expectation. They first attracted attention soon after leaving the twelve-mile house from Chicago, and appeared to form a belt between a quarter and half a mile in width, whose direction was N.W. and S.E. In crossing this belt it was uncommon to pass many rods without encountering a boulder. In general they were rather more than half buried in the soil. They varied in diameter, from ten inches up to three feet, and belonged to the following species of rocks; granite, granitic gneiss, and trap. A few detached boulders only were noticed between this deposit and Ottawa. Soon after leaving this place, another batch of them was passed; they were scattered over the Illinois bottoms so plentifully, as to prove objects of no inconsiderable annoyance in the road. . . . What serves materially to heighten one's interest in these boulders is the consideration that they must have been transported over a distance of between two and three hundred miles, since the southern shore of Lake Superior is the nearest region affording rocks of a similar character, *in situ*." [5]

From a report on the geology of Indiana, published in the same year, we gather that boulders of primitive rock are

[4] Silliman's Journal, xxxvi. 153-4. [5] Id. 141, 142.

numerous in that State, where they are called by the expressive name of lost rocks, also grey heads, and negro heads.[6]

In 1839 Dr. Hays published a letter in Silliman's Journal, which accompanied some specimens of rocks showing grooves. In regard to one of these he says: "You will perceive, that wherever a nodule of chert projects above the surface, a ridge of the softer limestone has been protected, in some measure, from friction, which invariably at this locality, as at the Black Rock quarry, one and a half miles distant, points in a southerly direction . . . . the nodules of chert often have a semicircular depression worn into the rock on their northern sides, opposite to the projecting ridge . . . . . Can proof be more conclusive, that these marks and scratches were produced by gravel stones and boulders, swept over the surface of the rocks by currents, tides or waves, which flowed from the north?"[7]

Reverting once more to Europe: on the 10th of August, 1840, M. Durocher presented a famous memoir to the French Academy, entitled "Observations sur les phénomène diluvien dans le nord de l'Europe." In this he summed up the results of long and careful researches in Russia, Poland and Finland. He especially examined the furrows and fissures on the northern rocks, from Altea on the Arctic sea to the south of Finland, and from the gulf of Bothnia to Lake Ladoga. These striæ, he says, are found everywhere in the north, where the rocks are sufficiently hard and fine-grained to have preserved them. They vary in Finland from a length of a few inches to a foot or more. They are often parallel and the larger grooves are often themselves traversed by finer ones. Sometimes on the same rock two sets of striæ are visible with an angle to each other of ten to twelve degrees. Sometimes there are deep striæ in one direction and fine ones in another. Sometimes the deep striæ intersect one another. The direction of the striæ is quite independent of the stratification. They occur on unstratified as well as stratified rocks and often make an angle with the dip of the strata. They occur less frequently and less regularly on irregular ground than when the surface is more or less regular. On perpendicular faces of rock the grooves are sometimes horizontal, as

[6] Silliman's Journal, xxxiv. 194.   [7] Id. xxxv. 191.

on the borders of Lake Ladoga, a fact previously observed by De Saussure on mount Salève. The mammillated rocks of Finland are entirely covered with scratches, those on the summits of the hillocks having a normal direction from N.N.W. to S.S.E., while the others have a tendency to follow the sinuosities of the surface. The surface of these rounded rocks is always polished, and they generally have a curved outline elongated in the direction of the striæ. It is only on the low hills of Finland that the whole surface is striated. On the higher hills above 100 feet in height, as Sefstrom and De Lasterye showed, the faces on the S.S.E. have preserved their original roughness. Although the normal direction of the striæ is pretty constant they occasionally and in certain localities, as notably between Brakestad and Gamle Carleby in Finland, diverge considerably, a fact also noticed by Böhtlink. The normal direction of the striæ is very nearly that of the valleys and lakes, and this both in Sweden and Russia. When the erratic detritus ceases to lie in continuous mounds and is broken up into hillocks, etc., the alignment of these in Russia corresponds in direction with the normal direction of the striæ, as Brongniart had shown that they do in Sweden. In Finland and Lapland the loose deposits are mostly unbroken and spread in a continuous mantle, in many places thirty and forty feet thick.

In regard to the erratic blocks, he remarks that all through the north of Europe they occur not only on the surface but buried in the sand, and when so buried their axes are often inclined to the horizon or even vertical, as if they had fallen all at once and been buried in the sand below. They occur most frequently on hillocks or high ground ranged in lines or bands. The larger blocks are never rounded; some of them have their angles exceedingly sharp, and when they are blunted it may be due to atmospheric erosion. In Finland and Lapland the blocks are of local origin, but north of Lake Ladoga there are blocks of a rock not occurring *in situ* at less than a distance of fifteen leagues further north. M. Durocher discusses the provenance of a large number of blocks found in Poland and North Germany agreeing with the conclusions already cited. He says that as we travel westward from Poland the blocks from Finland decrease and those from Sweden increase, and that on the zone between Tilsit and Berlin they are much mixed. In Denmark the blocks are generally of Norwegian and Swedish origin. M. Durocher remarked that everywhere large blocks of granite exceed greatly in number those of gneiss, although gneiss is the prevailing rock; and he accounts for the fact by the tendency of gneiss to split up into fragments. He remarked further that, taking the original site of each kind of stone out of which the erratics are formed as a centre, the present distribution of the latter occupies a third of the circumference of the generating circle, or even one half, so that some of the stones have moved at right angles to others. He quotes as an example that the *rapaikiwi* (a peculiar granite) of Wiborg is found from Kostroma in Russia as far as Pomerania.

The granite blocks have travelled further than the rest, and the calcareous blocks the least. The blocks are ranged not only in a fanlike way but also in concentric zones. Generally speaking, the range of the blocks may be measured by a section of a circle with a line joining Stockholm and Moscow as a radius and the former place as the centre. As we near the circumference the blocks diminish in number both in Russia and Poland, while in Saxony and Silicia, where the curve is broken and indented, they stop short more suddenly. This Durocher accounts for by the fact that in the former district the transporting agent gradually exhausted itself, while in the latter it was abruptly stopped by the mountains. The blocks over the vast area of Russia, North Germany and Poland are not all far-travelled erratics, some are derived from the rocks *in situ*.

When the travelled blocks cease the local blocks continue, as do also the beds of sand and soft material in which they are embedded, so that there is a zone encircling the outer margin of the transported drift composed of local materials otherwise continuous with it.

In regard to the cause of the varied phenomena thus described, M. Durocher distinguishes between the agent which transported the boulders and that which made the striæ and gave the äsar their shape and alignments. The former he attributes to ice, the latter he assigns to the operation of a diluvial current starting to the north of Scandinavia and sweeping over Lapland, Sweden and Finland in the direction marked by the normal direction of the striæ, etc., and explains the local variations by the deflecting influence of mountains and other barriers. His reason for putting the focus of disturbance to the north of Scandinavia is, that the striæ which further south have somewhat different directions, when we reach the 70th parallel in Lapland and in Finland take almost the same direction, and that they can be traced not only on the uplands but on the northern face of the rocks which border the polar sea.[8]

I do not propose to carry this part of my story any further, nor, in fact, would it be easy to do so, for at the epoch we have reached there arose another school of writers, who would have nothing to say to diluvial theories, and whose vigorous blows and acute rhetoric caused a very rapid revolution in the whole course of geological thought.

It will be noticed that among the names quoted in this and the previous chapter there occur nearly all those who are remembered as the fathers of geology; yet in the course of half a dozen years the views they had in various ways maintained in regard to the drift phenomena were almost entirely displaced, and one at least among them, the very standard-bearer of the diluvial school, went over to the enemy and became an advocate of the new views. Among those who remained staunch to the old creed the names of Sedgwick, Phillips, Von Buch, Durocher, and Sefstrom are the best known, but they preached in vain to the younger students of geology. The champions of water had to give place to the champions of ice. How this came about will be described in the succeeding chapters.

[8] Comptes Rendus, xiv. 78-110.

CHAPTER IV.

THE PHENOMENA OF THE DRIFT.

The Champions of Floating Ice.

"Saxa singula, petris primordialibus constituta, per Poloniam septemtrionalem dispersa, incolis *lapides Suecicos* appellari; quam notitiam auditori studiosissimus L. Zeuschner, Polono, debeo. Utrum Poloni in Suecia, an Sueci in Polonia Saxorum illorum cum petris Suecicis similitudinem agnoverint, decernere non possum; non improbabile autem videtur, denominationem illam, a veritate geologica non abhorrentem, e bellis, inter nationes illas gestis, originem suam traxisse."—HAUSMANN, Ac. G. A. Comm. Rec. VII. 15-16.

Tilas, 1740-2—Bergmann, 1769—St. Pierre, 1784—Dolomieu, 1793—Mierotto, 1790—Wrede, 1794—Von Buch, 1808-1811—Bakewell, 1815—Sir James Hall, 1815—Venturi, 1817-18—Dick, 1819—Horace H. Hayden, 1820—Bigsby, 1821—Eaton, id.—Maclure, 1822—Hitchcock id.—Dobson, 1825—Fischer, 1828—Lariviere, 1829—Kloden, id.—Lyell, 1830-1845—De la Beche, 1833—Murchison, 1835-1841—Bayfield, 1837—Cordier, 1837—Darwin, 1839-1841—Wissman, 1840—Hays, 1843.

SCIENTIFIC men naturally shrink from appealing to abnormal causes. It is of the nature of science that we should endeavour to explain the operations of Nature by invoking phenomena and laws which we see at work around us rather than by remote or occult causes. When it began to be seen, therefore, that it is impossible to account for the transport and distribution of the boulders and to explain their associated phenomena by a reference to water acting in its usual way, and that especially in North Germany and in other countries remote from mountains and impetuous rivers we must, if we appeal to water at all, have recourse to its action in a quite abnormal way, men began to turn elsewhere to find some explanation which should not involve other than forces actually at work and working in a normal way. It was natural that those living near the Baltic and in North America, who had opportunities of noticing the tremendous force exercised by shore ice and by floating icebergs, should turn to these causes and supplement them by theories of wide-spread submergence, which seemed also warranted by the finding of marine shells at a considerable elevation in many places.

Among the first to record such observations was the Swede, Tilas, who suggests that some of the larger stones found in Sweden were probably carried by floating ice which came from the Pole.[1]

T. Bergmann, writing in 1769, speaks of the way in which ice being frozen hard to the ground detaches from it, occasionally, earth and stones, and when itself loosened bears them away and deposits its burdens elsewhere. Thus he mentions how a big stone, eight to ten feet high and nine feet thick, which lay near the church of Kalajoki a foot above the water, was taken up by the ice, moved over the land, and then so deposited on some smaller stones that a man could creep under it. In the year 1760, the ice turned over a stone twelve to fourteen feet in length, at Breitgrund, and the year following it was moved still further.[2] While he points out how ice can occasionally move great stones, he urges that we cannot explain by its action the raising up of huge blocks and their deposition on the summits of high mountains.

St. Pierre, in his "Etudes de la Nature," published in 1784, argues that the tides of the ocean are largely due to the alternate thawing of the ice at each pole of the earth,[3] and he urges that if this melting were to take place suddenly or rapidly it would cause a general inundation, just as rapid melting of the snows on the Andes gorges the Amazon and the Orinoco, and the similar melting of the snow on the Mountains of the Moon floods the Senegal, the Nile, etc. He argues that at the time of the great deluge, the sun left the ecliptic and advanced further north and south, and in consequence the cupolas of ice at either pole and the great

[1] Ante, p. 9. [2] Phys. Besch. der Erdkugel, 399. [3] Op. cit. i. 240.

glaciers on the high mountain chains rapidly melted, and innumerable torrents, escaping by the various outlets of the polar basin, swept into the Atlantic charged with masses of ice, carried with them wreckage from the land and sea, broke down mountains and transported rocks, sand and gravel. To continue in our author's own words, "Ce fut alors que tous les plans de la nature furent renversés. Des iles entières de glaces flottantes, chargées d'ours blancs, vinrent séchouer parmi les palmiers de la zone torride, et les elephants de l'Afrique furent roulés jusques dans les sapins de la Siberie, où l'on trouve encore leurs grands ossements," etc.[4]

Dolomieu, in discussing the question of the transport of erratics, tells us how he had seen ice transport to great distances stones which running water could not move.[5]

De Luc reports Mierotto as saying he "had had opportunities of seeing in winter, on the coast of Pomerania, that while the sea is at its lowest level, the ice found on its shores encloses some of the blocks. Afterwards the level of the water rising and the ice breaking on the borders, detached masses of it containing these blocks are cast up on the strand, where, when the ice melts, the blocks remain. He argued from this that formerly the whole north of Europe was covered by a sea ,which was kept at a high level by a succession of barriers of rocks, forming a kind of dike around it . . . . that in consequence of ruptures in the dike, portions of this sea rushed down with impetuosity into lower basins, the dikes of which, in their turn, were also broken and that the blocks in question are fragments of these dikes . . . . the fragments were brought down by the rush of waters, having been carried to the borders of each basin by ice in the manner above described."[6] Elsewhere he reports the same geologist as urging that "the masses of stones scattered in low situations were fragments of dikes, whereby superior seas had originally been retained in successive basins, one lower than another, each of which had been emptied in its turn by the rupture of its dikes; that this having extended all the way into Sweden, masses of the granites and porphyries of that country had

[4] Id. 245-248. [5] Obs. sur la Physique, xiii. 59, note.
[6] De Luc's Travels, 1813, i. 420, 421.

been brought over to the lands on the south of the Baltic, by the ice of that sea, at its successively lower levels.[7]

Mierotto was presently followed by Dr. Wrede, the author of two remarkable works, one published in 1794, entitled "Geologische Resultate aus Beobachtung ueber einen Theil der Südbaltischen Lander," Halle, 1794; and a second, entitled "Geognostiche Untersuchungen ueber die Südbaltischen Lander."

His views were based chiefly on the facts observed in Scandinavia and north Germany. He was of opinion that a change has taken place in the centre of gravity of the earth which has been moving southwards and has caused a corresponding movement of a large part of the watery envelope of the globe to the Southern Hemisphere. He urges that formerly the Baltic was a much more extensive sea than it is at present, and quotes with approval the notion of Von Buch that the flat country between the Oder and the Volga, between the Baltic and the Carpathians, and between the Black Sea and Finland was once a vast sea.[8] Having postulated this sea he goes on to discuss in great detail the various modes in which the erratics reached their present position in the country south the Baltic, and continues on page 71 of his first work in these words: "In our experience we know of only one means of by which these effects can have been produced, namely by ice, in which granite blocks of this size were frozen, which were then carried away until the ice was pounded into fragments and eventually sank with its load." In support of this view he quotes a statement of H. Götze, vol. ii. p. 237, that annually blocks of granite, sometimes of twenty hundredweights, are detached in the Hartz mountains, etc., and are carried in the ice, in which they are fast, and removed for several miles. Plouquet, in his Swiss Travels, 1786, tells us the district between the lake of Brienz and that of Thun is strewn with débris similarly deposited by the Lütschine, and argues that both those lakes will eventually be filled up; the Rhine similarly carries blocks into the Boden lake and the Rhone into the Genser lake.[9] Dr. Wrede urges that if it be argued that the erratics are too large to have been carried by such icebergs as are now detached from the mountains of the Hartz, of

[7] Id. ii. 43. [8] Geog. Unter., etc., 39. [9] Op. cit. 71, 72.

Switzerland or Norway, that the higher mountains of former days would have greater torrents which would bear larger masses of ice. He calculates that inasmuch as the specific gravity of ice to salt water is as 6 to 7, and to fresh water as 8 to 9, the mean being as 7 to 8, that a block of granite weighing 389,078 lbs. might have been carried by a raft of ice 55,683 cubic feet in extent, and he shows that a depth of water of forty-four feet would amply suffice to float it.[10] If the stones, instead of being more or less cubical in shape, were spread out so as to be about the same thickness as the ice, then a much smaller depth would suffice to float the bergs with their loads. He urges that if only one of each group of blocks of ice which crowd the present streams in spring were to carry a stone, there would speedily accumulate a considerable deposit. He also attempts to show how, under the conditions he names, large stones would sometimes be rounded and polished. His argument does not imply that the erratics south of the Baltic came from Scandinavia specially, but from the various districts in the north where rocks like those from which the boulders were separated are found.

Similar views, it would seem, had also been propagated elsewhere. Von Buch tells us expressly that, according to those most worthy of credit in Switzerland, the blocks of stone had fallen from their original position upon icebergs which floated in the sea then covering that country, which had been driven by currents to the Jura, where they deposited their loads. Others deemed this view of immense icebergs inadmissible, and invoked natural rafts of trunks of trees like those sometimes seen on tropical rivers as the transporting agents. Von Buch urges the impossibility of supposing that such icebergs or rafts would all float in the same direction. They again would have deposited their loads at one level, namely, that of the then sea-level, and not at different elevations.

Bakewell, in his "Introduction to Geology," second edition, published in 1815, says vast "masses of rock near the sea-shore are sometimes enveloped in fields of ice, and raised up and transported to distant countries. Ice is specifically lighter than water; every cubic yard will support a stone of 100 lbs. weight;

[10] Id. 71, 72.

hence we need not be surprised at the insulated rocks of granite, that are sometimes found in situations far remote from primary mountains. These blocks have been floated over the ocean, and their angular points and edges defended from attrition during their passage by the surrounding ice. In this manner large fragments of granite, and other primary rocks, may have been brought upon our coasts from Norway and Greenland."[1]

Sir James Hall, writing in the same year, says he is induced to retain part of the system proposed by Wrede, involving the transport of blocks in some cases by ice. Having referred to the case, to be presently cited, of the Pitly stone, he goes on to say that a wave sweeping over the Alps would meet with many stones thus embedded in ice, which it would move along.[2]

He elsewhere urges that the fact of the blocks in Germany and the Baltic islands being found on hills of sand and not on the low ground, points to "these hills then constituting shallow sandbanks, which would afford the first resting-place for the floating bergs of ice which, grounding upon them, would accumulate to very numerous assemblages, and there deposit their granitic charges, while all other similar blocks flowed onwards in the deep water between."[3]

In 1817 and 1818 the Cavaliere Venturi published some papers in Brugnatelli's "Giornale de Fisica," in which he discussed the mode of transport of the Alpine blocks, and after quoting the views of Wrede and Hall with approval, he says that the main cause of the transport of the greater number of them was ice. When the sea was several thousand feet higher and bathed the feet of the Alps, the glaciers coming down the Alpine valleys carried stones; when these glaciers reached the sea they broke off into icebergs, which were blown along by winds and taken by currents, and thus deposited primitive blocks from the higher Alps on other strata."[4]

Mr. Thomas Lauder Dick read a paper in 1819, before the Wernerian Society, on a famous boulder known as the Travelled stone of Pitly, measuring about four to six feet in height and six to seven in thickness and height, and weighing about eighty tons, which, up to the 19th of February, 1790, served as the

[1] Op. cit. 55
[2] Trans. Roy. Soc. Ed. vii. 157, 158.
[3] Op. cit. 160.
[4] Op. cit. decad. 1. vol. x. 297.

march stone between the properties of Castle Stuart and Culloden. It lay just within the flood mark, near where a little stream empties itself into a shallow sandy bay. There had been a long-continued frost, which had frozen the upper part of the bay, and a thickness of about eighteen inches of ice clasped fast the stone. On the evening of the 19th of February, a hurricane arose and broke up the ice. In the morning the miller at the mill of Pitly missed the stone, and he called out to his wife that the "*meikle stane was awa,*" and there remained but a hollow pit in the sand with a long shallow furrow stretching from the pit towards the sea. Presently, when the tide ebbed, the stone was found lying out on the sands, 780 feet from its old position, and around it was still a cornice of ice. "In its new position," says Hugh Miller, writing in 1859, "the stone still lies; and only a few years ago, mayhap still, a wooden post which marked the point where the two contiguous properties met, marked also the spot from which, after a rest of ages, it had set out on its short voyage." [5]

The same graphic writer refers to a second instance of a similar kind which occurred at the beginning of the century on the eastern coast of Sutherlandshire. "Near the small hamlet of Toridal, on the upper part of Loch Fleet, there stood," he says, "about fifty years ago, a rude obelisk of undressed stone, generally regarded as Danish. It stood in a swampy hollow, which was frozen to the bottom. On the evening before the incident to be related, a drunken Highlander, returning home put his bonnet on the top of it and left it there. The next day the thaw came, the river rose, the ice-cake round the obelisk floated high above the level, wrenching up the obelisk along with it, and both floated down the loch to the sea. In the morning report spread that there was a man standing in the middle of the loch on the floating ice, being swept out to sea. A boat put out from the shore, when the supposed man was found to be the obelisk with a bonnet on it, and bonnet and obelisk were left to find their way to the German Ocean, where both now lie." [6]

Horace H. Hayden, in his "Geological Essays," published in 1820, while assigning a great current as the cause of the

[5] Geological Sketch-book, 69, 70. [6] Id. 71, 72.

distribution of the American drift, also invokes the assistance of ice. Thus, he says, "it is well known that large masses of rocks of different kinds are often enveloped in ice, and by the freshes of rivers or tides of the sea are raised and transported to great distances, before they are disengaged." Tilloch had observed that masses of stone are sometimes transported by cakes of ice in which they happen to be enveloped, and Hayden goes on to say that just as white bears have been known to be transported to Iceland by icebergs, so might great masses of rock be similarly transported.[7] He elsewhere describes the floods which took place on the Connecticut after the severe winter of 1780, when great masses of ice were scattered over the country on either side of it. "When this was accomplished, the following season large pieces of rocks and heaps of rolled pebbles were left exposed to view on an alluvial surface, on which, before, a stone could not be found for its weight in gold. These rocks and stones, from their characters, were known to be the same as those which composed the bed of the river many leagues above." And in discussing the great blocks found on the surface of Massachusetts, he urges that they were borne along by water when encased in ice, "for such is their form that if embraced in a mass of ice of sufficient size, and elevated from the earth by water, they might be floated over the ocean without the chance of escape, except by the melting of the ice."[8]

The intervention of ice in carrying the blocks he urges as another reason for bringing his diluvial currents from the north, "for," he says, "it would be absurd in the extreme, to suppose that masses of ice could have been found in the torrid zone, or even within the temperate zone, of sufficient magnitude to have transported those rocks *from the south* to their present position. . . . This opinion receives additional support from the following circumstance: the huge fields and mountains of ice, which every spring float down the Atlantic Ocean, seldom reach the latitude of from 39° to 40° (which is about the latitude in which those rocks lie), before they become so *weak and rotten* as to be incapable of supporting or retaining any considerable weight that might be attached to them. It therefore follows as a natural consequence, that by the time the masses of ice

[7] Op. cit. 86. [8] Op. cit. 89-91.

had reached those latitudes, they must necessarily have discharged the entire balance of their weight, if they contained any."[9]

"In the spring," says Dr. Bigsby, writing in 1821, "the ice occasionally removes fragments of great size. During the winter it surrounds those which are placed in the shallows, and on being broken up in April by mild weather and a casual rise of water, it carries them to some other shore. Remarkable instances of this are found on the islets near the south end of St. Joseph, where a few yards from the water, and little above its level, are deposited rolled stones *some yards* in diameter, with a furrow extending from them to the water, most probably tracing the last steps of the route to their place of rest."[1]

Professor Eaton mentions how on one occasion the ice of the Hudson, with gravel on it, in breaking up, pressed against the shores, and a large cake of it, pushed by others from behind, was squeezed up to a height of thirty-four feet above the water, when it melted and "left the gravel which it had transported from the northern countries on the bank."[2]

Dr. Hitchcock states how he had known masses of rock, from one to ten feet in diameter, raised from the bottom of a river by ice and removed down stream one or two miles.[3]

Maclure writes, in July, 1822: "The large masses of granite, some of them weighing tons, scattered over the secondary rocks between Lake Erie and the Ohio, while there is not an atom of granite in place nearer than the north side of the lake, would seem to point at the only mode by which they could have been transported; by supposing the lake extended thus far and that the large pieces of floating ice from the north side might carry those blocks attached to them and drop them as the ice melted in going south; few or none being found south of the Ohio, shows that the southern sun melted the ice before it got so far."[4]

In Hitchcock's paper on the geology of the district adjoining the Connecticut river, I find him saying: "At the outlet of the Connecticut through the mountains below Middletown, and

[9] Id. 95.
[1] Silliman's Journal, iii. 256.
[2] Id. v. 21.
[3] Id. vi. 37, 38.
[4] Id. vi. 102.

600 or 700 feet above the present bed of the river, I saw rounded masses of old red sandstone, several inches in diameter, mixed with the fragments of the rocks in place. Such a fact I never noticed at any other place in the primitive region along the river; certainly not on the east side of it. I was led irresistibly to the conclusion, that they were conveyed thither by the ice of the ancient lake, which would be floated to the ocean through this inlet."[5]

In a letter from Peter Dobson to Professor Silliman, written on Nov. 21st, 1825, we read: "I have had occasion to dig up a great number of boulders of red sandstone and of the conglomerate kind, in erecting a cotton manufactory, and it was not uncommon to find them worn smooth on the under side, as if done by their having been dragged over rocks and gravelly earth, in one steady position. On examination, they exhibit scratches and furrows on the abraded part. . . These boulders are found not only on the surface, but I have discovered them a number of feet deep in the earth, in the hard compound of clay, sand and gravel. One block, of more than thirty hundredweight, washed and worn as above described, was dug out of a well, at the depth of twenty-four feet.

"Boulders with these marks upon them I have observed not only in this town (i.e. Vernon, Con.), but in Manchester, Ellington and Wilbraham. I think we cannot account for these appearances, unless we call in the aid of ice along with water, and that they have been worn by being suspended and carried in ice, over rocks and earth, under water."[6]

Fischer, in a paper published in 1828, entitled, "Ueber die Gerollen des Nordens," attributes the portage of the erratics largely to ice. He cites Scoresby's observations on icebergs, and urges that in former times, when the climate was colder, such bergs were floated in a great flood of water, from the north or otherwise, on to the flat lands of Germany, and he also mentions how ground ice moves great stones in the Eastern Baltic.[7]

Engelspach Lariviere, in a memoir entitled "Considerations sur les Blocs erratiques," published in 1829,

[5] Id. vii. 18. [6] Id. x. 217, 218.
[7] Kastner's Archiv. xiv. 421-423.

describes at some length the erratics found in the Low Countries. He attributes them, and those found elsewhere, to more than one cause, but the principal one, he says, which brought the blocks from foreign countries to Germany and Holland, was enormous icebergs; and he mentions how, when he was at Memel in 1821, his companion, Baron Chaumont, called his attention to a mass of ice twenty-eight to thirty feet long, stranded on the shore, embedded in which was a triangular mass of granite like that of Finland, and on returning to see it again it had disappeared, having doubtless moved on.[8]

In the same year K. F. Klöden published a work, entitled "Ueber die Gestalt und die Urgeschichte der Erde," in which he endeavours to explain the contour of the present surface of the world by a theory of a change in the position of its axis. This, he goes on to say, would cause, if effected suddenly, a great rush of water, and he thus accounts for the rush from the N.E., which Bruckner had postulated as necessary to account for the surface deposits of Mecklenburg, Brandenburg, Pomerania and Prussia. This great current would bear along with it masses of ice which would carry great blocks of stone. Hence he accounts for the transport of blocks from the Scandinavian mountains, the icebergs dropping some of them in Sweden, Finland, Denmark, Prussia and Germany, as far as the Polish mountains, Lusatia, the Erzgebirge, the Thuringenwald and the Hartz, while some were also carried to the Netherlands and to Eastern England.[9]

In the first edition of his "Principles of Geology," published in 1830, Lyell says: "Marine currents are sometimes instrumental in the transportation of rock and soil, by floating large masses of ice to great distances from the shore. When glaciers in northern latitudes descend the valleys burdened with alluvial débris, and arrive at the shore, they are frequently detached and float off. Scoresby counted 500 icebergs in lat. 69° and 70° N. . . . Many of these contained strata of earth and stones, or were loaded with beds of rock of great thickness, of which the weight was conjectured to be from 50,000 to 100,000 tons. As the

[8] Op. cit. 29 and note.

[9] Op. cit. 347-350.

mass of ice below the level of the water is between seven and eight times greater than that above, these masses may sometimes take the ground in great numbers in particular parts of the sea, and may, as they dissolve, deposit such masses of matter on particular parts of the bottom of the deep, or on the shores of some isles, as may offer perplexing problems to future geologists."[1]

Applying these facts to Switzerland, Lyell says: "Those naturalists who have seen the glaciers of Savoy, and who have beheld the prodigious magnitude of some fragments conveyed by them from the higher regions of Mont Blanc to the valleys below, to a distance of many leagues, will be prepared to appreciate the effects which a series of earthquakes might produce in this region, if the peaks or 'needles,' as they are called, of Mont Blanc were shaken as rudely as many parts of the Andes have been in our own times. The glaciers of Chamouni would immediately be covered under a prodigious load of rocky masses thrown down upon them. Let us then imagine one of the deep narrow gorges in the course of the Arve, between Chamouni and Cluse, to be stopped up by the sliding down of the hillside (as Rossberg did in 1806), and a lake would fill the valley of Chamouni, and the lower parts of the glaciers would be laid under water. The streams which flow out of arches, at the termination of each glacier, prove that at the bottom of those icy masses there are vaulted cavities through which the waters flow. Into these hollows the water of the lake would enter, and might thus float up the ice in detached icebergs, for the glaciers are much fissured and the rents would be greatly increased during a period of earthquakes. Icebergs thus formed might, we conceive, resemble those seen by Captain Scoresby far from land in the Polar seas, which supported fragments of rock and soil conjectured to be above 50,000 tons in weight. Let a subsequent convulsion then break suddenly the barriers of the lake, and the flood would instantly carry down the icebergs, together with their burden, to the low country at the base of the Alps."[2]

"Ice," says De la Beche, writing in 1833, "seems to offer a possible explanation of the transport of many masses, for the glaciers which descend the valleys of high northern regions

[1] Op. cit. i. 299.

[2] Op. cit. iii. 1833, 149, 150.

are, like those of the Alps, charged with blocks and smaller rock fragments, which have fallen from the heights. Waters rushing up or down such valleys would float off the glaciers, more particularly as northern navigators have shown that they project into the sea. It is considered that the huge masses of ice known as icebergs, are the projecting portions of these boreal glaciers, which having been detached from the parent mass, are borne into more temperate climes, in some cases transporting blocks and smaller fragments of rock. The débris will, as Mr. Lyell has observed, be deposited at the bottom of the seas over which they pass; and therefore, if such bottoms were raised so as to become dry land, blocks might be discovered scattered over various levels of that land, presenting appearances that might be mistaken for the action of diluvial currents. If the present continents bore evident marks of long submergence beneath an ocean immediately previous to their present appearance, and if the blocks were merely scattered here and there, this explanation would by no means be without its weight; but there are too many circumstances tending to other conclusions to render it probable. The supposition of masses of ice, covered by blocks and smaller rock fragments, borne southwards with violence, though it may account for some appearances, does not, it is confessed, seem applicable to all, more particularly where blocks can be traced to their sources at comparatively small distances. Supposing a wave or waves discharged over Europe and America from the northwards, many phenomena would depend on the time of the year at which the catastrophe or catastrophes took place; for in the winter, waters rushing from that quarter would transport a larger quantity of ice, and many superficial blocks and gravels, bound by ice together, might be torn up and carried considerable distances, from the possible small specific gravity of the mass; for even in the case of rivers, it has been found that large masses of rocks have occasionally been transported from places, when encased in ice and acted on by the stream. In Sweden and Russia it is more than probable that many blocks would be thus encased during winter, and therefore a flood of waters passing over them would cause them to rise, to float, and to be borne onwards until, the ice melting, the blocks would sink and be finally brought to rest. Upon the hypothesis of a convulsion or convulsions, in the North, the effects would become less as we receded from the centre of disturbance, and finally, all traces of them would be lost.[3] . . . It has been frequently remarked that the Alpine erratic blocks frequently occur in groups. . . . It may be asked, as a mere conjecture, whether masses of floating ice charged with blocks and other detritus, rushing down the great valleys, into the more open country of lower Switzerland, might not be whirled about by the eddies, and that groups of rocks would afterwards be found where the whirlpools had existed. Masses of ice charged with blocks and pent up for the moment within such basins as might be formed between the Alps and Jura, might also be carried at certain levels against the sides of the opposite mountains, such as the Jura, and be there deposited in groups and in lines of level."[4]

Murchison, writing in 1835, says: "Once let it be granted that large frozen masses, like those now periodically liberated from the polar regions, were drifted to certain distances and in given directions by currents dependent on former configuration of the land, and we are furnished with an adequate agent, each ice-floe as it dissolved, might have dropped its load of stones, at intervals, upon a submarine surface of gravel, sand and shells." And he goes on to suggest that when England and Wales were separated by a strait, the distribution of land and sea may have been such as to have permitted the production of icebergs, which, being dislodged from the shores of Cumberland, might have been drifted into the straits of the sea then existing to the south of Bridgnorth. And in order to explain the generally rounded state of the boulders in Central England, he suggests that they may have been carried down by streams to the shores, and have been long bowldered there "previous to their insertion in the ice. Secondly, they may have been fragments falling from the adjacent rocks which were exposed to the action of water on the shores before their transport by ice. Thirdly, that granite is so prone to desquamation, that nearly all granitic chains are topped with rounded masses which, though *in situ*, have often the appearance of being boulders, and these, if dislodged from cliffs and embedded in ice-floes,

[3] De la Beche, a Geological Manual, ed. 1833, 192, 193.
[4] Id. 197.

would at once present the appearance objected to, though they had never been rolled under water. Lastly, if transportation by ice be supposed, we can account rationally for the blocks occupying for the most part the surface or upper portions of the drift, for we know by modern analogies cited by Captain Bayfield, that ice-floes, in narrow bays or straits, are generally stranded on coasts or shallow shores."

Speaking of the numerous icebergs referred to by Bennett, Murchison says: "The multitude of these ice-floes sailing together for such a great distance from the sources of their origin before they are dissolved, not only teaches us that any stones of the polar regions which they might be transporting, must now be part of submarine deposits, even in inter-tropical regions, but also explains how a vast number of such stones *might* be collected in *one* track very remote from the parent rock, as in the south of Staffordshire." [5] In a later paper he urges that in some cases "the different heights at which boulders occur may be explained by the fact that the bed of the ocean on which the stones were dropped was itself full of inequalities, which, when desiccated, would present the undulating outline of our hilly districts, and in other cases it might well be that great dislocations and disruptions took place when the land was raised above the water, when the boulders would be raised with the beds on which they lay." [6] He illustrates his position by the case of the Tyrolese mountains thus. "In one of the loftiest *courbes* of the *Alpine Limestone,* called Hogelsand, at a height of more than 3000 feet above the valley of the Inn, there is an accumulation of blocks of gneiss, granite, porphyry, chlorite-slate, etc., materials which can only have been derived from the Central chain of the Alps, from which they are now completely cut off by a deep valley." He explains this by postulating that there was a period when the blocks in the valley and those on the summits of the limestone mountains were deposited under water, when the limestone peaks were probably at no very different level from those now lying on the banks of the Inn. After the drift was deposited, powerful movements took place, heaving up the outer shores and raising

[5] Silurian System, 541, note, in which he says the preceding remarks are printed from a paper presented to the Geological Society in 1835.

[6] Id. 544, 545.

the pre-existing bottom of the sea, estuary or lake, to great heights, the blocks and loose materials being thus left at different heights, and in the detached positions we now find them.[7]

Lyell, writing again in 1835, and speaking of the boulders covered with lichens on the coast at Gefle, in Sweden, and on the Isle of Lofgrund close by, points out their size and sharp edges. One of the fishermen remarked that the ice might have brought them, and he undertook to show him much larger blocks which had been recently stranded on different parts of the skär. He accordingly went to a small island called Hustgrund, where he "saw blocks of red granite five or six feet in diameter, perfectly free from lichens, amidst other blocks of various sizes which were coloured grey, white and black, by a coating of these plants. The sailors named other spots where stones similarly bare or only beginning to be covered were to be seen among thousands which, having probably lain for a great many years at the same height above the sea, had entirely changed their colour. They declared they knew the exact date of the arrival of some of these blocks.[8]

Again, Mr. Westbeck, of Marstrand, for thirty years employed by the Swedish diving company, had opportunities of witnessing the extraordinary power of ice to lift up from the bottom of the sea and remove to a distance very heavy masses. In two instances the ice collected round sunken vessels which were under his charge, and having frozen round them, floated them off with their cargo and ballast from shallow into deep water.[9] Again, "I saw some considerable boulders overlying the deposits of recent shells at Capellbacken near Uddevalla, a phenomenon analogous to that described near Upsala, where these huge erratic blocks repose upon the sand hills, characterized by fossil shells of Baltic species. These rocky fragments continued therefore to be transported after the period when the modern shelly formations of both coasts were accumulated, and it may be inferred that the drifting of such blocks may be going on by means of ice every year. The largest blocks are found on the highest part of the ridges, and if these ridges were originally sand-banks in the sea, as the sea-shells found in some of them incline me to

[7] Id. 545, note. [8] Phil. Trans. 1835, 20, 21. [9] Id. 30, 31.

believe, the summits of such banks would have arrested the ice islands, which might transport fragments of rocks as suggested."[1]

In his address to the Geological Society in 1836, Lyell mentions, *inter alia*, how Captain Belcher, in 1815, fell in with an ice-field near St. John's Harbour in Newfoundland, in which there were muddy streaks, gravel, and even stones; and he goes on to suggest that in Scandinavia the blocks having been separated from their source by fissuring, one of them after another might be buoyed up and floated off on the rise of the Baltic. Turning elsewhere, he says: "Let us imagine a sunken reef in Baffin's Bay, in about 75° north, divided into fragmentary masses, and these masses becoming year by year involved in packed ice. In a few months they may be drifted more than 1800 miles to the southwards, through the Straits of Belle Isle, to lat. 48° N. After a repetition of these operations for thousands of years, the uneven bed of the ocean far to the south may be strewed over with drift fragments which have either stranded on shoals or have dropped down from melting bergs. Suppose the floor of the ocean, where they alight, to be on the rise as gradually as the bottom of the Baltic in our own times. . . . At length a submarine ridge, covered with the travelled fragments, emerges, and first constitutes an island, which at length becomes connected with the main land, in time perhaps the site of a university of Upsala."[2]

In a letter to Sir Charles Lyell, dated November, 1837, Captain Bayfield says: "The bed of the St. Lawrence below Lake St. Peter is full of immense boulders of primary rocks, most of them (but not all) rolled or water-worn or with their edges worn by atmospheric agency. . . . I have seen a granite boulder fifteen feet in length by ten in width and depth, transported many yards along a meadow by the ice, and last spring I watched the lake ice (lake St. Peter), which takes two or three days to pass Quebec every spring, and had the pleasure of observing several boulders of considerable size and many small stones, sand, earth, reeds and plants, on their way down the river, drifting at a rate measurable by the excess of every ebb tide over the

[1] Id. 32. [2] Proc. Geol. Soc. 1836, 384-5.

preceding flood. . . . Any boulders thus transported are liable to be dropped at various points along the bed of the river, as the ice gives way to the increasing temperature of the air and water in the spring of the year." He then goes on to report how large boulders, and also deep stakes which he had caused to be placed in certain positions, were removed and replaced by other boulders, while in the same season a mass of granite containing 1500 cubic feet, which he had particularly marked, was transported several hundred yards from its observed position. From these facts he infers that the older erratic blocks, from whatever source derived, have been dropped from time to time (from ice-floes) on the bed of the tertiary sea.[3]

M. Cordier, in his instructions to the voyage of the *Astrolabe*,[4] referring to the Anglo-American expedition of 1830, tells us that the reports of that voyage state that the shores of New Shetland are covered with blocks of granite, and consequently of a rock foreign to the country, and Mr. James Eights, naturalist to the expedition, does not hesitate to attribute the transport of these blocks to the portage of masses of ice which annually come there and deposit their loads.

In the year 1832 Darwin set out on his famous travels on board the *Beagle,* and during the next four years he accumulated a large part of the materials for his later works. It was during this voyage that he for the first time described in detail the erratic phenomena of South America. I shall quote largely from his notices.

He says he met with no boulders on the eastern plains of South America until he arrived on the banks of the Santa Cruz, in lat. 50° 10′ S., nor did they occur near the coast, but were first noticed in ascending the river about 100 geographical miles from the Atlantic, and sixty-seven from the nearest slope of the Cordillera. Twelve miles further west, in long 70° 50′ W., they were extraordinarily numerous, consisting of compact clay slate, feldspathic rock, a quartzose chloritic schist and basaltic lava. They were generally of angular form, and many of them resembled fragments at the foot of a precipice. Many were enormous; one, which was

[3] Silurian System, 733. [4] L'Institut, 1837, p. 283.

rounded, was sixty feet in circumference, and stood six feet above the ground. The place where they lay was about 1400 feet above the sea level. They lay on the surface of, and embedded in, a shingle certainly, says Darwin, of submarine origin, and within the period of existing sea shells which are found with them.

"I met," he says, "with erratic boulders nowhere else in Patagonia." Captain King, however, states in his "Sailing Directions," that the surface of Cape Gregory, a headland of about 800 feet in height, on the northern shore of the Straits of Magellan, is strewed with great fragments of primitive rocks.

In the Straits of Magellan and on Tierra del Fuego, boulders both rounded and angular abound. One of them, composed of syenite, and shaped somewhat like a barn, was forty-seven feet in circumference. There were many others half this size, and they must all have travelled at least ninety miles from their parent rock. The boulders there have all travelled from mountains to the west, which is in accordance with the present currents about Cape Horn. In some situations there is no trace of stratification in the long lines of cliff sixty feet high, while the clay contains boulders of all sizes heterogeneously mixed, and similar rocks do not exist *in situ* nearer than sixty miles off. In other places they are sometimes regularly interstratified with beds of shingle. "I entered in my note-book," says Darwin, "that a vast *débâcle* appears to have been suddenly arrested in its course." And he goes on to urge against Lyell's view of the deposition of sand and blocks from melting drift ice, in a tranquil sea of mud, how, near the eastern mouth of the Strait of Magellan, where the finest sediment has been arranged in horizontal laminæ, and the coarse shingle in beds, that it appears strange that stratification should be so entirely and often suddenly absent in the till alone. It should be remarked that boulders both rounded and angular occur both in the till and in the finest laminated matter.[5]

Darwin tells us that going north from the extremity of the continent along the west coast he did not land south of lat. 47°, but inasmuch as a fragment of granite was seen floating on an iceberg between lat. 49° and 50°, we may be sure that

[5] Journ. Geol. Soc. vi. 417-423.

erratic blocks occupy this space. Between lat. 47° and the south extremity of the island of Chiloe he landed several times, but saw no boulders, which he accounts for by the coast being separated from the Cordillera by intervening high land. Chiloe is an island 100 miles long between lat. 41° 46′ and 43° 26′, about thirty miles from the foot of the Andes. Boulders occur on it in extraordinary numbers on both the eastern and western beaches and up to a height of 200 feet on the land. They also abound on the islets between it and the coast. The boulders are of syenite and granite, the latter alone occurring on the south of the island. They all probably came from the Cordillera. The larger boulders were all angular, and resembled fragments at the foot of a steep mountain, one block at Chacas was a rectangular oblong, 15 feet by 11, and 9 feet high, another was pentagonal, quite angular, and 11 feet on each side, it projected 12 feet from the sand, with one point 16 feet high. There were many others smaller in size. The smaller ones were often more rounded. The land has risen here in post-pliocene times, and there are remains of numerous old straits along which the boulders were probably floated. Most of the boulders were in stratified gravel, but in some cases in a hardened unstratified mud like the Scotch till. This till often passes into finely laminated beds with shells. Darwin says that north of Chiloe in lat. 41° 47′ he met with no fragments of far-transported rock which can be classed with the erratic boulders just described, their great size, frequent angularity and complete separation by wide valleys or arms of the sea from their parent source, being taken as their distinctive characters.[6]

In regard to the origin of this phenomenon, Darwin, writing with the views of Agassiz and others before him, says: "Neither the till beds of Tierra del Fuego, which pass into and are regularly interstratified with a great formation of horizontally laminated sandstone containing marine remains, nor the stratified gravel and till which form low plains on the shores of Chiloe, and cross in regular beds the tertiary strata, can have been produced by ordinary moraines. I am led to the same conclusion with respect to the till of southern Tierra del Fuego. . . . The boulders on the lower levels at the head of

[6] Id. 423-426.

the Santa Cruz River are strewed on land which certainly has been modelled by the action of the sea. Those on the 1400 feet plain are sixty-seven miles from the Cordillera, of which the highest pinnacle is only 6400 feet and the general range considerably lower; this little inclination of the surface, with the absence of mounds or ridges on it, and the angularity of the fragments are opposed to the notion that the blocks have been pushed to this great distance by glaciers. Hence I conclude, that in the two first mentioned districts it is quite certain, and in the latter three highly probable, that the boulders were transported by floating ice. The fact of many of the blocks on the northern end of Chiloe being different from those thirty miles southwards, where there must anciently have been a channel across the island, he explains by the direction of the tidal currents as controlled by the seaward channels. According to the situation of the spot whence the iceberg with its cargo of rock was first launched, so would it be swept towards one or the other channel." The circumstance of the boulders on the higher and lower plain of Santa Cruz being of different kinds of rock, he explains by the fact that as the land rose some of the glaciers which formerly debouched on the coast would cease to do so, and that rocks hitherto submerged beneath the sea would become exposed and their fragments, falling on the glaciers, would be transported with the icebergs. "Masses of ice by which fragments of rock are conveyed are produced in two ways, by the breaking off of icebergs from glaciers descending into the sea, and by the actual freezing of the surface of the sea or its tributary streams. Great boulders can be enclosed in ice by this latter means, only (with rare exceptions) where the winter is extremely cold, as in the Gulf of Bothnia and on the shores of North America. A large proportion of the fragments thus enclosed will generally have been exposed to the wearing influences of the sea beach, and from the ice being in a sheet they will be liable to be repeatedly stranded in shallow places, and thus to become still more worn. The other method of transportal, namely, by the descent of glaciers to the sea-level and the production of icebergs, is far from necessarily requiring an extremely cold winter. Blocks are not embedded in glaciers, except high up at their sources, but rest on their backs, and from being exposed to much abrasion remain angular, hence only loose angular blocks of rock, as was the case with those on the floating ice in Sir G. Eyre's Sound, can be transported by icebergs detached from the glaciers of temperate climates, and such icebergs must be floated off perpendicularly and in large masses or the loose fragments would be hurled into the sea. . . . The angularity of the blocks at Chiloe and Santa Cruz accords with their transportal by icebergs, but it is not improbable that in Tierra del Fuego and the southern parts of the continent, where the boulders frequently show signs of attrition, as if they had been worn on a sea beach, the other agency, namely, the freezing of the sea, may also have come in; the shape of the fragments and their position enable us to discover whether they had been embedded in sheet ice or carried on the surface of deeply floating icebergs."[7]

In the Journal of the Geographical Society for 1839, p. 528, Darwin describes an angular block of rock embedded in the perpendicular face of an iceberg twenty feet above the sea level, in lat. 61° S. and long. 163° E., which was seen by Mr. Macnab, mate on board the *Eliza Scott*, when in the Antarctic seas. The iceberg in question was between 250 and 300 feet high. The same man about a week later saw on the summit of a low flat iceberg, a black mass which he thought was a fragment of rock. He had also frequently seen at considerable heights on the bergs, both reddish brown and blackish brown ice. The first of the icebergs above mentioned was 450 miles distant from Sabrinaland (if such land exists), and 1400 miles from any certainly known land, although land may possibly exist 100 miles due south, as the seas have not been explored there. An iceberg with a considerable block on it was met to the east of South Shetland by Mr. Sorrell, when in a sealing vessel, and it is to be noted that Cordier reports the shores of South Shetland as covered with erratic boulders which were supposed to have been brought there on ice.[8]

"Although," says Darwin, "the glaciers I saw were quite clear, many of the icebergs described by Mr. Kirke in Sir

[7] Trans. Geol. Soc. vi. 429-30. [8] Journ. Geog. Soc. ix. 528-9.

G. Eyre's Sound were dark coloured, and on the surface of one several blocks of granite and serpentine were found."[9]

In the appendix to his "Journal of Discovery," published in 1839, the same writer says, "As the numerous masses of ice, which fall from the glaciers at the head of the sounds on the South American coast, are slowly drifted outwards (owing to the fresh water flowing in from the foot of the glaciers), and in the more open channels are left to be acted on by the winds and currents; so must it have been with the icebergs from the glaciers of the Alps, situated in the same latitude, and under similar conditions. These icebergs would in most cases be driven on some part of the surrounding shore; but from floating deep they would ground a little way from the beach, and then being packed together, and driven to and fro, as the winds changed, and as the tides rose and fell, would they not, like a glacier on the land, though in a lesser degree, bruise and grind down everything and polish the surface on which they reposed? In the rapids of the North American rivers, over which large bodies of ice are driven, carrying with them pebbles and fragments of rocks, I am informed by Dr. Richardson, that the primitive rocks are scooped and hollowed, and have their surfaces polished and glossy. Dr. Richardson is not prepared to say, however, whether this is caused by the passage of the ice or of the pebbles. In the Arctic seas, he tells me, great icebergs are packed together and are driven with such force against the shore, that they push up before them, to the height of several feet, every pebble and boulder which lies on the bottom; and consequently the submarine ledges of rock are kept absolutely bare. If a fragment were to be wedged beneath one of these mountain-masses of ice, when forced upwards with such overwhelming power, it is impossible to doubt that the underlying surface of solid rock would be deeply scored. As it is known that the shingle on most beaches has a tendency to travel in one direction, so must the icebergs, and hence we may conjecture, that the grooves would generally be slightly oblique to the line of coast, and parallel to each other.

"Although the icebergs might be drifted from side to side of the sound, if they were moved after having grounded, it would

[9] Trans. Geol. Soc. vi. 428.

be along the shore by the set of the currents or winds, and perhaps slightly up and down by the tidal changes. Would not the necessary effect of this be, that the scratches formed by the sand grating between the rocks and the bottom of the icebergs should be, with some irregularities, longitudinal or (from the effect of the tidal movement) oblique? And as the mountains slowly emerged during ages, every part would be thus acted on; and consequently the whole surface would be marked by longitudinal scratches. As the icebergs on the South American coast sometimes transport angular fragments to the distance of many miles from the glacier whence they were detached, and as the winds and tides generally are strong enough to drive them ashore . . . . the blocks would be *generally* landed on the shores of the channels between the Alpine ranges and not dropped in the intervening spaces. Occasionally, however, these loads would be launched into the deep, and this would account for the extraordinary position of immense single blocks, as referred to by Charpentier, which are sometimes found planted vertically in the soil, in the valleys as on the sides of a mountain, and split up through their whole length from top to bottom, which would go to show, as Charpentier urges, that they had fallen perpendicularly from a certain height on the spots we now find them, and had been rent asunder by the fall." Charpentier argues that they fell down crevasses, but Darwin claims with justice that his theory explains them as well.[1] "Again, if a berg were arrested on its way by some pointed rock so near the surface that when thus charged it would ground on it, the block would, when the ice melted, be there left. . . . From what I have seen when passing in boats through the channels of Tierra del Fuego, and from frequent examination of the armings of lead used in sounding, I feel nearly sure that *absolutely bare* submarine rock is not very common. Where matter is depositing near a shore, the finer the particles are the further they are drifted. Thus, when the coast is rising, a layer of sand will be covered with a layer of small pebbles, and these by larger pebbles, and as icebergs would seldom be driven up on the beach of a sheet of water if, like the channel between the Jura and the Alps, it were protected from the open sea, any fragments of rock

[1] Id. 620, note.

transported by them would have been dropped some way outside, and therefore, when upraised with the whole country, they would be found in most cases reposing on beds (where the loose matter had not subsequently been removed), characterized by the order of superposition just described."

In replying to Sir James Hall's diluvial theory, Darwin says that "the boulders embedded in the blue clay near Edinburgh are, according to Mr. James Hall and Mr. Smith of Jordanhill, themselves marked with parallel lines having one direction, which shows that they were held fast while drifting across the country, and were not rolled over and over, like a pebble in a stream . . . . What then would be the effect of tides and gales of wind, driving packed icebergs with irresistible force, through channels, and over rocky shoals; each part of the surface being exposed for centuries, as the country was elevated, to this action? Would not the fragments of rock embedded in the ice grate in a direct path over the surface, regardless of *minor inequalities*, and would not the fragments themselves be grooved and scored in one direction? Can we for one moment believe it possible that boulders, either in water or in the thickest mud, could be driven over a rugged surface, or along a perpendicular face of solid rock with such enormous velocity as with their points to groove and scratch it, and nevertheless not to be rolled over and over, like a stone descending a mountain, but be marked with parallel lines of abrasion, equally with the fixed, underlying mass? . . . Travellers in arctic regions tell us that the drift ice, with its irresistible power, can force up the gravel and sand into mounds,[2] and drive before it great boulders, and even ships and masses of ice, high and dry on the beach. What then would be the effect of a few pebbles, or a single fragment, between such masses of ice and a steep coast-wall of rock? Would not scratches, 'horizontal, or nearly so,' be formed, indicating (to use Sir James Hall's words) that 'grinders had been pressed against the rock,' as if 'independently of their gravity'?"[3]

In a paper by Wissman, on the erratic blocks of Switzerland, etc., published in 1840, he devotes several pages to a de-

[2] See Geog. Journal, vii. 221. [3] Darwin's Journal, Addenda, 615-625.

scription of the phenomena, and concludes that they can be best explained by invoking floating ice.[4]

In 1840 Sir Charles Lyell wrote a paper in the "Philosophical Magazine," on the Boulder formations of Norfolk. In this he describes the drift on the East Anglian coast as attaining a thickness of over 300 feet of clay, loam and sand, in some places stratified, in others wholly devoid of stratification, and as having interspersed in it pebbles and large boulders of granite, porphyry, greenstone, lias, chalk and other transported rocks. Pure and unmixed white chalk rubble, and even huge fragments of solid chalk, being also associated with it in certain localities. Lyell says in this paper: "In no other parts of our island, or perhaps of Europe, are there evidences of local disturbance on so great a scale, and of an equally modern date, for there are proofs of the movement both downward and upward of strata several hundred feet thick for an extent of many miles, together with most complicated bendings and foldings of the beds and the intercalation of huge masses of chalk, and what is no less perplexing and difficult of explanation, the superposition of contorted upon horizontal and undisturbed strata."[5]

He goes on to say that the boulder formation of Norfolk is strictly analogous to that of Denmark and Sweden, Holstein, Hamburg, Bremen, Osnabruck and Holland, and he urges that "it has all been accumulated almost exclusively on ground permanently submerged beneath the waters. . . . The entire want of stratification in the till, whether of Scandinavia, Scotland or Norfolk, implies some peculiarity in its mode of origin; yet in all these countries some of the till has accumulated contemporaneously, and apparently in the same body of water as much of the accompanying stratified gravel, sand and clay. Moreover, the stratified drifts are often identical in composition with the till, the distinction consisting merely in the mode of arrangement. I have seen no kind of deposit now in progress precisely similar in character to the till, except one, namely, the terminal moraines of glaciers which are accumulated without the influence of currents of water, by which the materials would be sorted and arranged according to their relative weight and size. . . . There can be no doubt that

[4] Neues Jahrbuch, 1840, 314-325. [5] Op. cit. 3rd ser. xvi. 346.

similar accumulations must take place in those parts of every sea, when drift ice, into which mud, sand, and blocks have been frozen, melts in still water and allows the denser matter to fall tranquilly to the bottom. The occasional intercalation of a layer of stratified matter in the till, or the superposition or juxtaposition of the same, may be explained by the existence or non-existence of currents, during the melting of the ice, whether successively in the same place or miscellaneously in several places.

"It is, I believe, a common error of those who are not unwilling to admit the agency of ice in reference to the larger fragments of transported rock, to forget that what carries heavier masses from place to place must unavoidably convey a much larger volume of lighter and finer materials."[6]

On a later page he reverts to "that most perplexing phenomenon, the superposition of bent and folded beds upon others which appear to have undergone no dislocation." Speaking of the cliffs between Bacton gap and Mundesley, he says: "The curves of the various coloured beds of loose sand, loam, and gravel are so complicated, that not only may we sometimes find portions of them which maintain their verticality to a height of ten or fifteen feet, but the replication is often such that a continuous seam of fine loose sand between two layers of gravel or loam might be pierced three times in one perpendicular boring. As it is clear that some of the underlying horizontal beds, and apparently the till also, of which the surface is so even, have not participated in these movements even in the smallest degree, we are compelled to suppose that some lateral force has been exerted against the upper masses of drift which has not been applied to the lower ones. Yielding beds having a thickness of at least fifteen or twenty feet must in some cases have been subjected to this sideway pressure and moved bodily; and it is impossible to conceive that any original irregularity in the mode of depositions, or any shrinking or settling of the materials, or anything in short but mechanical violence, could have produced such complicated folds."[7]

In discussing the well-known outliers of chalk at Trimingham, and the drift in which they stood, he says he was

[6] Id. 348. [7] Id. 351-2.

fully confirmed in his opinion that "both the chalk and incumbent formation, for the thickness of several hundred feet, must have been subject to some common movement, whether sudden or gradual, by which the strata of both have been tilted."[8] Not only are the beds of drift tilted, but the drift is in certain places found to underlie the chalk. Lyell urges that the outliers were brought into their present position after the deposition of the greater part of the drift which has been subjected to precisely the same movements and abuts in some cases in vertical beds against the wall of displaced chalk, and that the derangement is a very modern event.[9] Near Cromer the cliffs contain much more chalk, and sometimes huge masses of it are intercalated in a remarkable way. Some of the enormous masses of transported chalk not only contain layers of undisturbed flints, but also sand-pipes in the middle of thin or cylindrical cavities filled with sand and gravel, such as are found penetrating the chalk at various depths from the surface in the interior of Norfolk. These pipes seem to imply that such masses of chalk were once at or near the surface of emerged land and have been brought bodily into their present position. The intercalated masses of unregenerated chalk are sometimes horizontal and sometimes vertical.[1]

The boulders which are contained in the Cromer cliffs, some of large size, seem to imply, says Lyell, "that while a great proportion of the mass of the till may have been derived from neighbouring regions, part at least has come from a great distance. . . . In different parts of the interior of Norfolk, boulders weighing several tons have been found in blue clay or till. . . . Professor Sedgewick informs me that in the unstratified brown clay or till of certain parts of Cambridgeshire, large angular blocks of lower greensand and chalk, with fossils of the Oxford clay and lias, occur. The till alluded to attains at some points a thickness of 300 feet; it resembles that in the Norfolk mud cliffs and has been traced over many of the adjoining counties."[2]

In regard to the origin of these deposits, Lyell says: "As I believe coast ice and icebergs to have been instrumental in transporting much of the Scandinavian drift, so I presume

[8] Id. 356. [9] Id. 359-60. [1] Id 365-6. [2] Id. 374-5.

that at the same period, the effects of the same agency was extended to the British seas, although on a smaller scale. But while some of the Norfolk erratics may be of northern origin, other portions of the associated drift may have been brought from neighbouring regions, and perhaps in an opposite direction, just as we now observe that some granitic boulders are floated on ice from the distant shores of Labrador into the Gulf of St. Lawrence, while other fragments of rock, together with much gravel and sand, are firmly frozen into ice and carried down every winter by various rivers into the same gulf."

Lyell's views would favour Dr. Mitchell's notion, that many of the boulders may have come from the destruction of strata which once occupied the site of the German Ocean. "We ought always," he says, "to bear in mind that fragments of chalk, greensand, oolite and lias embedded in the drift of Norfolk and other countries may not have come from the westward, where those formations now crop out, but possibly from the N.N.E., like the erratic blocks, if some of these be of Scandinavian origin."[3] Lyell argues that the present contorted and dislocated condition of the strata in the Norfolk cliffs can be explained by ordinary subterranean movements, by landslips and slides, and lastly by the stranding of icebergs and large masses of packed ice, a view to which he inclines. This, he urges, may explain the great foldings and contortions of the upper beds while the lower ones are undisturbed. He says coiled and folded beds of loam, gravel and sand are frequent in different parts of Scotland, Sweden, Norway, and probably everywhere in Europe where drift occurs between lat. 50° and 70° N." He mentions how Dean and Simpson, in lat. 71° N. and long. 156° W., found a low spit named Point Barrow, composed of gravel and coarse sand, in some parts more than a quarter of a mile broad, which the pressure of the ice had forced up into numerous mounds, that, viewed from a distance, assumed the appearance of huge boulder rocks;[4] and he cites what he had himself seen in the Baltic and the Gulf of St. Lawrence, as proving that lateral pressure, exerted under favourable circumstances by drift ice on banks of stratified and incoherent sand, gravel and mud, is an

[3] Id. 376. [4] Journ. Geog. Soc. viii. 221.

adequate cause for producing considerable flexure and dislocation. "The banks on which icebergs run aground occasionally, between Baffin's Bay and Newfoundland, are many hundred feet under water, and the force with which they are struck will depend not so much on the velocity as the momentum of the large floating islands. The same berg is often carried away by a change of the wind, and then driven back again upon the same bank, or in other cases it is made to rise and fall by the waves of the ocean, and may thus alternately strike the bottom with its whole weight, and then be lifted up again until it has deranged the beds over a wide area. On these beds new and undisturbed strata may be afterwards thrown down. In other cases, where banks of mud and sand forming the top of a shoal have been made to assume various shapes by the lateral pressure of icebergs, the bed of the sea may subside, and then the disturbed beds may be overspread by horizontal strata, which may never afterwards be deranged by similar mechanical violence."[5]

In a paper by Murchison and Verneuil on the Geology of Russia, published in the Edin. New Phil. Journal for 1841, those authors urge that "the Russian detritus, the blocks of which have all been derived from the north, is shown to have been deposited under the sea or on a sea bottom, since it covers beds with marine shells. . . . In endeavouring to account for the immense distances to which these blocks had been transported, the authors express their belief that they had been floated on former icebergs; these, breaking loose from ancient glaciers which they suppose have existed in Lapland and the adjacent tracts, were dislodged upon an elevation of the northern chain, and impelled northwards into the sea of that period in which the post pliocene shells were accumulated; and they compare the conditions to those in Central England, where a similar collocation of blocks and shells was accounted for, by supposing that the northern blocks were borne thither in vessels of ice, which in melting dropped them upon what was then a sea bottom."[6] Later in the same paper, after criticizing Agassiz's views, they go on to offer as a probable explanation of the chief phenomena of the

[5] Phil. Mag. 3rd ser. xvi. and 379, 380.
[6] Op. cit. 136.

north of Russia, that currents strongly determined in given directions by the elevation of the northern continental masses might dislodge and set in motion ice-floes and detritus, which, grating upon the bottom of a sea, may have produced the parallel striæ. This is confirmed by the fact that the longer axes of the lakes and stony ridges in north Russia have generally the same direction, which is consistent with the supposed icebergs and land detritus having been borne in that direction also.[7]

Durocher, in his admirable report on the Geological work done during the voyage of the *Recherche*, argues that there is evidence of two causes, distinct in date, for the erratic phenomena. (1) The breaking up of the Northern or frozen ocean, by which a current loaded with ice was sent over the partially submerged country, abrading the rocks, producing striæ and boulders, and carrying them to the south-east. (2) The forming of icebergs on the coast which carried off each summer their load of rocks and earth, and deposited them to the south-east. In the meantime the land was gradually rising and brought the detritus above the surface.

He says that he had very seldom noticed stones on the Spitzbergen icebergs, and never large blocks, only gravel;[8] and he urges that the greater portion of the blocks from the north have been derived from low lands where glaciers cannot exist, and cannot become the mothers of icebergs. He rather invokes ice, acting as it does on the shores of the Baltic, where it often holds masses of granite, etc., in its grip when it breaks up, and transports them when it moves off in the spring. He cites Dr. Beck as having frequently seen such occurrences on the coasts of Denmark, and he refers especially to two instances communicated by Baer to the St. Petersburg Academy, where blocks had been raised up on the Gulf of Finland. The ice of rivers has a similar carrying power; thus at Uleaborg there was a block of stone, three or four feet in size, which was carried down the Ulea Elf on an ice-raft. He further cites M. Dubois, who described how, at Rabelle in Kurland, the low ground by the River Neiss had, according to

[7] Id. 137.

[8] M. Eugene Robert tells us that on the 18th of June, 1838, every one on board the *Recherche* saw floating ice covered with pebbles and sand. (Bull. Soc. Geol. xi. 299.)

the testimony of some old men, been once strewn with great boulders, but they had been taken up by the winter ice and deposited high above their former place on the banks. A similar thing had been noticed on the low ground near Pokroie. Durocher argues that at this moment the floating ice in the Baltic is depositing on its bottom various erratics, and reproducing in some measure the drift beds of north Germany and Russia.[9]

Darwin did not limit his geological speculations to South America. In a paper published in the Edin. Phil. Journal for 1842, while furnishing numerous proofs of the former existence of glaciers in North Wales, he argues that the till between Bethesda and Bangor was accumulated in a sloping sheet beneath the waters of the sea, and he explains the irregularly stratified gravel and boulders, with layers of coarse yellow sand, on Moel Tryfan in the same manner. As, he says, "there are no marks of the passage of glaciers over this mountain, we must suppose the boulders were transported on floating ice; and this accords with the remote origin of some of the pebbles, and with the presence of the sea shells. . . . May we not conjecture that the icebergs, grating over the surface, and being lifted up and down by the tides, shattered and pounded the soft slate rocks as they appear to have contorted the sedimentary beds of Eastern England? . . . The drifting to and fro and grounding of numerous icebergs during long periods near successive uprising coast lines, the bottom being thus often stirred up and fragments of rock dropped on it, will account for the sloping plain of unstratified till, occasionally associated with beds of sand and gravel, which fringes to the west and north the great Carnarvonshire mountains. . . . "As far," he says, "as my very limited observations will serve, one of the best criterions between the effects produced by the passage of glaciers and icebergs is boss or dome-shaped rocks." [10]

A report by Mr. Hays on the probable influence of icebergs upon drift, read before the Association of American Geologists and Naturalists in April, 1843, comprised a large mass of information collected from over eighty people who had seen icebergs. *Inter alia*, the report states that

[9] Op. cit. 132. [10] Edin. New Phil. Journ. xxxiii. 352-363.

blocks of rock and earth are found on the surface and in the interior of the glaciers in the Arctic and Antarctic regions, and instances were cited where some of these glaciers or fixed icebergs in the Antarctic regions were strewn with stones transported from a distance, which stones had afterwards been covered by new deposits of snow and ice; where large rocks were found in the perpendicular face of the glacier overlooking the sea; and where they had been covered with piles of sand and volcanic scoriæ. The motion of icebergs was shown to be very slow and steady, and in the direction of the great undercurrents of water tending from the poles to the equator. Facts were also adduced to show the great depth of water in which icebergs sometimes ground and remain grounded many years, and the limit of transport of icebergs was shown to be about 40° of N. lat. and 36° of S. lat. Many original facts were stated as to the transportation of boulders, from a small size to the diameter of many feet, and from the examination of the positive and negative evidence it was inferred that icebergs are rarely seen charged with foreign materials except near their source. From the facts exhibited the following inferences were drawn.

1. The steadiness of the movement in icebergs from great undercurrents in our northern hemisphere, causes which must have prevailed in ancient as well as modern seas, favours the theory that icebergs with gravers of rock in their lower portions, on pressing the sand and gravel with their immense weight, along the surfaces of the rocks in the bottoms of the ancient oceans, might have scored and grated along the rocks, grinding off their salient points, and leaving their surfaces smoothed and striated in the fixed southerly direction in which they now occur.

2. The immense magnitude of the icebergs in our present seas, and the evidence as to their present mechanical power, when moved by powerful currents, warrant the conclusion that they must have exerted a powerful influence in pushing and crowding along the sand and gravel which formed the bottoms of the ancient seas, and in this way forming accumulations analogous to the moraines of the glaciers.

3. The length of time during which icebergs may remain aground, even when swept by rapid currents which might

surround them with sand and mud, or sweep away the loose materials, leaving hills or banks upon spots protected by the stranded bergs, favours the idea that this agency had an influence in giving the present form to our drift.

4. The formation of glaciers upon the present coasts under such circumstances that fragments of rock and detritus from the land become attached to them, the constant advance and separation of the glaciers from the land, and their floating into the sea as icebergs with loads of earth and rock, lead to the conclusion that icebergs breaking off from the shores of ancient seas were important agents in the transportation of rocks and earth from their parent beds. The existence of immense fragments of rock in situations where they could not have been carried by water alone, as on the sides of hills with valleys intervening between them and their parent beds, but where they might have been stranded by icebergs, favours this conclusion.

5. The fact that icebergs are for the most part broken to pieces and dissolved near their source, and that they are seldom seen with blocks on them far away, leads to the inference that the same cause limited the transportation of the boulders and larger fragments of the drift to within the comparatively small distance from the parent rocks at which they now occur.[1]

Mr. Redfield urged at the same meeting, that the natural and the deflected course of the currents issuing from the Arctic Ocean and carrying icebergs into the North Atlantic, correspond severally in direction with each of the two systems of striæ found in North America. While he argued that the phenomena of single icebergs in the open sea would fail to produce conviction of their efficiency as a cause of drift phenomena, those who had attentively considered the narrative of Back's last voyage might be satisfied that the movements in mass of such vast *packs* of ice and icebergs as those in which his ship was enclosed for many months would be an efficient cause. They were probably moved slowly a great distance to the southward by the force of the great current; while the agitation produced by the vast floes and the base of the bergs, upon the incoherent portions of the shores and the

[1] Silliman's Journal, xlv. 317-319.

subaqueous topography, and the grooving of the faces of the coherent rocks, were causes which, acting without stint of time, were sufficient to produce most of the phenomena of the drift formation.[2]

In describing the American drift in his American Travels, published in 1845, Lyell urges that it was deposited during the successive submergence of a region which had been previously elevated and denuded, and which had already acquired its present leading geographical features and superficial configuration. "At South Brooklyn," he says, "I saw a fine example of stratified drift, consisting of beds of clay, sand and gravel, which were contorted and folded as if by violent lateral pressure, while beds below of similar composition, and equally flexible, remained horizontal. These appearances, which exactly agree with those seen in the drift of Scotland or the North of Europe, generally accord well with the theory which attributes the pressure to the stranding of ice islands, which when they run aground are known to push before them large mounds of shingle and sand, and must often alter greatly the arrangement of strata forming the upper part of shoals or mud banks and sand banks in the sea, while the inferior portions of the same remain unmoved."[3]

Referring to Mr. Mather's observations, that the boulders on Long Island are methodically arranged and occur opposite the places on the other side of the sound where the parent rocks are respectively found *in situ*, Lyell remarks that this reminds one of the similar arrangement of the Jura boulders, which have crossed the intervening great valley of Switzerland at right angles. "The sound," he says, "which separates Long Island from the mainland, is from five to twenty-five miles broad. The fragments have doubtless been transported by ice, but we must suppose them to have been floated by ice islands in the sea, as there are no high mountains in this part of North America from which glaciers can have descended after the continent had acquired nearly its present shape and altitude."[4]

In regard to the origin of the polished and grooved surfaces near lakes Erie and Ontario, he urges that "the land was submerged and that large islands and bergs of floating ice

[2] Id. 326. [3] Op. cit. i. 239, 240. [4] Id. 241-2.

came from the north, which, as they grounded on the coast and on shoals, pushed along all loose materials of sand and pebbles, broke off all angular and projecting points of rock, and when fragments of hard stone were frozen into their lower surfaces, scooped out grooves in the subjacent solid strata. The sloping beach, as well as the level bottom of the sea, and even occasionally the face of a steep cliff, might all be polished and grooved by this machinery; but no flood of water, however violent, or however great the quantity of detritus, or size of the rocky fragments swept along by it, could produce straight, parallel furrows such as are everywhere visible in the district under consideration. . . . After the surface of the rocks had been smoothed and grated upon by the passage of innumerable icebergs, the clay, gravel and sand of the drift were deposited, and occasionally fragments of rock, both large and small, which had been frozen into glaciers, or taken up by coast ice, were dropped here and there at random over the bottom of the ocean, wherever they happened to be detached from the melting ice. During this period of submergence the valleys in the ancient rocks were filled up with drift, and finally, during re-elevation, the ridges so conspicuous between lakes Ontario and Simcoe were formed and certain valleys were re-excavated." [5]

Again, speaking of the valley of the St. Lawrence, Lyell remarks that "wherever the drift or superficial clay and gravel have been removed, the rocks are worn, smoothed and furrowed, the furrows being clearly defined on the sandstone. The direction of all the straight and parallel grooves is nearly N.E. and S.W., differing uniformly from those traced by Hitchcock and Percival through New England, where they usually run from N.N.W. to S.S.E. In both places the blocks have been transported southwards along the same lines as are marked out by the direction of the furrows. . . . I know of no theory that can account for both of them with any plausibility, except through the agency of large islands of floating ice, which by their buoyancy and enormous weight supply the carrying power and pressure required to scratch, polish and groove the solid floor of the ocean, and to convey stones of all sizes, firmly fixed and frozen into the ice, to great distances. . . . At the bridge above the falls, at Montmorenci, over which a large

[5] Op. cit. ii. 98-101.

quantity of ice passes every year, the gneiss is polished and kept perfectly free from lichens, but neither there nor elsewhere, where ice is now at work in this loose way, are any long, straight grooves observable;" and Lyell urges that any degree of freedom of motion in the rocky fragments forced along by small pieces of ice, or by a flood of water, would be quite incompatible with the mechanical effects exhibited in what are called glacial furrows.[6]

On a later page he describes some remarkable striæ in Nova Scotia, occurring on a recently formed ledge of rock, straight and about half an inch broad, some nearly parallel. On inquiring, he learnt that the place in winter was covered with packed ice in which fragments of hard stone coated with quartz crystals were frozen; and he adds, "I have no doubt that the hardness of these gravers, firmly fixed in masses of ice, which, although only fifteen feet thick, are often of considerable horizontal extent, have furnished sufficient pressure and mechanical power to groove the ledge of soft sandstone.[7] In our speculations, therefore, on the carrying power of ice, we ought always to remember that besides gravel and large fragments of rock, it transports with it the finest mud. Dr. Harding informed me that the surface of mud banks along the estuaries near Wolfville, are often furrowed with long straight and parallel ruts, as if large waggons had passed over them. These conform in their general direction to the shore, and are produced by the projecting edges of irregular masses of packed ice, borne along by the tidal current."[8]

Elsewhere Lyell tells us that in Canada boulders are usually found on the surface; at some points, they have been found in sinking wells, thirty feet deep in the clay and sand. He also says the American drift is more stratified than that associated with large boulders in Scotland. It is generally barren, but in some places molluscs abound, and, as on the other side of the Atlantic, they are of Arctic types. Captain Bayfield, he says, had called his attention to the fact that boulders accompany the shells in such a manner as to imply that they had been dropped from melting icebergs to the bottom of a sea in which the mollusca lived and died.[9]

[6] Id. 135-137. [7] Id. 173-175. [8] Id. 175. [9] Id. 138 and 146.

One more reference to the work of floating ice will suffice. "Off the western part of Shippagan Island within the bay of Chaleur, there is an extensive shallow flat, extending nearly two miles from the land, called the 'Grand Batture.' On this flat are numerous large blocks or boulders firmly imbedded. These boulders are brought over from the wild and mountainous shores of Caspe, directly across the bay, by the huge masses of floating ice driven over by the northerly gales, which ground and melt, adding the rocks they bring to those already deposited. Mr. Wilson mentioned that there was formerly a very large rock directly in front of his landing-place, at Miscow, which was much in the way of his boats, but the severe winter of 1848 and 1849 caused the ice to attain an unusual strength and thickness near the shore of Miscow; and when it moved off in the spring, it carried off this large rock to deposit it where he hoped it would be less troublesome."[1]

I do not propose to carry this part of my subject further at present. Unlike the diluvial theory, that of floating icebergs as an explanation of the so-called glacial phenomena still maintains its hold upon some inquirers, notably upon those of America and Northern Europe, who have to face the obvious evidence of submergence involved in the stratified drift and the presence of sea shells.

It will be convenient, however, to stop at this point while we trace the history of another ice theory, which has acquired a wider fame and numbers a larger army of supporters, namely the theory of glaciers.

[1] Porley's Report on the Fisheries of New Brunswick, 26; Silliman's Journal, N.S. xi. 425.

[In writing the previous chapter I overlooked a passage which the fame of its author justifies me in adding in a note. Speaking of Scandinavia, Goethe describes masses of ice passing through the Sound and carrying with them pieces of rock torn from its shores, and he suggests that the blocks found in North Germany, and which have been traced to Scandinavia, may have been detached from rocks which once bordered the Baltic and have since been denuded. He goes on to say: "If a great cold covered the greater part of North Germany with ice, when the water was of a depth of 1000 feet there, one can imagine that the collision of glaciers would cause great destruction, and storms coming from the north-west and the east would carry icebergs with blocks on them further south."[2]]

[2] Göthe, Gesamm. Werke, i. 129.

CHAPTER V.

THE PHENOMENA OF THE DRIFT.

The Champions of Glaciers.

"Zuletzt wollten zwey oder drei stille Gäste sogar einen Zeitraum grimmiger Kälte zu Hülfe rufen, und aus den höchsten Gebirgszügen auf weit ins Land hingesenkten Gletschern, gleichsam Rütschwege für schwere Ursteinmassen bereitet, und diese auf glätter Bahn fern und ferner hinausgeschoben im Geiste sehen. Sie sollten sich, bei eintretender Epoche des Aufthauens, niedersenken und für ewig in fremden Boden liegen bleiben."—GÖTHE, "Wilhelm Meister," 2nd ed. (1829), vol. ii. ch. x.

Playfair, 1802-1816—Perraudin, 1815—Venetz, 1822—Esmark, 1826—Göthe, 1829—Charpentier, 1835, etc.—Agassiz, 1837, etc.—Le Blanc, 1838, etc.—Renoir, 1839-1841—Hogard, 1838—Studer, 1839—Conrad, 1839—Buckland, 1840—Murchison, id.—Whewell, id.—Lyell, id.—Fargeaud, 1839—Durocher, 1841—Boubée, 1842—Max Braun, 1843.

So far as I know, the first person who appealed to glaciers to explain the distribution of the old erratics was Playfair. In 1802, before he had visited Switzerland, he published his "Illustrations of the Huttonian Theory." On page 126 of that memoir, as reprinted in the first volume of his collected works, speaking of the transporting power of water, etc., he says: "Hence fragments of rock, from the central chain, are found to have travelled into distant valleys, even where many inferior ridges intervene: hence the granite of Mont Blanc is seen on the plains of Lombardy or on the sides of the Jura, and the ruins of the Carpathian Mountains lie scattered over the shores of the Baltic." Later on, in the same volume, he has the following pregnant sentence. "For the moving of large masses of rock, the most powerful agents, without doubt, which nature employs are glaciers, those lakes or rivers of ice which are formed in the highest valleys of the Alps, and other mountains of the first order. These great masses are in perpetual motion, undermined by the influx of heat from the earth, and impelled down the declivities on which they rest, by their own enormous weight, together with that of the innumerable fragments of rock with which they are loaded. These fragments they gradually transport to their utmost boundaries, where a formidable wall ascertains the magnitude, and attests the force, of the great engine by which it was erected. The immense quantity and size of the rocks thus transported have been remarked with astonishment by every observer, and explain sufficiently how fragments of rock may be put in motion, even where there is but little declivity, and where the actual surface of the ground is considerably uneven. In this manner, before the valleys were cut out in the form they now are, and when the mountains were still more elevated, huge fragments of rock may have been carried to a great distance; and it is not wonderful if these same masses, greatly diminished in size and reduced to gravel or sand, have reached the shores, or even the bottom of the sea."[1] This was written in 1802, before Playfair had visited Switzerland. In the notes of his journey thither in 1816, we find some interesting sentences. Thus we read that, "On entering the Val de Travers in Mont Jura, I met with a phenomenon, curious in itself, which had often engaged my attention, namely, the existence of loose blocks of granite, gneiss and mica slate, on the surface of a chain of mountains entirely calcareous. They first appeared within the French frontier, on the western declivity of the Jura, and numbers are scattered all the way through to Neuchatel. The largest and most striking of them, the Pierre a-Bot (so named from the farm in which it is situated), lies concealed in a wood upon the rapid slope of a hill, at an elevation of at least 700 feet above that town, and measures 64 feet in length, 32 in breadth, and 16 in height. It contains therefore 32,768 cubic feet, which (allowing 13 cubic feet of granite to a ton) gives a weight of 2520 tons. When we consider that the

[1] Huttonian Theory, Playfair's Works, i. 385.

nearest point where granite is to be found in its native place, is at a distance of seventy miles, it will appear no easy matter to assign a conveyance by which this block could have performed such a journey over intervening hills and valleys without considerable injury. A current of water, however powerful, could never have carried it up an acclivity, but would have deposited it in the first valley it came to, and would, in a much less distance, have rounded its angles, and given to it the shape so characteristic of stones subjected to the action of water. A glacier which fills up valleys in its course, and which conveys rocks on its surface free from attrition, is the only agent, we now see, capable of transporting them to such a distance, without destroying that sharpness of the angles so characteristic of these masses."[2]

This conclusion, which Playfair was the first scientific man to publish, had probably been reached by the unsophisticated natives of Switzerland before. Thus, Charpentier tells us that in 1815 he passed the night in the hut of a famous chamois hunter, named J. R. Perraudin, who lived in the hamlet of Lourtier, in the valley of Bagnes. The conversation having turned upon glaciers, he told Charpentier that in the mountains where they were, the glaciers had formerly been much more extensive than at present. "The whole valley," he said, "as far as a great height above the Drance (i.e. the torrent running in the valley), was once occupied by a vast glacier which extended as far as Martigny, as is proved by the blocks of stone which are found near that town, and which are too big for water to have carried them."[3]

M. Charpentier reports another conversation he had had with a woodman from Meiringen, when travelling along the valleys of Hasli and Lungern in 1834, when, having passed a great block of granite, his companion said to him, "There are many of these stones here, but they have come a long distance. They all came from the Grimsel, for similar stone is not found nearer," and he went on to explain that the glacier of the Grimsel had brought them and deposited them on both sides of the valley, for, he said, that glacier once extended as far as Berne. Water could not have carried them without filling up

[2] Playfair's Works, i. p. xxix; Forbes's Travels, 42, 43.
[3] Charpentier, Essai sur les Glaciers, 241, 242.

the lakes (i.e. the lakes of Brienz and Thun). M. Brard[4] also mentions a person at Chamounix who attributed the transport of the blocks of protogine which occur on certain heights in that valley, and the parallel scratches there, to the agency of glaciers.

The induction of the peasants was assisted no doubt by the fact, now perfectly ascertained, that glaciers have periods of extension and periods of shrinking, in consequence of which serious damage has often occurred. The Glacier de Getroz, in the upper Rhone valley, having outstepped its limits in April, 1818, precipitated a large portion of itself down the valley and formed a barrier which dammed back a new lake. In fear that this dam would break down suddenly and that the valley would be flooded, an engineer named Venetz was appointed to devise some means of escape. This led him to study glaciers more closely, and in the course of the year 1822, he read a paper before the Helvetic Society of Natural History, in which he showed that when we first begin to have historic records about Switzerland, the glaciers were not so extensive as now, and that they have gradually grown to their present size during the last eight centuries. This he showed from the evidence of old roads and Alpine passes, bridges, chapels, pastures and forests which once existed in places now covered with ice, also from the records of old vineyards, cherry orchards and walnut plantations. This evidence was supplemented by that of tradition and by the names of places,

On the other hand, he went on to argue that long before the historic time the glaciers, instead of having been smaller, had been much larger than now, and for evidence he turned to a large number of cases of old moraines existing much below the present glaciers in the valleys of Piedmont and Valais, which pointed to the temperature having once been much lower.[5] Forbes tells us that although M. Venetz did not publish any more memoirs on the subject, he was the first person publicly to maintain in Switzerland the former extension of glaciers to the Jura, as the transporting agents of

[4] Dictionnaire des Sciences Naturelles, tom. xix. p. 16.

[5] Bibliothèque Universelle des Sciences, etc. Section of Sciences and Arts, vol. 21, Geneva, p. 77.

the erratics; and he adds that he was introduced to him in 1832, as the originator of a speculation which, although it had not then probably another advocate, was acknowledged to be novel, ingenious and bold.[6]

In the Edinburgh New Philosophical Journal for October, 1826, is the translation of a paper in the Christiania Journal by Professor Esmark, in which he discusses the erratics of Norway and attributes their portage to ice. In this paper we read, *inter alia,* "There are not wanting many facts to confirm the conclusion that ice has worked prodigious changes on mountain masses, and conveyed from them large rocks into regions where now no perennial ice is to be found, at a great distance from the mountains from which they must have come." Going on to speak of the erratics on the Jura and elsewhere, our author says, "A few suppose that these masses have been brought to their present situation by ice. According to Von Buch, this is a very general opinion in Switzerland. It is not only in mountainous regions we find this phenomenon, but also in flat alluvial districts, where these rocky masses lie upon gravel and sand, a circumstance which cannot be explained in any other way than by their having been brought thither in combination with masses of ice. It cannot be admitted that they could be brought to such situations by torrents of water, for the same torrents which could have been capable of bringing such masses of rock, must at the same time have carried off the gravel and sand on which they rest. One needs only travel through the plains of Denmark to perceive how improbable the supposition is, that such masses could be placed where they are by water. In short, in every country, whether it be mountainous or flat, we shall find similar traces of the operation of masses of ice. The prominent conglomerations to be found in many districts, may be easily accounted for in the same manner. But it is particularly in Norway I have found many proofs of the operation of immense masses of ice which have now disappeared." [7]

1. "As in other countries, we find large loose rocky masses lying spread over pretty level plains; for example, in travelling from Marstuen on the Miosen to Lenton in Hedemark.

[6] Forbes's Travels, 44.

[7] Op. cit. 114, 115.

These must have been brought from a great distance, for there are in the neighbourhood no mountains of the same character as these masses.

2. "In no other satisfactory way than by the operation of ice can we explain how those prodigiously large loose stones, sometimes with sharp corners, have been brought up to the ridges and tops of mountains, which are found in such numbers in the province of Christiansand. . . . By the assistance of immense masses of ice it is easy to conceive how they could have been brought from a great distance, and pushed high up on the mountains.

3. In travelling over our mountainous districts, especially in Osterdal, it will be frequently found that the slope of the mountain towards the valley is covered with very large loose stones, mixed with a great quantity of loose sand and gravel, and that this covering extends to a considerable height over the bottom of the valley." Our author urges that this "could not have been deposited by a current of water, which would have first deposited the big rocks and have carried the gravel and finer sand to places where the influence of the current was less powerful." Nor could it have been the effect of two currents, for if a mighty current had first brought down the big stones, they would have rested on each other without anything interposing, and the gravel, etc., brought by the subsequent current would merely have filled up the cavities, etc., whereas the big stones are separated from one another, and surrounded by sand and gravel, "which cannot be explained in another way than by supposing that the whole has formerly been filled up with ice which has pushed the whole mixed mass up the slope of the mountain. The water of the ice, afterwards thawing, carried off by its rapid streams a part of the stones and gravel, which were then heaped together, deeper down in the valleys: these heaps resemble entirely those which glaciers carry before them."

4. Our author refers to the evidence presented by a large level sandy plain near Fossan in the Stavanger Fiord. "Its upper end," he says, "is bounded by a glacier-dike or rampart, extending across the whole valley. This dike is remarkable, and," he continues, "so far as I know, the only one of the kind lying close to the level of the sea, in a district where you find

only a few heaps of perpetual snow in hollows of the mountains. The length of the dike is 2250 feet, and its perpendicular height 100; a break about 200 feet long, where its highest point is only 12 feet above the plain, exists at the end nearest the mountain. It consists of coarse gravel and sand mixed with a great number of immense blocks of gneiss, which is the prevailing kind of rock in the mountain. This gravel and sand is not only heaped up across the valley, but pushed up in great quantity on the opposite side of the dike to the length of 1400 feet towards the mountain. The bottom of the valley is filled with a lake, which is diked in fact by the rampart. . . . From this description," says M. Esmark, "it will be easy to see that this dike could have been formed only by masses of ice, which must have filled up the whole valley, and by their spreading and pressure have hollowed out its bottom. . . . Not only the dike itself, but the whole horizontal surface, exhibits proofs that there has been a glacier here, for the plain exactly resembles those which I found adjoining to the glaciers at present existing between Londfiord and Lomb in Gudbrandsdal." Esmark says so striking is the resemblance that a young mineralogist, named Tank, who accompanied him to both places, on seeing the latter, immediately exclaimed, without a hint from himself, that the dike they had seen at Stavanger must be a glacier dike.[8]

Esmark goes on to say that these facts prove the Norwegian mountains must have been covered with ice down to the level of the sea, and he adds that inasmuch as the containing mountain precipices are often 3000 to 5000 feet high and the valleys also several thousand feet wide, it is a matter of astonishment to think of such valleys being filled with ice to the extent of several miles. This ice in lower districts must have stretched a long way out into the sea, and, on its thawing, large masses must have broken loose, and gone out to sea, as we find takes place now in the polar regions.

Our author goes on to say that the part of Norway separated from the islands of Gulé and Eastern Lulé, consists of a solid conglomerate, composed of boulders from the size

[8] Id. 115-118.

of a pea to that of a man's head, consisting chiefly of gneiss, quartz, and clayslate, which are involved and bound together in a mass so solid that it is difficult to find out what the binding material is in the interstices between the large stones. This rock, he says, presents proofs of the powerful operation of ice. The precipices on the side of the mountain next the sound are several feet high and perfectly perpendicular, and though composed of boulders cemented together are perfectly smooth. If the precipices had been caused by rents attended by successive masses falling down, each boulder would have adhered to one mass and left a cup or cavity in the corresponding mass. This is not the case, however, but the face of the rock is worn perfectly smooth and has the appearance as if these boulders had been cut across by a knife, and he urges that he can only explain the fact by the supposition that large masses of ice have pressed through the Sound and worn down the boulders and their matrix to a common level.[9]

Our next witness is Göthe. I will translate the heading of the present chapter, which comes from his famous novel. "Finally two or three quiet guests called to their aid a period of intense cold, and saw in their mind's eye glaciers descending from the highest mountain like gliding roads, far into the low country, upon which, as on an inclined plane, heavy primary blocks were slid forward and farther onwards. So that, at the period of thawing of the ice, they sank down and remained permanently on the foreign soil."[1]

In vol. ii. of his collected Works, page 129, after rejecting the theory of a violent transportation of the erratic blocks in Savoy and Switzerland, he says: "We state that there was a period of great cold when the sea covered the continent to a depth of 1000 feet, and the Lake of Geneva was in communication with the North Sea. The glaciers of Savoy descended much lower than now, and the moraines and long lines of boulders extended as far as the valleys of the Arve and the Drance, where the rocks were transported with their angles intact."

In 1835, M. Jean de Charpentier published a memoir, entitled "Notice sur la cause probable du transport des blocs

[9] Ed. Phil. Mag. ii. 107-120.
[1] Wilhelm Meister, 2nd ed. (1829), vol. ii. ch. x.

erratiques de la Suisse." In this memoir he argued strongly in favour of the conclusions of M. Venetz. He urged that since the erratic blocks are found mixed together of all sizes, from that of a grain of sand to that of several thousand cubic feet, and this on the Jura as well as the Alps, the transporting cause could not have been water, which would have sorted them according to their weight. Water, again, would stratify its deposits, while the great masses of erratic débris are unstratified, and it is only very sporadically and exceptionally that slight beds of sand, gravel, etc., are found thus arranged. The fact, again, of the angles of the boulders in many cases not being rounded, and the absence of traces of rolling, precludes an appeal to running water, which would have had the effect of so rounding them. Then, again, the deposits in question are found ranged in elongated lines like ramparts or in pointed hillocks, sometimes isolated, sometimes in rows, and never in continuous layers or mantles. These alignments are planted horizontally to the feet of the mountains and along their flanks, generally one behind the other in parallel rows, and in the direction of the valleys. Occasionally they approach one another and unite, terminating in one or several ramparts. The highest point where they have been traced in the valley of the Rhone is about 1100 or 1200 feet above the river in the environs of Bex, and about 2400 feet in those of Sion. The ground on which they are placed is always the solid rock *in situ*. Water, urges M. Charpentier, could not have arranged its deposits thus. Nor, again, could water have carried the stones across the lakes which so often are found at the feet of the valleys; nor would it have left them in many cases planted vertically on the flanks of the mountains or on the plains, and occasionally broken or split from top to bottom as if they had been dropped upon their present sites.

Again, the blocks coming from the lateral valleys seldom or never mix with those of the main valley, or with those coming from the valley opposite. Thus the feldspathic and talcous rocks of the valley of Hereus, which form considerable depôts near Sion, are not found mixed with the calcareous stones coming from the valleys of the Sionne and the Lierne. The ramparts or lines of blocks coming from these valleys, which join the main valley of the Rhone nearly opposite to one another, are perfectly separate and distinct. Water flowing rapidly would have mixed these stones, and especially would it have done so when the current abutted with its load upon the Jura.

Again, the valleys in which the blocks occur have their unweathered surfaces polished. If water carrying a load of stones rushed along those rocks, it would scratch and plough them up like some gigantic emery paper, but could not polish them; and this is confirmed by the fact that the polished surfaces do not occur above the line where the highest ramparts of erratics are found. Above that line the rocks are scored and weathered and rough, but not polished. The same conclusion follows from the fact that the vaults of the so-called *barmes* or *balmes* of the mountaineers are similarly polished, and also the surfaces under rocks which jut out, and which would preserve them from a torrent bearing stones. The fact also that the higher we mount in the Alps, the more pronounced does this polishing become, while it becomes less and less as we get into the lower valleys, is precisely opposite to the effect which running water with a cargo of stones would produce.

M. Charpentier goes on to confess that at first his prejudices were strongly aroused against the view of M. Venetz, which in lieu of water summoned glaciers on a great scale to explain these ancient moraines, glaciers which once filled all the Alpine valleys, and spread out like huge fans at their feet, which spread as far as the Jura, and even climbed that range and went over to the other side of it; but he was obliged to concede that glaciers and glaciers alone could explain the facts. The deposits left by existing glaciers are ranged tumultuously and without stratification, great blocks and minute particles all mixed together and all carried to similar distances. The moraines of existing glaciers frequently form dykes at the bottoms of small lateral valleys, which intercept the drainage and cause small lakes to be formed, into which sand, gravel, etc., are washed and arranged in beds, thus accounting for occasional patches of stratified materials found in the ancient moraines. Glaciers still bear along on their backs great blocks whose angles are intact. The form of moraines which glaciers are still making is precisely that of the ancient moraines; they are either like ramparts or conical hills, sometimes isolated, sometimes in lines, and when several occur they are always parallel to one another. So that both the interior and the exterior configuration of the moraines is precisely that of the deposits of erratic blocks. Glaciers, again, never deposit moraines in the shape of mantles and continuous beds. In advancing they plough through the soft materials down to the bare rock, and push before them the débris which they move; hence why the lakes at the feet of glaciers are not choked by débris, which, if water had carried it, would have filled them. Hence M. Charpentier accounts for the extremely few erratics found in the plains at the foot of the Alps, where the ancient glaciers formerly rested, and for their abundance in places which once formed the edges and borders of the glaciers. They abound on the steep flanks of the Rhone valley, and are much less frequent in the plain of Valais and those at the feet of the Alps.

Water would have deposited them in a different fashion, and spread them out on the valleys. The stones found split vertically and still standing upright, he explains as having fallen down the crevasses.

As M. Saussure showed, the moraines of converging glaciers do not mix when both reach the main valley, but continue to march separately, which explains the fact already named, that among the deposits of erratics the same separation into different kinds of stone is found. Occasionally a mass of angular fragments, which once apparently formed parts of one great block, are found together. These M. Charpentier explains as resulting from the fall of some stone avalanche on the hard back of the glacier where it was shattered. Glaciers, again, rub, wear and polish the rocks with which they are in contact, and struggling to dilate, they follow all their sinuosities, and press and mould themselves into all the hollows and excavations they can reach, polishing the surfaces, even those which are overhung, which a current of water could not accomplish. This, again, accounts for the evidence of rubbing and polishing being more frequent in the upper valleys, where the glaciers have been longer at work than lower down. Finally, it is possible to follow lines of erratics from the Jura right to the feet of modern glaciers, where similar blocks are being laid down in the modern moraines. M. Charpentier urges that it is not necessary, in explaining the presence of erratics on the Jura by means of glaciers, to suppose that the whole intervening valley was filled up with ice to the height of the Jura. He affirms that when a glacier deploys into a valley at right angles, or nearly so, it is sometimes known to cross it, and, without becoming thicker, to climb the opposite mountain to a certain height, which depends on the volume of the glacier and the steepness of the slope.[2]

I have quoted at considerable length from this excellent memoir, because I look upon it as having definitely applied inductive methods to this question, with results which are for the most part sound and unanswerable, and as having formed a notable epoch in the discussion. It finally established for the Alpine country and for Switzerland, the fact that glaciers were formerly much more extensive than now, and that the erratic blocks and débris are largely the wreckage of former moraines.

The next name which occurs in the history of the theory is that of Agassiz. In 1837, Agassiz opened the session of the Helvetic Society of Natural Sciences with a memoir on glaciers, moraines and erratic blocks. In this he accepted the results of MM. Venetz and Charpentier, and supported them by a number of fresh facts. Thus he mentions how, at various distances from existing glaciers, there are found at different altitudes moraines perfectly like those now being formed. These are also concentric and follow the inequalities on the flanks of the valleys. They are also found at various stages, the highest being at an elevation of several hundred feet above the bottoms of the higher Alpine valleys, where glaciers no longer exist. But on descending to the lower valleys they occur at 12 or 15 or even 18 hundred feet above the bottoms of the valleys. Near St. Maurice, in the Valais, in the Rhone valley, they occur at 2000 feet above the level of the Rhone, and may be followed on the two flanks of the lake of Geneva. They also occur very high above Vevey and in the neighbourhood of Lausanne, where they correspond with those on the south side of the

[2] Annales des Mines, 3rd ser. vol. viii. 219-232.

lake. These old moraines differ entirely from the *talus* formed by an avalanche, which is always angular, and from the mounds of débris left by rivers, in being unbroken and of one continuous form, in having a triangular section and occurring in parallel rows along the flanks of the valleys instead of being spread out in continuous sheets. They are also formed of rounded stones. The lateral valleys present the same evidence as the main ones. He also urges the fact of the existence of polished rocks, wherever the surfaces have been protected from weathering, at a long distance from the present glaciers; as, for instance, on the borders of the valley of the Rhone, and on the banks of the lake of Geneva.[3] He says the southern flanks of the Jura, facing the Alps, from the lake of Bienne to beyond the Orbe, present polished surfaces called *laves* by the natives, which are quite independent of the stratification of the beds and of the direction of the mountains. They traverse all kinds of rocks, and are found in the valley as they are on the highest summits, and are polished almost like a mirror wherever the protecting mantle of gravel or clay has been recently removed. Sometimes these surfaces are flat, sometimes undulating, and often traversed by fissures more or less deep and sinuous or marked by longitudinal bosses, much rounded, but which never occur in the same direction as the slope of the mountain but obliquely to it. The polished surface runs uniformly, irrespective of the hardness or softness of the rocks, and when they contain shells these are smoothed down like those in polished marble. Often the surfaces are finely scratched like a glass scratched by a diamond point. These scratches, again, are arranged obliquely. In many places there are deep excavations and pits. All these effects are consistent only with glacier action and quite different from the effects of water or of strata sliding over one another.

Agassiz cites several localities where these effects can be well studied, and points out that in some places there are deep pits and excavations which could only be produced by cascades falling down crevasses. He remarks that the polished surfaces are not found in the fissures which run longitudinally in the Jura, nor on scarped rocks facing the Jura, but occur only on the surfaces facing the Alps, and he shows how the *laves*

[3] Actes de la Société Helvétique, 1837.

differ from polished surfaces caused by faults in the rocks or by the sliding of strata on one another. The former are very local, while the latter show marks of erosion running along the slopes and not independently of them like the *laves*. He also distinguishes the polished surfaces caused by running water or by floating bergs, and concludes that those in question are consistent only with glacier action. He goes on to show that as the ancient moraines occur on the banks of the lake of Geneva, on both sides of it, at the same height, there must have been a time when that lake was frozen to its bottom while the ice rose considerably above its present level, and he concludes that these old moraines were left behind by the glaciers when they began to retire and to shrink to their present dimensions.[4]

In answer to the champions of water, Agassiz urges that since the blocks are found on all sides of the Alps and in all the Alpine valleys, if the Alpine chain was in existence when the currents were in motion, they must have flowed from their summits in different directions with sufficient depth and speed to have hurried the blocks along, so that they should not fall into the valleys but be transported right across to the heights where they are now found on the Jura. But how can we invoke currents several leagues in length, able to sustain such blocks at a height of 1000 feet? The fact that the blocks in the different valleys differ from each other and are found spread out below like a fan, also excludes the action of violent torrents on this scale or a great diluvial wave passing over Switzerland. If the torrents carried the blocks before the elevation of the Alps, how is it that when the mountains rose they did not destroy the continuity of the old moraines, which still exist at the same level on both sides of the valleys and follow their sinuosities? Water, again, has a different force according to the nature of its bed; ice alone has the same energy at all points.[5]

Agassiz then turns to Lyell's theory of floating icebergs, and shows that icebergs act generally at high water mark, and, if grounded, the marks they make trend in various directions, owing to the rocking or rotating move-

[4] Actes de la Société Helvétique, 1837, vii.-xviii.
[5] Id. xviii. and xix.

ment of the iceberg. Again, he urges, that before the drift phenomena and the polished and engraved surfaces can be attributed to floating ice and icebergs, it must be shown that these appearances have been produced by some agency moving from the seaboard towards the land, and extending up to the very summits of the mountains, or else that all the countries exhibiting glacial phenomena have been sunk below the ocean to the greatest height at which glacier marks are found, and have since gradually emerged to their present level . . . All the evidence of the kind we have, at the period under consideration, indicates only a comparatively slight change of relative level between sea and land within a narrow belt along the shores. . . As to the supposition that the motion proceeded from the sea towards the land, all the facts are against it, since the whole trend of these phenomena is from inland centres toward the shore, instead of being from the shore upward. Certainly no one familiar with the facts could suppose that floating ice or icebergs had abraded, polished, and furrowed the bottoms of narrow valleys as we find them worn, polished and grooved by glaciers.[6]

It was natural that these conclusions should, in the first instance, have been derived from the Alps and their attendant satellite, the Jura, where glaciers still remain close by and where it is only a small step to postulate their having been once much larger. It was not long, however, before the same conclusions were drawn from similar evidence elsewhere, where glaciers no longer exist, and notably was this the case in the Vosges.

In 1838, M. Le Blanc, in discussing Agassiz's views, noticed how in the valleys of Giromagny, Saint Amarin, and St. Maurice, in the Vosges, there are masses of débris, which he identified with old moraines, the work of glaciers which once descended from Giromagny and Drumond, and he added that the people of Wisserling, who are Swiss by origin, had eighty years before given the name of moraine to one of these masses of débris situated to the east of them.[7] A year later M. Renoir noticed that the unstratified heap of sand and rolled pebbles on which Wisserling, in the valley of St. Amarin, is

[6] Geological Sketches, ser. 2, 37-40.
[7] Bull. Soc. Geol. de France, ix. 410.

built, was in all probability a terminal moraine. Other similar deposits were noticed at the entry of the vale of Urbey (Urbés) as well as above Oderen, below Gruth, and near the castle of Wildenstein. M. Renoir also noticed polished surfaces on the right bank of the Thur, and striated ones on the southern slope of Giromagny, while ancient moraines are well preserved at the foot of this mountain, such as the three terminal moraines in front of the village of Puy. The upper part of the village of Giromagny is built on a similar mass of débris. Similar phenomena occur in the valley of the Moselle.

In regard to these morainic deposits of the Vosges, M. Renoir remarks that their absolute height is greater in the middle than towards the rocks encompassing the valley, a form peculiar to true moraines, and which is the very opposite of that which deposits formed by rapid currents would assume. In the valley where the village of Giromagny lies, the deposits lie in parallel mounds like lateral moraines which follow the sinuosities of the valley, as we see in modern moraines, and are not accumulated on transverse heights as they would have been if laid down by water.

In the Vosges as in the Alps the detritus is spread out in a fanlike shape from the mountain summits, which is contrary to the action of water and like that of glaciers.[8] M. Hogard, in the same year, published fresh observations confirming those just cited, and describing how the débris opens out fan-like from the mountain summits, and when it reaches the wide valleys is either spread out like a sheet, or in raised rounded dikes, or in long strings of débris. He argued that while the lateral moraines have been largely preserved intact, the terminal ones have often been broken up and re-arranged by subsequent torrents, to which the more or less stratified beds at the entrance of the valleys are to be also attributed. He also argued that the erratic blocks so often found isolated are the remains of old lateral moraines from which the gravel and sand have been washed away. He especially called attention to the phenomena about Rochesson, Gérardiner, Retournemer, etc.[9]

[8] Ed. Phil. Journ. xxix. 280-295.

[9] Ann. de la Soc. d'Émulation du Dep. des Vosges, iv. 91.

In another paper he refers to such lakes as those of Longemer, de Retournemer, de Foudromé and others now dried up, which, he says, have always one bank much steeper than the other. He urges that the débris found on the sides of these lakes lacks stratification, and forms barriers to the water in the way moraines do. If these dikes had been deposited by water, the lake basins would have been filled up.[1]

M. Constant Prevost mentions how he had seen on the way to Chambery calcareous surfaces deeply furrowed and boulders foreign to the country, notably a block of green schist fifteen feet in diameter, and which he attributed to the action of glaciers.[2]

M. Le Blanc, at the same meeting, reported that M. Fargeaud, professor of Physics at Strasburg, had discovered similar phenomena in the Black Forest and the Pyrenees, which he explained by the theory of glaciers.

In 1839 Studer, who had been a champion of the diluvial theory, read a paper before the French Geological Society, in which he confessed that after examining the glacier of Gorneren, the lower part of which is formed by the union of the glaciers of the Cima de Jazi, Monte Rosa, and Breithorn, and on climbing the Riffel, now raised about 500 feet above the glacier, he says: "The height of this place above the glacier prevents us supposing that the glacier ever rose so high, since the commencement of the present epoch; and yet we saw the surface of the serpentine rocks polished like a mirror, and covered with furrows and striæ nearly horizontal, and in every respect resembling those in contact with the glacier itself. The supposition of currents of water carrying along stones, to which this condition of surface may be attributed, is in like manner rendered a very improbable explanation, in consequence of the isolation of the crest between two very deep valleys of ice and the proximity of the summits of all this group of mountains. At a league distant from this place, below Zermatt . . . we again find on granitic gneiss the same polished surfaces, furrowed and striated, and those rounded

[1] Bull. Soc. Geol., 2nd ser., ii. 249; see also a paper by M. Collomb in the same volume, p. 506, etc.

[2] Bull. Soc. Geol. de France, xi., 65.

prominences which Saussure calls *moutonnées*. On the southern slopes of the Alps we have observed the same phenomenon, both in the immediate proximity of the glaciers and at continually increasing distances from them, to which they cannot reach in the existing physical conditions. On the route from Cormayeur to the Col de la Seigne, for example, we find one of these surfaces opposite the glacier of the Brenva, and if this glacier may have extended so far in present times, this cannot be admitted in regard to the polished and striated rocks which we have noticed between Macugnagna and Pesterena, nor in regard to those of Val-Quarrazza, at the northern base of Monte Turlo; and still less in relation to the distinctly marked ones which we observe near St. Vincent, in the vale of Aosta, on coming from the defile of Mont Jovet." Our author then goes on to describe as a remarkable feature the larger bank or range of hills separating the basin of Ivica from that of Biella, extending to the foot of the Alps over a length of five or six leagues, as far as Santja in the plain of Piedmont. "The height of this range, on the road from Mongrande to Bolengo, cannot be estimated at less than a thousand feet above the neighbouring plain; it then diminishes in height as it recedes from the ridges of the Alps, in such a manner that the latter, when seen at a distance, appear like a very uniform talus, which had been cut on its two sides by the rivers which it separates. . . . The entire mass of this hill seems to be composed of Alpine débris. Blocks of gneiss and other rocks of all sizes, many of them from fifteen to twenty feet in length, with their edges little worn, are scattered over the whole surface, and wherever the soil itself is disclosed, nothing is to be seen but unstratified gravel and sand, enclosing a great number of these same blocks. If we traverse the plain of Ivica and ascend Lessolo to the plateau of Vico and Brosso, we perceive a like number of Alpine blocks covering the whole of this rather steep declivity, which rises to a height nearly equal to that of the opposite edge above the soil of the plain." Studer goes on to say: "I am almost disposed to believe that the partisans of new doctrines respecting moveable rocks or boulders, will see in these hills a classical example of the colossal grandeur of ancient moraines, and that, if they happen to examine the *roches moutonnées*, and striated surfaces in the neighbourhood of Ivica, they will declare their conviction that the ancient glacier, of which we discovered traces at St. Vincent, had extended as far as the edge of the great plain of Piedmont."[3]

Studer, while to some extent a convert to the theory of glaciers, did not subscribe absolutely to it as explaining all the facts. Writing in 1848 about the glacial phenomena of Scotland, we find him still saying: "To the assumption of former glaciers, I feel quite favourably disposed; although it appears difficult to believe in so unlimited an extension of them as must be assumed in this case, as well as in many others, if we attribute the whole phenomena of erratics to this agency."[4]

Meanwhile the same theory had been applied to the American phenomena. In some notes on American geology by T. A. Conrad, published in Silliman's Journal for 1839, he writes: "The phenomena of the 'diluvial epoch' have long attracted peculiar attention, from the many curious and highly interesting facts which they embrace, and the great difficulty of reconciling them with existing hypotheses. Enormous angular masses, transported perhaps a hundred miles from the parent rock, and reposing on sand or gravel which even a mill-stream would have swept away, bid defiance to the mighty currents which so long flourished in the imaginations of certain geologists. Whence came these floods and whither did they go? Such gigantic movements would soon have restored the equilibrium of the waters; and truly they should have been busy during their short reign on earth, to grind down mountains into sand, roll into smoothness myriads of siliceous pebbles, plough deep trenches in the solid rocks, and polish their surfaces with sand.

"The boulders rest usually on sand, gravel, or the natural soil, which would naturally have been swept away, had currents transported these huge fragments, leaving them in every instance reposing on indurated strata. The hypothesis of ice floes bringing them from the north, floating on the waters of an ocean, and depositing them where they

[3] Bull. Soc. Geol. Feb. 1840; Ed. New Phil. Jour. xxix. 274-279.
[4] Ed. Phil. Jour. xlvi. 171.

are now found, has been supported by some of the geologists of the present day; but this is in direct opposition to another theory of these same geologists, that a higher mean temperature prevailed over the northern regions of that period than now reigns in temperate climates. This would not have been the case, all other things being equal, if the northern half of the continent had been nearly all formed by the ocean, notwithstanding the mean temperature is greatly modified in the same parallel of latitude, by the presence or absence of large bodies of water, rising with the former and falling with the latter condition of the globe. Whence then this immense body of ice, which has scattered boulders over so vast a country? Appearing too at an epoch subsequent to the extinction of the *mastodon* and other mammalia, which evidently lived in this region and enjoyed an equatorial climate anterior to the icy period. Nothing can reconcile this apparent contradiction but the admission of a fall of temperature far below that which prevails in our day, freezing the enormous lakes of that period, and converting them into immense glaciers, which probably continued undiminished during a long series of years. At the same time, elevations and depressions of the earth's surface were in progress, giving various degrees of inclination to the frozen surface of the lakes, down which boulders, sand and gravel would be impelled, to great distances from the points of their origin. This in some cases might result from gravity alone; but in others, during the close of the epoch when the temperature had risen, and avalanches began to descend from the mountain tops and from numerous less elevated places, there occurred on a vast scale the same phenomena which are now familiar to the travellers on the Alps. . . . M. Agassiz attributes the polished surfaces of the rocks in Switzerland to the agency of ice, and the 'diluvial scratches' as they have been termed, to sand and pebbles which moving bodies of ice carried in their resistless course. In the same manner I would account for the polished surface of the rocks in Western New York. . . . The scratches and grooves, Mr. Hall informs us, on the rocks bordering the Genesee river, have a direction N.N.E. and S.S.W., and they would probably therefore follow the dip of the stratum down which the ice moved. . . . If a glacier had previously occupied this area, the uplifts would have produced a symmetrical line in the ice, and pebbles and boulders thus brought from opposite directions. Mr. Hall has noticed this phenomenon, but attributes it to the agency of opposing currents.

"While granite boulders have been removed to surprising distances from the rocks *in situ*, those of transition limestone and sandstone seem never to have been far removed from the parent mass, a fact which harmonizes with the theory of refrigeration. The vast thickness of granite, and its corresponding uplift from the force of crystallization, has protruded its naked summits through the overlying strata, and from these peaks, rising to a great altitude, replete with parallel fissures, and split and rent by the upheaving power, large masses would necessarily fall, and when coming in contact with the surface of a glacier, however slightly inclined from the horizon, many of the boulders might of course traverse the extreme limit of the slope, and without losing their angular form; but the limestone fragments being imbedded in the bottom of the glacier would be only affected in position by contraction and expansion of the ice, and the more extensive movements caused by its breaking up in melting, which would have ample power to wear down the angles of these fragmentary rocks." Lastly, our author says, "I think it impossible to explain the breaking up of rocks to a distance of many feet below the surface except by the agency of intense cold, freezing the water which filled the fissures and thus forcing the rocks into tubular fragments and disturbing their position by the lateral and upward pressure."[5]

Early in the summer of 1840 Agassiz visited England with the express intention, as he tells us, of finding traces of glaciers in Great Britain, and was advertised in *Punch* as on a sporting tour in search of moor-hens (moraines). He travelled through the north of England, Scotland and Wales with Buckland, who had, on a previous visit to Switzerland, been much impressed with his presentation of the glacial phenomena of Switzerland. The result of this visit was certainly very remarkable.

"The first valley I entered in the glacial regions of Scotland,"

[5] Silliman's Journal, xxxv. 240-243.

says Agassiz, "was barred by a terminal moraine; and throughout the north of England, as well as in Scotland and Ireland, I found the hill sides covered with traces of glacial action, as distinct and unmistakable as those I had left in my native land. Not only was the surface of the country polished, grooved and scratched, as in the region of existing glaciers presenting an appearance, corresponding exactly to that described elsewhere, but we could trace the path of the boulders where they had come down from the hills above, and been carried from the mouth of each valley down into the plains below."[6] Again he says: "Some of the finest *roches moutonnées* I have seen in Scotland are to be found at the entrance of the valley of Loch Treig. . . . I have observed many other *roches moutonnées* in Scotland, especially about the neighbourhood of Loch Awe, Loch Fyne and Loch Etive. In fact, they may be found in almost all the glens of Scotland, in the lake-region of England, and in the valleys of Wales and Ireland."[7]

On November 4th M. Agassiz read his famous paper on the former existence of glaciers in Scotland, Ireland and England. In this paper he tells us he had arrived at the conclusion that great masses of ice, and subsequently glaciers, existed in these portions of the United Kingdom at a period immediately preceding the present condition of the globe. He urged that "the distribution of blocks and gravel, as well as the polished and striated surfaces of rocks *in situ*, do not indicate the action of a mighty current flowing from N.W. to S.E., as the blocks and masses of gravel everywhere diverge from the central chains of the country, following the course of the valleys. Thus in the valleys of Loch Lomond and Loch Long, they range from N. to S., in those of Loch Fine and Loch Awe from N.W. to S.E., of Loch Etive and Loch Leven from E. to W., and in the valley of the Forth from N.W. to S.E., radiating from the great mountain masses between Ben Nevis and Ben Lomond." Ben Nevis in the north of Scotland, and the Grampians in the south, are considered by the author to constitute the great centres of dispersion in that kingdom and the mountains of Northumberland,

[6] Geol. Sketches, 30, 31. [7] Id. 63-64.

Westmoreland, Cumberland and Wales, as well as those of Ayrshire, Antrim, the west of Ireland and Wicklow, to be other points from which blocks and gravel have been dispersed, each district having its peculiar débris, traceable in many instances to the parent rock, at the head of the valleys. Hence, M. Agassiz, says, "it is plain the cause of the transport must be sought for in the centre of the mountain ranges, and not from a point without the district." The Swedish blocks on the coast of England do not, he conceives, contradict this position, as he adopts the opinion that they may have been transported on floating ice.[8]

Agassiz attributes the parallel roads of Glenroy to the damming up of a main valley by glaciers coming from lateral valleys, thus forming lakes which gave rise to stratified deposits and parallel roads or beds of detritus at different levels.[9] He also seems to argue that similar barriers may have existed at the point of contact with the sea, whence the formation of salt marshes inhabited by animals the remains of which are found in the clays superimposed on the Scotch till.

In his paper on the parallel roads of Glenroy, Agassiz enlarged on these arguments with his customary graphic force. "Tides or currents," he says, "driven powerfully and constantly against a rocky shore, and bringing with them hard materials, may produce blunt, smooth surfaces, such as the repeated blows of a hammer on stone would cause; but they never bring it to a high polish because the grinding materials are not held steadily down in firm, permanent contact with the rocky surfaces against which they move, as is the case with a glacier. On the contrary, being dashed to and fro, they strike and rebound, making a succession of blows, and never a continuous, uninterrupted pressure and friction. The same is true of all the marks made on rocky shores against which loose materials are driven by water currents. They are separate, disconnected, fragmentary; whereas the lines drawn by the hard materials set in the glacier, whether light and fine or strong and deep, are continuous, often unbroken for long distances and rectilinear. Indeed, we have beneath a glacier a

[8] Proc. Geol Soc. iii. 328, 329. [9] Id. 332.

complete apparatus adapted to all these results. In the softer fragments, ground to the finest powder under the incumbent mass, we have a polishing paste; in the hard materials set in that paste, whether pebbles, or grains of sand, we have the various graving instruments by which the finer or coarser lines are drawn. Not only are these lines frequently uninterrupted for a distance of many yards, but they are also parallel except where some change takes place in the thickness of the ice, which may slightly modify the trend of the mass, or where lines in a variety of directions are produced by the intermittent action of separate glaciers, running successively at different angles over the same surfaces. The deeper grooves sometimes present a succession of short *staccato* touches, just as when one presses the finger vertically along some surface where the resistance is sufficient to interrupt the action without actually stopping it, a kind of grating motion, showing how firmly the instrument which produced it must have been held in the moving mass. No currents or sudden freshets carrying hard materials with them, even moving along straight paths down hill-sides or mountain slopes, have ever been known to draw any such lines. They could be made only by some instrument held fast as in a vice by the moving power. Something of the kind is occasionally produced by the drag of a wheel grating over rocks covered with loose materials." [1]

The 18th of November, 1840, was a memorable day in the history of geology in England. On that day Dr. Buckland read a memoir on the evidences of glaciers in Scotland and the north of England. In it the great champion of diluvian methods and the author of the "Reliquiæ diluvianæ," confessed that he had been won over by the arguments of Agassiz, who, in October, 1838, had urged upon him that the polished, striated and furrowed surfaces on the S.E. slope of the Jura near Neuchatel, and also the transport of boulders to the Jura were the effects of ice. Buckland tells us that, after devoting some days to the examination of actual glaciers in the Alps, he had acquiesced in Agassiz's theory relative to Switzerland. He goes on to say that, as early as 1811, he had observed on the head rocks on the left side of the gorge of the Tay, near

[1] Geological Sketches, 2nd ser. 35, 37.

Dunkeld, rounded and polished surfaces; and in 1824, in company with Lyell, grooves and striæ on granite rocks near the east base of Ben Nevis. About the same time Sir George Mackenzie pointed out to him, in a valley near the base of Ben Wyvis, a high ridge of gravel laid obliquely across, in a manner inexplicable by any action of water, but in which, after his examination of glaciers in Switzerland, he recognized the form and condition of a moraine. In his present memoir Dr. Buckland brought together a number of effects which he had noticed in Scotland, and which he assigned to glacier action. *Inter alia,* he mentions (1) a long terminal moraine, like the *vallum* of an ancient camp, covered with grass, about 400 yards long, 100 feet broad at the base, and from 20 to 30 feet high, and chiefly formed of a mass of unstratified pebbles, in the ravine of Creekhope Linn, near Dumfries. (2) He considers the gravel and sand covering the greater part of the granite table-land from Aberdeen to Stonehaven, as the detritus of moraines, the larger insulated tumuli and tortuous ridges of gravel occupying 100 acres near Forden, a mile east of Achinbald, he looks upon as terminal moraines, and the blocks, large pebbles and small gravel spread over the level portions of the valley of the North Esk, after emerging from the Grampians, as the residue of moraines re-arranged by water.

To glacier action, modified subsequently by water, he assigns the cones and ridges of gravel at Cortachy and Piersee in Forfarshire; the polish and striæ on a porphyritic rock immediately above these; the vast longitudinal and insulated ridges of gravel extending for two or three miles up the valley east of Blair Gowrie, and the transverse barriers forming a series of small lakes in the valley of Lunanburn; the lofty mounds comprising the ornamental grounds near Dunkeld Castle, and the detritus in the valleys of the Tay, the Tummel and the Garrie. He refers to the vast congeries of gravel and boulders on the shoulder of the mountain exactly opposite the gorge of the Tummel; the mammillated, polished and striated slate rocks about a mile above the falls of the Tummel; and the rounded outline and polish on veins of quartz, which project eight or ten inches above the weathered masses of mica slate near the same locality. Similar mammillated masses of mica slate retaining striæ and flutings are visible, he says, at Bohaly, one-and-a-half miles east of Tummel Bridge.

Dr. Buckland pointed out the rounded, polished, and striated surfaces, many of them recently exposed by the making of roads, on the shoulders of Schiehallion, and also polished and striated surfaces on dykes of porphyry recently exposed in the valley called the Braes of Fors descending from Schiehallion and on the intervening space, where the slate and quartzite are rounded, polished, grooved, and striated. The striæ, as in the case of the porphyry dykes, are parallel to the direction a glacier would naturally take.

Two lofty ridges of gravel crossing the park at Taymouth Castle, Dr. Buckland considers to be moraines or the detritus of moraines, and, like the deeply scored and fluted boulders of hornblende near Fortingal, to have been distinctly due to glacier action.

On the highlands separating the Tay and the Bran are a remarkable group of moraines, and between the sixteenth and the fourteenth milestone are thirty or forty round-topped moraines, from thirty to sixty feet high, crowded together like tumuli. These mounds, composed of unstratified gravel and boulders, cannot be referred to the action of water since they are placed precisely where a current descending from the adjacent high lands would have acted with the greatest velocity, and they exactly resemble, according to Dr. Buckland, some of the moraines in the valley of the Rhone between Martigny and Löck.

In Strathearn are ridges and terraces of gravel, the detritus of moraines; near Comrie in the same valley hard slaty rocks of the Devonian era have been rounded and striated.

Near there Dr. Buckland marked down spots where, if glaciers formerly existed, traces ought to occur, and he found them accordingly on a hill above the so-called Devil's Cauldron, where there are rounded surfaces of greenstone covered by moraines. At Kenagart, near the same place, is a small cluster of moraines, easily separable into lateral and terminal. Two miles higher up the valley is a medial moraine. He mentions numerous other traces in the same valley.

At Loch Earn Head, at the junction of Glen Ogle and Loch Earn, is a group of conical moraines. They are at the very point where, had they been brought by a rapid current, they would have been propelled into the loch, but nevertheless in the exact position where the terminal moraine of a glacier would be deposited.

Near Callender, the Dean says, "Moraines are stated to cover more or less the valley of the Teith from Loch Katherine to Callender, and the lofty terraces flanking the valley from Callender to Doune are considered to be the detritus of moraines, modified by the great floods which accompanied the melting of the ice. One of them has been mapped as the *vallum* of a Roman camp, and the greater part of the first table-land on the right bank of the Teith between Callender and Doune is composed of re-arranged glacial detritus.

Having thus surveyed the Highlands, Dr. Buckland moved into the lowlands. He points to striæ and furrows still existing on the hills of Stirling and Edinburgh, and assignable only to ice action. Some of these may have been caused by stones projecting from the side or bottom of floating masses of ice, but this could not be the case at the Blackford Hills, two miles south of Edinburgh. On the south face of this hill, at the base of a cliff of trap, is a natural vault filled with gravel and sand, cemented by recent infiltration of carbonate of lime. The sides and roof of this vault are highly polished and covered with striæ, irregularly arranged with respect to the whole surface, but in parallel groups over limited extents. He shows that these marks could not be referred to pebbles moved by water, or fixed to floating ice. It is, however, easy to explain the polish by the long-continued action of fragments of ice forced into the cave laterally from the bottom of a glacier descending the valley, on the margin of which the vault is placed, and the irregular grouping of the parallel striæ by the unequal motion of different fragments of ice, charged with particles of stone firmly fixed in them, like the teeth of a file. The cave is not more than 300 feet above the level of the sea, and justifies the belief that glaciers may also at that period have covered Calton Hill and the Castle Hills of Edinburgh and Stirling.[2]

South of Edinburgh Dr. Buckland drew attention to a

Proc. Geol. Soc. iii. 332, 337.

longitudinal moraine in the valley of the North Tyne, about a mile east of Haddington, to others near Dunbar, and others again dispersed in terraces in the high valleys at the eastern extremity of the Lammermuir Hills. While some round and oblong mounds of gravel are to be seen on a slope of a hill 300 to 400 feet above the sea-level, on the left bank of the Tweed, three miles north of Berwick.

Mr. C. Trevelyan had pointed out to the Dean in 1831 a tortuous ridge of gravel at North Charlton, between Belford and Alnwick, supposed to be an inexplicable work of art, but which he in 1838 became convinced, from an examination of the upper glacier of Grindelwald and that of Rosenlaui, is a lateral moraine. Enormous moraines are said to cover a tract four miles from N. to S. and two from E. to W., immediately below the mountains of the eastern valleys of the Cheviots. The high road winds among cultivated ones near Waller, through North and South Middleton, and by West and East Lillburn to Rosedean and Wooperton.

Turning to Cumberland and Westmoreland, he pointed out extensive moraines loaded with enormous blocks of porphyry and slate near the junction of the Eden with the Eament and the Lowther, as the remains of glaciers which formerly descended from Helvellyn, Martindale, etc. Similar moraines exist near Eden Hall, four miles east of Penrith. Immediately below the gorge through which the Kent descends from the valleys of Kentmere and Long Sledmere, many hundred acres of the valley of the Kendal are covered with large and insulated piles of gravel, and smaller moraines or their detritus nearly fill the valley from Kendal to Morecombe Bay. A group of moraines occupies a space of about 200 acres at an elevation of 500 feet on the shoulder of a mountain in front of the valley of Long Sleddale, on the high road from Shap to Kendal. South of Kendal the high roads from Burton and Milnthorpe to Lancaster pass for the greater part over moraines or their detritus.

Dr. Buckland considers that many of the conical hills laid down on Fryer's large map of Cumberland, in the valley of the Duddon, at the south base of Harter Fell, are moraines; that some of the hillocks on the same map on the right of the Esk, at the E. and W. extremities of Muncaster Fell, are also moraines formed by a glacier which once came down from Sca Fell, and that many of the hillocks near the village of Wastdale were formed by moraines descending westwards. Extensive medial moraines exist on the shoulder of the hill called the Braw Top at the junction of the valley of the Greta with that of Derwent Water.

He found polished and striated surfaces on a recently exposed surface of greywacke at Fox House near Ambleside, also in a slate quarry at Rydal, and on newly-bared rocks by the side of the road ascending from Grasmere to the pass of Wythburn. Mány of the round and mammillated rocks at the bottom of the valley leading from Helvellyn to Windermere, he urges, also owe their form to glacial action.

The remarkable assemblage of boulders of Criffle granite at Shalkbeck, between Carlisle and Cockermouth, he thinks, were conveyed across the Solway Firth on floating ice, as similar boulders are said to have come from Scandinavia.

Dr. Buckland concludes his paper by an examination of the critical case of Shap and its granite boulders. The fact that these boulders have spread north, east and south seems only consistent with glaciers descending the valleys radiating from Shap Fell. "There are abundant proofs," says Dr. Buckland, "of the existence of such a glacier in large mud and boulder moraines, in the ascent of the gorge between Shap Fell and Birbeck Fell, and in the furrows and striæ as well as the mammillated forms of the rocks at the portals of the gorge, particularly on the northern side." In the physical structure of this neighbourhood, Dr. Buckland pointed out other conditions which would have facilitated the accumulation of glaciers, as the lofty mountains of Yardale Head, which overtop Shap Fell on the N.W., and the still higher mountains to the W., whose snows must have nourished enormous glaciers; and he concludes by stating that Professor Agassiz, during an independent tour, arrived at similar conclusions respecting the mode by which the Shap boulders were distributed.[3]

The reading of this paper by Buckland led to a dramatic scene. Murchison at once took up the cudgels. "The day will come," he sneeringly said, "when Highgate Hill will be regarded as the seat of a glacier, and Hyde Park and Belgrave

[3] Id. 345, 348.

Square will be the scene of its influence." He protested against having the question prejudged by giving the name of moraines to the diluvian mounds. "There ought to be a co-ordinate relation in the phenomena. But in the Highland mountains, not one-third the elevation of the Alps, we have moraines two or three times the magnitude of any known in Switzerland. . . . These grooved and striated surfaces and heaps of boulders are also to be found in Scandinavia, on the east of the Gulf of Bothnia, all proceeding from the north and north-west. Have these crossed the gulf on ice? In Russia too we shall find them where there are no mountains, and if we look to the remains of marine shells found in beds elevated, differing in no respect from those in our present seas, except that they are called Pleistocene, we have proof of a lower elevation at the very time when these glaciers should be introduced. On these accounts I am still contented to retain our old ideas, that when a mountain was elevated or a body of water passed over a series of elevations, the diluvium would descend with the strike, and be disposed in mounds and terraces according to the direction of currents, etc.

On the same evening Dr. Whewell objected to the glacier explanation, *inter alia*, (1) Because of the extent of the diluvium over such wide areas in countries of opposite physical structure, surface, climate, etc.; (2) From the marine remains of the glacial period showing the continents to have been submerged; (3) Because of the physical conditions under which glaciers now exist. "We find them universally stretching out from lofty mountain chains which take their rise in *warm* countries, so as to allow of the downward motion and the retiring in summer. . . . Where shall we obtain mountains as *fulcra* for glaciers, stretching many leagues into the plains, producing such results as are ascribed to their action in Scotland?"

In reply Agassiz said that grooves are to be found in the valley of the Aar glacier seven leagues (22 miles) below the end of the present glacier. On soft rocks the grooves are annually renewed in winter and removed by the atmospheric action in summer. He said he had been for many hundred feet under the glacier of Monte Rosa, and found the quartzose sand forming a bed beneath, and acting like emery on the rocks. "Moraines differ from all other accumulations, the lateral moraines are exposed to constant friction with the rocks with which they are brought in contact, and their terminations are passed over by the whole mass of the glacier, so that they become rounded and striated; while the medial moraines, remaining on the surface, continue angular. When the glacier retreats in the summer, the medial moraine, composed of angular fragments, is spread out over the surface of the lateral and terminal moraines, composed of rounded fragments. There are moraines in the Alps 200 feet wide, composed of boulders several feet in diameter."

Lyell, at the same meeting, said a glacier had been known to retire half a mile in a single summer; a number of moraines might have been left in succession, and in severe winters might be driven successively into one by the downward motion of the glacier.[4]

At the same time that Dr. Buckland was giving up his diluvian views in favour of the glacier origin of the phenomena of the drift, etc. etc., Lyell was similarly surrendering to some extent his theory of drifting ice. This partial change of view was embodied in papers read before the Geological Society on November the 18th, and December the 2nd, 1840, on the former existence of glaciers in Forfarshire, in which he goes over very much the same geological reasoning which was pursued by Buckland. I do not think it necessary to state his views in detail.

Lyell's final conversion to the glacial theory did not take place, however, till his visit to Switzerland in 1857.

As was to be expected, the adherence of two such doughty champions as Dr. Buckland and Mr. Lyell to the glacier theory caused a great revolution in geological opinion in this country, and the champions of water, as the agent by which the so-called diluvial phenomena had been caused, had to beat a retreat. Some still held out, but virtually the battle was over, and the army of enthusiastic workers in the still green and largely fresh pastures of geology betook themselves with wonderful patience and pertinacity to map out and to survey in detail the effects of former glaciers in many countries.

In January, 1841, M. Leblanc mentioned having seen

[4] Notes of a discussion at the Geological Society, taken by Dr. S. P. Woodward, "Midland Naturalist," October, 1883.

erratic blocks in the Atlas range in Africa, notably those on the mamelon of Minirch, near Blidah, where they were arranged in the manner they ought to be if a glacier had occupied the valley of Wed Sidi el Kebir, and if Mirnich had been a kind of projecting boss on the glacier forming a so-called "jardin" on its *mer de glace*.[5]

In the Bulletin of the French Geol. Soc. for the same year, M. Renoir illustrates the erratic phenomena of northern Russia by an appeal to glacier ice. He thus explains the fact mentioned by M. Robert that while on the left bank of the Neva as it comes from lake Ladoga and on the margin of the lake at the same point primitive blocks rolled and rubbed occur in great numbers, none occur on the other bank. He explains them as forming a moraine which directed the course of the Neva and formed a dyke for it on its issuing from the lake. He similarly accounts for the line of blocks between Wol-Raculskaia and Copachefskaia on one of the banks of the Dwina, while the other bank is free from them. A great flood, he says again, spreads débris evenly and has not a tendency to create small hills. The numerous mounds and small hills occurring in north Russia, he urges, are the remains of moraines; the fact of their occurring in parallel lines is another evidence of the same fact. M. Robert had described the whole coast of Finland from Helsingfors to Abo with the adjacent islands, as being perfectly polished for some distance into the interior. M. Renoir urges that this perfect polish is clearly the work of masses of ice moving immediately over the surface of the rocks.[6]

The first to explain the erratics of the Pyrenees by means of glaciers was, as I have said, Fargeaud, who, however, seems to have published nothing on the subject.[7] In 1841 M. Durocher mentions having found in the Pyrenees all three types of phenomena which characterize the Alpine drift, namely polished and striated surfaces, erratic blocks and stretches of gravel on the flanks of the valleys. The striæ follow the valleys on both sides of the chain, while the polished surfaces occur wherever the rocks are sufficiently hard to preserve them, as in the valleys

[5] Bull. Soc. Geol. de France, xiii. 134.
[6] Op. cit. p. 68.
[7] Vide ante, 178, see Bull. Soc. Geol. de France (1839) xi. 65.

of La Tet, Audoire, Segre, Ariège, Vicdessos, Luchon, Essera, Gédre, Baréges, Cauterets and Assan."[8]

On the 2nd of May, 1842, M. Boubée read a paper before the French Geological Society, in which he confessed that against his will he had been obliged to appeal to glaciers to explain the erratic phenomena which he had observed in the Cevennes and the Pyrenees. He had spent nearly two years in making observations in the latter chain and found on both flanks of it polished and striated rocks, undoubtedly due to the action of glaciers long before the historic period, and also great moraines beyond the limits of the mountains themselves. He noticed polished and striated surfaces in the valleys of la Pigne, Lys, Larboust, Aran, Venasque, Louron, Gavarine, etc. Moraines existed in these valleys and in others besides, as those of Garen, near Bagnères de Luchon, and those met in descending the valley of Venasque on the Spanish side of the chain. Some are extraordinarily preserved as that of Labroquère, near St. Bertrand de Comminges, that of Tibiran near the cave of Gargas, those of Gripp, of Argelis, of the valleys of Gavarine and Heas, that of Arudy near Pau, that at the meeting of the Taule, Tarascon and Orgeix in the Ariege, that of Vernet at the foot of the Canigu, etc.; and he is very emphatic in the conclusion that precisely the same evidences of former ice action which are found in the Alps occur also in the Pyrenees. Although he had not examined the Cevennes with the same care, he felt that the evidences he had there seen, proved the same fact.[9]

Lastly, Max Braun[1] found in all the Pyrenean valleys polished and striated rocks, *roches moutonnées* and moraines, but he found no erratic blocks on the southern flanks of the Corbières of the Black mountain, nor of the Agénais, which points to the fact that the Pyrenean glaciers did not extend beyond the chain itself.

The new theory was now making rapid way and was the subject of fierce discussion in every scientific circle in Europe. Von Buch, Sedgwick, Sefstrom, Durocher, the younger De Luc,

[8] Compte Rendu, xiii. 902, and Voyage of the *Recherche*, 140.
[9] Bull. Soc. Geol. de France, xiii. 317-18.
[1] Neu. Jahr (1843), p. 80.

and others fought it bravely, but the close reasoning, the careful and minute observation and the generally sober judgment of its two great advocates, Charpentier and Agassiz, were irresistible.

In 1841 Charpentier published a second memoir on the dispersal of the erratics by glaciers, entitled "Essai sur les Glaciers," in which he, in the most methodical manner, proceeded first to demolish every other theory up to that time suggested to explain the dispersal of the erratics, and then stated in greater detail the positive case in favour of his own theory of the former great extension of glaciers. I have already incorporated some of his criticisms.

In this paper he sums up in several pages his general objections to the transport of the blocks by water. He urges that in the famous flood in the Vallée de Bagnes in 1818, only five large blocks of stone were moved, and then only for a distance of 1800 feet, in the narrowest part of the torrent, and they were borne along its bottom and not carried along the top or the borders, while the other materials were deposited directly the torrent began to widen and spread out. The transport in this case was also assisted by the great masses of débris in the shape of chalets, trees, etc., etc., which were also borne along. Along the streets of the village and the town of Martigny, where the current was nine and a half feet deep and sufficiently powerful to carry along houses, etc., it left no gravel or even sand, but only fine mud. He urges that this being so, it is impossible to conceive how a block 100,000 cubic feet in extent could be deposited on the mountain Plan-y-Beuf near Ferret, 2700 feet above the level of the valley, a valley, be it remarked, only three and a half leagues in length. If a great current, as in the case of Bagnes, deposits its débris when it spreads out, why did that coming down the Valais not leave its burden in lower Switzerland instead of carrying it along to the flanks of the Jura?

2. The fact that the rocks are not sorted according to their size and weight, requires not only that the current should have been exceedingly powerful, but should also have been suddenly arrested. If it abated gradually it would sort its contents, but the débris are not so sorted.

3. The fact that on the Jura the greater number of the blocks are found opposite the outfall of the great Rhone

valley, shows, according to M. Charpentier, that they were not deposited by a current, for, he urges, when a torrent suddenly comes upon a transverse barrier it throws its contents to the right and left and does not leave them directly in front of it.

4. If the blocks were transported in great trajectories the ruin they would cause in striking such a barrier as the Jura would be visible in many places, and the blocks themselves would be broken to fragments, but this is not the case, and the exposed beds of Chasseron, which must have borne the chief shock, are no more weathered than those of the rest of the Jura, while the blocks which lie on it are no more broken than those elsewhere.

Again, if, as M. Studer urged, the current was so charged with mud as to make it more easily carry the great blocks, how could it acquire the necessary speed? One necessity here seems to destroy another. As a matter of fact, blocks which sometimes appear to be floating along a muddy current are in reality rolling over other débris which supports them, but there are no remains of such débris left in the valleys on which the great blocks could thus roll. If there had ever been such débris it must have filled up the lakes.

Certain valleys peculiarly situated, like that of Vallorbe, ought to have contained large quantities of débris if it had been deposited by water, but they are singularly free.

It is also impossible to understand how great blocks of 60,000 cubic feet in extent should have preserved their angles sharp if they had been rolled by water. How, again, could water deposit the débris in the form of bands and continuous dykes running in parallel lines along the flanks of the mountains and corresponding to one another on each side of the valley? Why are these dykes only found on the flanks of the mountains and not in the open valleys? How could a torrent, again, deposit its débris in some places heterogeneously and in others in a stratified way? How, again, by this theory can we explain the groups of similar rocks found together in so many places, and which occur as the granite groups above Monthey, those of Plan y Beuf, Steinhof, etc., and the calcareous group of Devens and the Bois de la Chaux, unmixed with other kinds of rock? If, as has been supposed, these groups are the débris of vast masses broken to fragments, we must suppose, as in that of Monthey, that such a mass was three-quarters of a mile long and several hundred feet wide. How, by such currents, can we explain the suspended and poised blocks, and those which are split from top to bottom? [2]

In regard to the champions of icebergs, Charpentier says those who have invoked icebergs in Switzerland have done so in one of two ways, in one they postulate a vast sea or lake as enveloping the whole country, in the other that the bergs were brought by currents. The former views are referred to by Von Buch, and also formed the main position of Darwin. Charpentier says frankly that this theory accounts for the moving of great blocks without rounding their angles, for the occurrence of erratics in the valleys and in Lower Switzerland, as well as on the flanks of the mountains, and for the non-occurrence of blocks above a certain level; but he urges the great improbability of a great lake or sea occupying Switzerland after the general elevation of its surface so as to reach the feet of the glaciers. If we admit such a sea as possible, its border ought to correspond with the limit at which the erratics occur, but such a limit must have been horizontal, unless the surface of the sea was not horizontal, whereas the upper limit of the débris is in some cases, as on the Jura, marked by a curve, and in others the upper limit is many hundreds of feet higher in some spots than in others. The notion of a curved or irregular surface to such a sea or lake is contrary to the laws of physics.[3]

Again, if Darwin's view were right, we must admit one of three things, (1) either Switzerland acquired its present relief under the level of the sea, if so, glaciers could not be formed, for they cannot grow under water, and without glaciers there can be no icebergs; (2) after the present configuration of the country was attained, the sea must have invaded Switzerland, until its depth between the Alps and the Jura was at least 4428 feet, which is incredible; or (3) Switzerland and the rest of Europe must have been buried under water up to the level of the highest erratics, which is also incredible.

But granting the possibility of any one of these solutions,

[2] Charpentier, op. cit. 214-228. [3] Essai sur les Glaciers, 178-181.

where do icebergs deposit loads simulating moraines so closely as do those in many parts of Switzerland? How is it that the only bergs which reached the Jura were those which came from the valleys of the Rhone and the Arve? How was it none reached there from the valleys of the Sarine, Aar, Reuss, Limmat or Rhine? or how is it that the bergs coming from any one of these valleys never invaded another? Why did no bergs take the contrary direction and remove loads from the Jura to the Alps? How is it the bergs coming down the lateral valleys of "the Valais" did not proceed right across the main trunk and deposit their load on the flank of the mountains immediately opposite? Why are not the largest accumulations found where the valleys change their course, and where, consequently, the majority of the bergs would be stranded? How explain the mounds whose position in the valleys is exactly that of terminal moraines, being perpendicular to the enclosing mountains? How explain the vertical furrows and erosions? How is it this sea has left no traces of its recent sojourn in Switzerland, no traces of wave action, no marine débris, no shells, no sand or shingle like that we find on the sea shore? [4]

Another theory urges a vast current as coming from the upper Alps, which detached and carried off masses of ice-bearing rocks like ice-rafts. This theory overlooks, Charpentier says, the very sinuous character of the Alpine valleys and also the fragile nature of ice. The current must have been at least 2500 feet deep, with a fall of one in three. How could bergs on such a torrent arrive safely in lower Switzerland or on the flanks of the Jura? Many of them coming from the upper Valais must have been broken to pieces against the Col du Trient near Martigny, and left their loads there, but we may look in vain for them; not a trace of rocks from the upper Valais has been found at that very frequented place. The same argument applies elsewhere. How could such rafts, again, have safely navigated defiles such as that by which the valley of Binnen opens into the Rhone valley, or the defile of Entremont, etc., without being broken and depositing their loads? When the lower glacier of Geetroz was destroyed by

[4] Id. 181-183, note.

M. Venetz, blocks of ice 10,000 to 40,000 cubic feet in size were detached which fell into the Drance, but they were broken to pieces so quickly that when they reached the village of Lourdier, two leagues from the glacier, only small fragments remained.[5]

Another theory still remains, namely that by which it is sought to explain the dispersal of the erratics by the supposition that the ice covering the Alpine lakes (in which fragments of rock were embedded) was broken to pieces, which floated away with their loads. This theory invoked considerable lakes as formerly filling the Alpine valleys, one of them occupying the plain of lower Switzerland, and opening a communication between the Alps and the Jura. These lakes, frozen in the winter, were melted by the summer sun, which broke their covering into fragments that floated away with their loads of rock from one lake to another. Charpentier urges that there is no evidence of the former existence in Switzerland of such a system of lakes at different levels, nor would such a series explain the contour-line formed by the erratics on the flanks of the mountains; nor, unless we invoke transcendental cold, can we suppose that in one winter the lakes would be sufficiently frozen to bear great masses of stone, several thousand cubic feet in extent, and to create ice rafts sufficiently large to bear them away.[6] Having thus negatived other theories, M. Charpentier proceeds to press reasons in favour of his own view that the erratics were distributed by huge glaciers.

In addition to the arguments already quoted, he showed that the erratic fragments deposited in each of the great Swiss valleys comprise fragments of all the rocks found in the mountains bordering that valley, just as is the case in a modern moraine. The blocks are mixed, some with sharp angles and some rounded also, as in modern moraines, where the former have been carried down on the back of the glacier, and the latter rolled by its underlying stream. Hence why the largest blocks are the best preserved, because when they fell on to the glacier they would travel nearer its medial line and thus escape the rubbing they would have met with if

[5] Id. 184-188.

[6] Op. cit. 189, 190.

smaller and deposited near its flanks.[7] No block is so big that a glacier would not move it. Thus the Blaustein, a mass of serpentine containing 244,000 cubic feet, has been moved by the actual glacier of Matmark or Schwarzberg, and there were old men living at Saas in 1821 who stated how they had heard their fathers tell they had seen this block on the glacier. The largest erratic known to Charpentier was that of Devens, containing 161,000 cubic feet, and therefore considerably smaller than the Blaustein.[8] The mode in which the débris is arranged, again, is exactly that of modern glaciers. In some cases the fragments are strewn sporadically. We may, in illustration of this, compare the abandoned bed of the recent glacier of l'Arpellaz or Salène, near the hamlet of Pratdefort, in the valley of Ferret, with the surface of certain sterile districts of the canton of Fribourg, of the plateau of Jorat, of the environs of Cossonay, la Bretonnière, Premier, etc. On the other hand, when the deposits are in masses on a flat surface, they occur generally in the form of long mounds or dykes, just as a glacier deposits its moraine. If the slope is very great it forms bands, single hills or rows of hills.

The occurrence of long and parallel bands of deposit at the same height on both sides of the valleys is exactly paralleled by modern glaciers, and proves that the retreat of the glacier was not continuous but intermittent and by a series of oscillations. M. Charpentier accounts for the absence of piled-up deposits in the Valais, except in the neighbourhood of Lausanne and of the plateau of Jorat, by the almost horizontal character of the valley from which this great glacier retired continuously and without oscillations.[9] The occurrence of occasionally stratified portions of sand and gravel he explains by the occasional forming of lakes in the lateral valleys where the glacier in the main valley barred them and prevented drainage, as is the case now. In regard to the finding of fragments of one kind of rock together, he mentions that if the famous landslip of the Diablerets, which occurred in 1749, and that of the Rorsberg in 1806, had fallen on glaciers they would have formed two groups of fragments, one entirely composed of calcareous rock and the other of gompholite.[1] In some

[7] Id. 250, 252. [8] Op. cit. 252, 253. [9] Id. 253-256. [1] Id. 261.

cases blocks are found singularly perched one on another, as in the case of the well-known Pierre à Dzo, which is perched on another block, and the Pierre à Bot, resting on one of its smallest faces; others again, like the Pierre à Mourguets, split horizontally by having struck against another rock when rolling, the Pierre Bassa and other stones similarly split, but vertically. These are paralleled by similar rocks in modern moraines formed by retreating glaciers, and are apparently explainable in no other way than by an appeal to glacier action. Charpentier shows how a glacier descending a main valley intercepts all the débris coming down the lateral valleys and prevents it from traversing the main trunk, thus accounting for the fact that when two lateral valleys abouch opposite one another the débris of one valley is not deposited in the mouth of the other. The size of former glaciers is to be measured by the height at which they left their débris in any place. Thus from Aernen to Brigue, in the Haut Valais, the glacier must have reached 2800 feet above the Rhone. At Brigue the valley widens and the glacier must have spread out, and consequently we find the débris only reaches 2500 feet above the Rhone. For seventeen leagues the valley remains fairly even in size and the débris remain at about 2500 feet. Between Martigny and St. Maurice, however, it contracts and the glacier must also have contracted, and we consequently find the débris between these towns occurring as high as 3000 feet. Below St. Maurice the valley again widens and keeps its new width as far as the basin of Geneva, consequently we find the height at which the débris occurs falling to 2300 feet. Beyond this the glacier would meet the plateau of Jorat, which would impede it and cause it to spread out laterally until blocked towards the south by the chain of Verreaux. Hence why we find débris on the mountain of Playau, 2600 feet above the Rhone. Towards the west there was no obstacle, for there is the lake, hence we find that beyond Thonon the line of débris comes down to the level of the lake itself. Having traversed the Jorat, the glacier would meet with no barrier until it reached the Jura, where an immense accumulation of ice naturally followed, and hence we find the remains of its frontal moraine at a height of 3000 feet, at the point where the pressure was greatest, but as it relieved itself laterally, the depth of ice would diminish, and thus leave a curved line of débris on the flanks of the Jura, just as we find to be the case; and inasmuch as a glacier moves fastest along its medial line, it would deposit more débris where this medial line abutted on the barrier, as we in fact find to be the case. The effect produced by a mountain barrier would be imitated where another glacier intervening prevented the forward flow of the ice and caused it to thicken vertically. At the Châlets du Boule, near Givrins, the débris is found at a height of 2300 feet, for there the glacier coming from the Rhone valley must have encountered that coming from the Arve.[2] Just as the beds of the torrents flowing under modern glaciers are free from moraines, so we must not expect to find, and do not find, that kind of deposit where we have reason to believe the beds of the ancient glaciers lay, of which that of the Rhone valley was much the largest in Switzerland.[3]

Perhaps it will be well to state Charpentier's general conclusion in his own words:—"It goes without saying that not only all the valleys of the Valais were filled with ice up to a certain height, but that all lower Switzerland, in which we find the erratic débris of the Rhone valley, must have been covered by the same glacier. Consequently all the country between the Alps and the Jura, and between the environs of Geneva and those of Soleure, has been the bed of a glacier."[4]

While Charpentier battled bravely for his theory, he was helped by a still more famous champion, namely Agassiz. To use the graphic words of the latter, "Every terminal moraine is the retreating footprint of some glacier, as it slowly yielded possession to the plain, and betook itself to the mountains. Wherever we find one of these ancient semicircular walls of unusual size, there we may be sure the glacier resolutely set its icy foot, disputing the ground inch by inch, while heat and cold strove for the mastery. . . . By these semicircular concentric walls we can trace the retreat of the ice as it withdrew from the plain of Switzerland to the fastnesses of the Alps. It paused at Berne, and laid the

[2] Id. 269 273. [3] Id. 275. [4] Id. 266.

foundations of the present city, which is built on an ancient moraine; it made a stand again at the lake of Thun, and barred its northern outlet by a wall which holds its waters back to this day.[5] The great mountain chain of the Alps has been a central axis from which immense glaciers at one time descended in every direction, not only to its base, beyond which the lowlands extend in flat undulations, but to a greater or less distance over the adjoining plains, while at present they are confined to the higher valleys."[6]

Elsewhere, speaking of the glacial striæ in the State of Maine running invariably due north and south, and never being deflected except by local causes, he asks: "How is it possible to suppose that floating icebergs would advance over such an uneven country with this steadfast, straightforward march? Instead of ascending the hills, they would be caught between them in the intervening depressions, or, if the land were completely submerged, floated over them. . . . Equally impossible is it to suppose that anything so unstable as water has produced such straight and continuous lines.[7] . . . In the State of Maine I have followed, compass in hand, the same set of furrows, running from north to south in one unvarying line, over a surface of 130 miles, from the Katahdin Iron Range to the sea shore. These furrows follow all the inequalities of the country, ascending ranges of hills varying from twelve to fifteen hundred feet in height, and descending into the intervening valleys, only two or three hundred feet above the sea, or sometimes even on a level with it. I take it to be impossible that a floating mass of ice should travel onward in one rectilinear direction, turning neither to the right nor to the left, for such a distance. Equally impossible would it be for a detached mass of ice, swimming on the surface of the water, or even with its base sunk considerably below it, to furrow in a straight line the summits and sides of the hills, and the beds of the valleys. It would be carried over the depressions without touching bottom. Instead of ascending the mountains, it would remain stranded against any elevation which rose greatly above its

[5] Agassiz, Geol. Sketches, 6, 7. [6] Id. 20.
[7] Geol. Sketches, ser. ii. p. 150.

own base, and if caught between two parallel ridges would float up and down between them.[8]

Elsewhere Agassiz points out that we cannot attribute the so-called terminal moraines extending across the valleys to the action of water. "Any current," he says, "powerful enough to bring the boulders and débris of all sorts of which these walls are composed to the places where they are found, would certainly not build them up with such regularity, but would sweep them away or scatter them along the bottom of the valley. That this is actually the case is seen in the lower course of the valley of the Rhone where there are no transverse moraines, while they are frequent and undisturbed in the upper part of the valley." He goes on to argue that the lake which formerly occupied the ground here, swept away and totally obliterated the traces of former glaciers.[9]

Agassiz urges that the longitudinal moraines found in the Alpine valleys differ from glacier-detritus remodelled or spread out by water, in being disposed in ridges with a double talus, one flank of which is presented to the glacier and the other to the side of the valley; and their continuity and parallelism at the same height easily distinguish them from the débris disposed along the bottoms of valleys by currents.[1] He contrasts the stratified deposits made by water with those of glacier detritus. The materials of the former are much smaller, the pebbles also are elongated, and fine gravel and sand generally form the upper beds. In the latter, large and small materials are arranged without order, the largest blocks being often in the upper part; and where very large angular blocks occur, they rest on the surface. Referring to polished and striated surfaces, without denying the power of water to produce such results, he says he has sought for them in vain on the borders of rivers and lakes, and on sea coasts; and that the effects produced by water are sinuous furrows, proportioned to the hardness of the rocks; not even, uniform, polished surfaces, such as those presented by the rocks under discussion, and which are independent of the composition of the stone. The parallel striæ in the general direction of the movement which the glaciers

[8] Id. 159. [9] Geol. Sketches, ser. ii. 55. [1] Id. 330.

must have taken, and the so-called *roches moutonnées* are other phenomena to which he refers as due only to ice, and which he says he has traced under the glacier of the Aar, in the valley of the Rhine and of Chamounix, and in Scotland on the banks of Loch Awe and Loch Leven, and he adds that they are very remarkable near Kendal. The most striking points in the distribution of the striæ, are their diverging at the outlets of the valleys, and their being oblique, and never horizontal on the flanks, which they would be were they due to the agency of water, or floating masses of ice.[2]

Agassiz explains the Alpine lakes of Switzerland and northern Italy as largely due to their lower extremities being closed by transverse moraines. "The greater part of them," he says, "have such a wall built across one end, and but for this masonry of the glacier there would have been nothing to prevent their waters from flowing out into the plain at the breaking up of the ice period. We should then have had open valleys in place of all these sheets of water." Turning from the plains to the mountains, he speaks of the traces of old lateral moraines as uncertain and fragmentary, depending on the existence of some ledge or terrace on the wall of the valley affording support for them, and thus marking the former limit of the ice; but, he continues, "thousands of feet above the present level of the glacier, far up towards their summits, we find the sides of the mountains furrowed, scratched and polished, in exactly the same manner as the surfaces over which the glaciers pass at present. . . . Above the lines at which these indications cease, the edges of the rocks are sharp and angular, the surface of the mountain rough, unpolished, and absolutely devoid of all marks resulting from glacial action."[3]

Later on he contrasts the effects of water and of ice in their abrading power, and says: "On any surface over which water flows we shall find that the softer materials have yielded first and most completely. Hard dykes will be left standing out, while softer rocks around them are worn away, furrows will be eaten into more deeply,—fissures will be widened, clay slates will be wasted,—while hard sandstone or

[2] Id. 331. [3] Agassiz, Geol. Sketches, 25-27.

limestone and granite will show greater resistance. Not so with surfaces over which the levelling plough of the glacier has passed. Wherever softer and harder rocks alternate, they are brought to one outline; where dykes intersect softer rock, they are cut to one level with it; where rents or fissures traverse the rock, they do not seem to have been widened or scooped out more deeply, but their edges are simply abraded on one line with the adjoining surfaces. Whatever be the inequality in the hardness of the materials of which the rock consists, even in the case of pudding stone, the surface is abraded so evenly as to leave the impression that a rigid rasp has moved over all the undulations of the land, advancing in one and the same direction and levelling all before it."[4]

Turning to the *roches moutonnées* (so-called, as we have seen, by De Saussure, from their resemblance to the rounded backs of sheep resting on the ground), Agassiz urges that "they are the result of the prolonged abrasion of rocks separated by deep indentations wide enough to be filled up by large glaciers, overtopping the summits of the intervening prominences, and passing over them like a river, or like tide currents flowing over a submerged ledge of rock." He notices two circumstances specially striking about this phenomenon. "The hummocks have frequently glacial marks on one side only, the other being rough and unequal like a hill slope not acted upon by ice. The polished side was clearly the one turned towards the advancing glacier which the ice pressed in its onward motion, while it passed over the other, the lee side as we may call it, without coming in immediate contact with it, bridging the depression and touching bottom again a little further on. The lee side of the knolls often has on it accumulations of the loose materials left by the glacier in its progress. These effects only occur when the knolls are sufficiently high and abrupt to allow a rigid body to bridge over the space in passing from summit to summit. When the knolls are low and slope gently downwards in every direction, they present the characteristic glacier surfaces equally on all sides.[5] Another feature of the *roches moutonnées* is that the scratches

[4] Id. 40, 41.

[5] Id. 41, 44.

upon them ascend the slope in a direct line on one side while they deviate to the right and left on the other side of the knoll, more or less obliquely according to its slope. The summits of the knolls are often again occupied by boulders which would have been swept away by currents, but which, on the supposition that the ice where they rested melted away, would remain poised on prominent peaks or ledges."[6]

Agassiz forcibly contrasts the stratification of deposits from water with the unstratified deposits from glaciers. The débris of which the drift consists is thrown together pell mell without any arrangement according to size or weight, larger and smaller fragments being mixed so indiscriminately that the heaviest materials may be on the very summit of the mass, and the lightest at the bottom in immediate contact with the underlying rock, or the larger pieces may stand at any level in the mass of finer ones. Impalpable powder, coarse sand, rounded, polished and scratched fragments of every size, are mixed together in a homogeneous paste, in which the larger materials are embedded, to use a homely but expressive comparison, like raisins and currants in a pudding. The adhesive paste holding all these fragments together is, no doubt, the result of the friction to which the whole was subjected under the glacier, and which has worked some of the softer materials into a kind of cement. The mode of aggregation of water-worn materials is very different. Examine the shingle along our beaches; we find it so distributed as to show that the fading tide-wave has carried the lighter materials farther than the heavier ones, and the successive deposits exhibit an imperfect cross-stratification, resulting from changes in the height of the tide and the direction of the wind. Moreover, in any materials collected under water we find the heavier ones at the bottom, the lighter on the top. . . . Another distinction between water-worn deposits and drift consists in the fact that the former are washed clean, while the latter always retain the mud gathered during its journey and spread throughout its mass."

I do not propose to carry this part of my subject further, nor, in fact, beyond applying the arguments to fresh localities,

[6] Id. 44, 45.

[7] Id. 51-54.

the accumulation of minute observations, such as those embodied in the works of Guyot, Desor, Maclaren, and many others, or such as Hooker's extension of the theory to the Lebanon and the Himalayas, Hector's and Haast's extension of it to New Zealand, etc., etc., could it be carried much further than it was carried by Charpentier and Agassiz.

It is plain to those who would look without prejudice that the rounded and mammillated surfaces, the scratched, polished and grooved rocks, and a great number of the phenomena which accompanied the distribution of the boulders and the drift, are consistent only with the fact that in the last geological age there was an immense development of glaciers which occupied not only the high ranges of the Alps and the Dovrefelds, but the secondary ranges and lower heights of the continent of Europe and North America. This conclusion seems supported by every form of converging evidence, and is apparently beyond the reach of cavil. So far there is no question at issue.

CHAPTER VI.

THE GROWTH AND CULMINATION OF THE GLACIAL NIGHTMARE.

Part I.—Europe.

"When the rivers that are called elivagar had flowed far from their sources," replied Har, "the venom which they rolled along, hardened, as does dross that runs from a furnace, and became ice. When the rivers flowed no longer, and the ice stood still, the vapour arising from the venom gathered over it and froze to rime; and in this manner were formed in Ginningagap many layers of congealed vapour, piled one over the other. That part of Ginningagap," added Jafnhar, "that lies towards the north, was thus filled with heavy masses of gelid vapour and ice, whilst everywhere within were whirlwinds and fleeting mists."—"The Prose Edda," Mallet's Translation, 401.

Bernhardi, 1832—Schimper, 1835-6—Agassiz, 1837-1843—Charpentier. 1842—C. Martins, 1846—R. Chambers, 1850-52—Murchison, 1853—Von Post, 1856—Kjerulf, 1876, etc.—Erdmann, 1866—J. Geikie, 1873—Torell, 1875—Penck, 1875—Belgrand, Collomb, and Tardy, 1870—Ramsay, 1862 — Jamieson, 1862-65 — A. Geikie, 1863, Tiddeman, 1872—J. Geikie, 1871—Croll, 1875.

The diluvial theories of the earlier geologists had now been largely abandoned and the struggle had lapsed into a fight between the champions of land ice and icebergs—a struggle which still continues.

It was not long, however, before men, while accepting the glacier theory of Charpentier or the iceberg theory of other writers as explaining a great deal in the distribution of the drift deposits, began to have serious doubts as to whether they explained everything, and these doubts began very early.

It was perhaps natural that students living in the plains of Germany, far away from the sea and from any mountain ranges which seemed capable of nursing large glaciers, should, while appealing to ice to account for the phenomena they desired to explain, have appealed to it in some other form than that of icebergs or glaciers. There arose an evident necessity of enlarging the induction which had been so ably made by Charpentier if the distribution of the erratics and drift over the plains of Russia and Poland, many hundreds of miles away from any mountain chain, the similar phenomena occurring in America, the existence of striæ and polished surfaces and great boulders not only in the valleys but on the very crests of the hills, etc., etc., were to be duly explained.

Hence there arose another form of appeal which took two directions. On the one hand, a resort to continental ice moving over the earth's surface in spite of gravity and by an inherent quality of the ice itself; and, secondly, the postulating of huge ice caps about the two polar areas extending far into the temperate regions and even into the tropics. It is not quite easy to separate the history of the two theories, and one great inquirer, Agassiz, successively urged them both. He has been treated as the inventor of the new views, but as a matter of fact he derived his inspiration from others. The first person known to me to advocate them was A. Bernhardi, professor at the Forest Academy at Dreissigacker, who, writing in 1832, modestly urges that he lived away from the centres of culture and that the latest scientific publications were therefore not accessible to him. He formulates his conclusion in these words: "Once the polar ice reached as far as the southern limit of the district which is still marked by the erratics. This ice in the course of thousands of years shrank to its present proportions, and these deposits of erratics must be identified with the walls or mounds of rock fragments which are deposited by all glaciers large and small, or in other words are nothing less than the moraines which this vast sea of ice deposited in its shrinkage and retreat."[1] This is assuredly a remarkable generalization to have been made by a virtually unknown student long before Agassiz published his theory.

A similar view had meanwhile occurred to another German student, and as the story is of some interest to others besides geologists, I shall enlarge somewhat upon it.

Carl Schimper, a botanist, and known as an authority upon mosses, was born at Mannheim on February 15th, 1803. He

[1] Leonhard and Brunn, Jahrbuch, 1832, p. 258, 259.

was a precocious boy, and at Munich became the favourite pupil of Schelling. In his various rambles in Bavaria he had been attracted, as many others have been, by the mysterious boulders scattered over the country, which were a double object of interest to him since they were covered by lichens and mosses which were strange to him. According to the testimony of Dr. Gustav. Bezold, he gave a series of lectures in the winter of 1835-36 at Munich to a society of old friends there, including Schinzlein, afterwards professor at Erlangen, Meyer the poet, Döbner, afterwards professor at Aschaffenburg, etc., etc. Of these lectures Dr. Bezold took notes. In them Schimper distinctly urged that the blocks found in the Isar and Würm valleys and on the Starnberg, "foundlings" he called them, had been transported thither not by rivers or water floods, but by ice, and he invoked a long period of cold during which they were distributed.

In July, 1836, Schimper took part in a gathering of Swiss naturalists at Solothurm, when he met Charpentier and Hugi, who had both written about the Swiss glaciers. Agassiz was also there, but he was busy with his work entitled "Poissons, Echinodermes, et Mollusques fossiles," and had merely gone to see Charpentier's collections. Schimper stayed at Bex with Charpentier from September to December, 1836. Agassiz, who was there also, was busy with his shells, and although he listened to the discussions of the two friends, he only once accompanied them on a glacier exploration to the Col de Balme and the Trient Glaciers, which was conducted by Schimper. During this time Schimper pressed upon Charpentier the fact that the ice action to which he had appealed was in no sense a *res domestica,* and limited in its operations to Switzerland, but was, in fact, a widespread effect of a mantle of ice covering a much wider area. On the 19th of December, 1836, Schimper discovered the famous glacier marks in the chalk of the Jura, near Landeron. This fact he communicated to Agassiz, whose imagination was at length fired by the impulsive and picturesque conversations of his friend, and he determined *for the first time to appear in the character of a geologist,* and advertised a course of lectures to be delivered to his fellow-citizens of Neuchatel which were duly announced in the *Courier Neuchâtelois* of January 24th, 1837. In order to carry out his purpose more effectually he asked Schimper to lend him the notes of the lectures the latter had given at Munich the winter before. As these were locked up, Schimper wrote to his friend and pupil Bezold to forward him his own notes.

These were duly sent. Schimper acknowledged their receipt in a letter dated the 5th of February, 1837, in which were enclosed some verses on the lake of Neuchatel. In this letter he again urged that the granite blocks had not been transported by rivers, but by glaciers before the upheaval of the Alps and the Bavarian mountains, and that, in the penultimate geological period, masses of ice several thousand feet thick covered, not only Switzerland, but all Europe; and that the then existing lakes were also filled with ice, whence it was that the same lakes were not filled up with alluvion and drift. He goes on to urge that the great blocks must have originally been scattered on the ice, and been carried along by it according to the physical laws which govern ice, and that they were thus carried over lakes, valleys, and mountains, and deposited on the high ridges bounding them, etc., etc.

The lectures above referred to proceeded the next month, but in the very first one, Agassiz's knowledge of physics, especially of the physics of ice, proved so slight, that Schimper offered to help him in the subsequent ones; which offer was gladly accepted. Schimper, to commemorate Galileo's birthday, issued a poem on the 15th of February, 1837, entitled "Die Eiszeit: fur Freunde gedruckt am Geburtstage Galilei, 1837." It was signed Dr. K. F. Schimper, Neuchatel, February 15th, 1837, and was distributed by Agassiz among his auditors. This is the first time the term "Ice Age" occurs, and Dr. Volger dates the birth of the glacial theory from its publication; so does Charpentier.[2]

In July, 1837, the Helvetic Society of Natural Sciences met at Neuchatel, and Agassiz addressed to it his famous "Discours préliminaire," which is said to have startled his hearers, and which has been generally accepted as the first

[2] See Essai sur les Glaciers, 228.

public pronouncement on the subject of the glacial theory. In view of this meeting, Schimper, who was then at Carlsruhe, wrote him a letter, in which he asked "his brother Agassiz" to bring, in his stead, before the meeting this vast and important truth more quickly to the notice of the learned world. In the letter he complains naïvely how his discovery had already brought him some troubles, since it had aroused the inveterate prejudices of the "desiccating-water" and the "cooling-fire" men, and ran counter to the traditional notion of a merely mechanically progressive diminution of the earth's temperature,[3] and he concludes that the polished surfaces in the Jura, and the erratic blocks, are the result of the action of a great field of ice, "ein grosses Eisfeld."[4] He extended his researches elsewhere than to the Jura, and tells us erratic blocks are scarce in the Black Forest, but he had found some of them, in 1826, near Lake Tili above Höllenthal, and he had also found them about Carrigou, in the Pyrenees, in 1825.[5]

This letter clearly shows who was the originator and who the mere disseminator of the new views. Agassiz himself, in his address, speaks of his theory as a fusion of his own views with those of Schimper, and refers to an unpublished memoir by his friend. Even this scanty acknowledgment, however, was itself very shortlived, and when Schimper wrote to complain that the press were attributing to Agassiz what had been the work of his own brain, he replied that he did not read the newspapers, and was not aware of their contents, but that the printed report of the society's proceedings would put everything right. When Agassiz published his well known "Etudes sur les Glaciers," in 1840, it will hardly be credited that he did not even refer to his friend, and the process of ignoring him, in spite of protests, he carried out to the end.

Braun was a friend both of Agassiz and Schimper, and cognisant of the facts. He was too timid to interfere, however, although a letter is extant which he wrote to Professor Röper, at Rostock, dated February 22nd, 1840, in which he says, speaking of Schimper, "Agassiz und Charpentier, die in

[3] Id. 5, see also Actes de la Société Helvétique, 1837-38, etc.
[4] Id. 46. [5] Id. 49, 50.

dieser Sache jetzt das Meiste thun, sind, wenigstens was die allgemeine Ausdenhung der Ansichten und die tieferen physiologischen Gedanken dabei betrifft, seine Schüler." Agassiz himself, in a letter to Braun, accompanying a copy of the work last cited, and subsequently included in Braun's "Life," published by his daughter, writes the following extraordinary sentence: "Du darfst Dich nicht wundern, den Namen von Schimper darin nirgends genannt zu finden. Ich wollte so seine Anmassung strafen. Alles, was er hätte sein nennen können, wenn auch nur von ferne, habe ich unberuhrt gelassen, selbst wenn ich beistimmen müsste." Mr. G. P. Evans says of this, "Wherein consisted this presumption which Agassiz wished to punish by a policy of utterly ignoring the achievements of a colleague, in a manner which, in the interests of true learning and to the honour of human nature, one would gladly think is rare in the annals of scientific research? Merely in the modest expression of a desire to have his name publicly mentioned in connection with a theory, of which, as is now clearly shown, he was the real and only author."[6] Well may Mr. G. Evans say, "Whatever glory emanated from the new doctrine haloed round the brow of Agassiz as its public expounder, and naturally enough he soon grew fond of the easily-won fame. The nimbus of the saint is a covetable head-gear, provided one is not compelled to win it by the thorny crown of martyrdom."[7]

The result has been what might be expected. Schimper's name has almost dropped out of geological literature, and is not even referred to in such works as Professor James Geikie's "Ice Age" and "Prehistoric Europe," whose main thesis is that first maintained by the German botanist.

While it is clear that Agassiz has got credit for inventing a theory which he only appropriated from Schimper, there can be small doubt that it was the persistent, picturesque and seductive preaching of the new doctrine by one so famous as a naturalist as himself that made it acceptable in many quarters where scientific idolatry is maintained chiefly by the size of the idol, and he certainly did not shrink either from disseminating his views or pressing them to their logical conclusion. He first had a polemic

[6] G. P. Evans, North American Review, cxlv. 96, 97. [7] Id. 96.

with Charpentier, against whom he argues that the remains could not have been those of the glaciers in the present Swiss valleys, however large we choose to make them, but a vast ice sheet covering Switzerland and reaching higher than the summits of the Jura. This is proved, he says, by the presence of old moraines 2500 metres above the sea level on the borders of the lake of Geneva; and he urges that since the polished surfaces have been found on the summit of "the Pilgrim," which dominates Vevey, and is 3301 feet above the sea level, the bottom of the ice sheet must have been at least at this height. He contests Charpentier's notion that the glacier ice from the Alps crept along the valleys and mounted the Jura to its summit, and urges that this is disproved by the fact that the blocks on the Jura are generally both larger and less rounded than those in the moraines which are being deposited at this moment, from which it follows that they cannot have been subjected to the very long process of trituration involved in travelling up and down the inequalities of the intervening land on a glacier since they would have been ground into a smaller size. This observation, however, about the relative size of the boulders on the Alps and Jura, seems at issue with that of M. Elie de Beaumont made in the valley of the Durance.[8]

Agassiz urged that the whole space between the opposite chains of the Alps and the Jura was once filled with ice; that this mass of ice completely covered the Jura, with the exception of a few high crests, rising island-like above it, and mounted to a height of some nine thousand feet upon the Alps, while it extended on the one side into the northern plain of Italy, filling all its depressions, and on the other down to the plains of Central Europe.[9]

He further argues that while Charpentier's theory may explain the portage of the angular blocks, it cannot explain the other facts without having recourse to currents, and as against currents he quotes the view of Schimper, that in traversing the valleys they must have filled up the lakes with débris and have worn down their angular and steep banks, and however

Actes de la Société Helvétique, 1837, vii.-xviii.
[9] Geol. Sketches, 28.

violent and powerful the currents, they must have eventually relaxed and left in their last efforts portions of their burdens behind, while very few blocks indeed are found between the Alps and the Jura. How, again, was it that the softer and more intangible débris, such as the sand, etc., which such currents would have carried just as much as the blocks, did not fill up the inequalities of the valleys? Sand does occur in small quantities on the polished surfaces of the Jura, so does gravel, and it is generally on a quantity of one or the other, from a few inches to a few feet thick, that the blocks are found, the sand being below the gravel and the latter topped by the blocks. Sand would have been swept off these polished surfaces by a violent rush of water, which again would have deposited the big boulders at the bottom and overlain them with the finer débris, and not deposited them in the reverse order;[1] and he concludes that the existence of the polished surfaces, the fact that these are immediately covered with gravel and sand, and that on these again rest the angular blocks, is inconsistent with every explanation forthcoming about the blocks, previous to his own.[2]

The dispersal of the mantle of gravel which is found so much developed in Switzerland, and of the loess which he attributes to the denudation of the tertiary rocks there, he accounts for by invoking *débâcles*, due to the melting of the ice sheet, which may also have borne along with them ice-rafts loaded with stones.[3] His general conclusion is formulated in the following sentence: "Ce serait donc une grave erreur de confondre les glaciers qui descendent du sommet des Alpes, avec les phénomènes de l'époque des grandes glaces qui ont précéde leur existence."[4]

In 1833 M. Godet made an excursion to Sweden, and in 1837 communicated to the Helvetic Society an account of his observation of the erratic blocks of Scania, upon which M. Agassiz made some remarks, attributing them to a similar ice sheet to that he had invoked in Switzerland.[5]

In his famous memoir, entitled "Notice sur les Glaciers," published originally in 1839, he says: "Je conclus de l'ensemble

[1] Id. xix.-xxii. [2] Id. xxii. [3] Id. xxviii. [4] Id. xxix. [5] Id. 122.

de ces phénomènes et des stries analogues observées en Suède par M. Sefstroem, qu'à une certaine époque l'Europe entière s'est couverte de glacé: que cette époque est celle de la disparition des grands mammifères que l'on trouve déposés dans les graviers glacés du Nord."[6]

Such was the beginning of what I choose to call the Glacial Nightmare. A far more experienced geologist than I can ever claim to be has said of this famous and epoch-making memoir: "Dans ce premier développement trop rapide de l'hypothèse, le célèbre zoologiste dépassa le but, et ce fut à la difficulté de le restreindre au domaine de l'observation rigoureuse des faits comme des inductions rationnelles que l'on doit une grande partie des recherches auxquelles on s'est livre depuis."[7]

In his paper on glaciers, read before the Geological Society of London on November 4th, 1840, Agassiz says that if the analogy of the facts which he had observed in Scotland, Ireland, and the north of England with those of Switzerland be correct, it must be admitted, that not only glaciers once existed in the British Islands, but that large sheets (*nappes*) of ice covered all the surface. He again urged that glaciers did not descend from the mountain summits into the plains, but that they were the remaining portions of the sheets of ice which at one time covered the flat country. "It is evident," he says, "that if the glaciers descended from the high mountains and extended forward into the plains, the largest moraines ought to be the most distant, and to be formed of the most rounded masses, whereas the actual condition of the detrital accumulations is the reverse, the distant materials being widely spread, and true moraines being found only in valleys connected with great chains of lofty mountains. It must then be admitted that great sheets of ice resembling those now existing in Greenland once covered all the countries in which unstratified gravel is found."[8]

Elsewhere he says: "Scandinavian boulders, scattered upon English soil and over the plains of Northern Germany, tell us that not only the Baltic Sea, but the German Ocean also, was

[6] Op. cit. 15, 16.
[7] D'Archiac Histoire des Progrès, etc., ii. 240.
[8] Proc. Geol. Soc. iii. 331.

bridged across by ice on which these masses of rock were transported. In short, over the whole of Northern Europe, from the Arctic Ocean to the northern borders of its southern promontories, we find all the usual indications of glacial action, showing that a continuous sheet of ice once spread over nearly the whole continent, while from all the mountain ranges descended those more limited glacial tracks terminating frequently in tranverse moraines across the valleys, showing that, as the general ice sheet broke up and contracted into local glaciers, every cluster or chain of hills became a centre of glacial dispersion, such as the Alps are now, such as the Jura, the Highlands of Scotland, the mountains of Wales and Ireland, the Alps of Scandinavia, the Hartz, the Black Forest, the Vosges, and many others have been in ancient times." [9]

It will be noted that Agassiz's conclusions as here formulated, although they were supported by a wider generalization, did not carry the theory a step further than it had been carried by Bernhardi and Schimper. The memoir last quoted was nevertheless a great landmark in the development of the Glacial Nightmare, which at this stage of its history involved an appeal to an ice sheet co-extensive with the distribution of the drift deposits.

The theory thus formulated still largely maintains its hold upon the scientific world, and soon attracted adherents who wrote with weight and authority. Among these none was more remarkable than Charpentier himself.

In the year 1842 he communicated to the Bibliothèque Universelle of Geneva a paper in which he examined the erratic phenomena of Scandinavia and the north of Europe. In this paper he contests the notion that mountains are necessary to the existence of glaciers, except that they sometimes favour the accumulation of snow by the wind. "It is only," he says, "their cold snowy and rainy climate which causes the formation, development and movement of glaciers, and if these conditions existed in a flat country, even at the level of the sea, there would be nothing to prevent glaciers being formed and developed." He also urges that, according to his theory of dilatation, glaciers can and do move over level plains. In

[9] Geol. Sketches, ser. ii. 76.

order to account for the former large development of glaciers, he postulates that the winters were more snowy and the summers more rainy and colder, with a smaller difference in the mean temperature of summer and winter than exists at present; such a climate now exists in Tierra del Fuego and the Straits of Magellan. He further urges that the ice of glaciers is only partially derived from the *nevés* whence they spring, a large part of it accumulating *en route*, and he urges that there is therefore "no need to suppose that all the ice of the diluvian glacier of the north came from one single point. On the contrary, that glacier would be constantly increased by the rain and the snow falling directly upon it, and its increase must have gone on augmenting in proportion as it acquired a larger surface."

Charpentier does not shrink, therefore, from the conclusion that a large glacier once occupied Finmark, Lapland, Norway, and the greater part of Sweden and Finland; and he adds, "as it crossed the Baltic, and extended to the north of Germany, Prussia, and the plains of Russia as far as Moscow, and perhaps the east of England, there is nothing extraordinary in supposing that the erratic formation really reaches to Moscow, Stezyka, Oppeln, Leipzig, etc."

He says the phenomena of Scandinavia are like those of the Alps. The widely scattered débris of rocks are sometimes spread out, sometimes collected in bands and mounds, fragments of all sizes being found mixed together pell-mell, many having their prominent surfaces well preserved, while the rocks, as in the Alps, have marks of wear and rubbing, smooth surfaces, striæ, furrows, and vertical erosions in the form of cauldrons.

In order to account for the true diluvium, that is, the deposits containing sea-shells and traces of stratification attributable to the action of water, he postulates the former great extension of the Baltic and the existence of icebergs supplementing the ice sheet.

Answering Durocher, he refuses to believe that the marks of rubbing and of wearing on rocks is due to a great current from the north, which must have risen on the coast of Finmark to the height of 2500 feet, since they occur on the summit of Raipas and on the high plateau of Norwegian Lapland, and since water by itself cannot polish and scoop out rocks, it must have been a flood of water charged with other materials. He urges that such a torrent would have filled up the Baltic and other hollows which it traversed and would not have allowed the quiet or stratified deposit of the true diluvium. He argues that a large proportion of the erratic débris are not the remains of what the glacier carried on its back, but that glaciers have the power of raising such fragments from their bed to their surface, a view to which we shall presently revert. He says further that the external shape of the äsar is much more consistent with ice action than with that of a current or with floating ice, and so with the fine striæ occurring on the surfaces of hard rocks and also the giants' cauldrons.

He argues that the moraines, etc., are never permanent in a glacier which is growing, since it must overthrow and destroy them in its progress until it reaches its maximum extension where a terminal moraine would exist, if not swept away by torrents or buried under true diluvium. Every oscillation of a glacier gives rise to new accumulation of débris. He finds the remains of such frontal moraines in the so-called Steindamme stretching E. and W. in Prussia. As the ice-sheet retreated beyond the Baltic and shrank into its original limits, it would be lacerated or broken up into fragments forming separate glaciers which would each deposit moraines. These he finds in the existing äsars. Thus also he accounts for the striæ sometimes crossing each other. "The localities where the striæ cross have been covered at two different times by the ice. The first time they have been invaded by the great ice-sheet, which has scratched them in the general direction from north to south; and the second time they have been so by partial glaciers, whose action has there produced striæ, in some degree anomalous, which cross the first in various directions."

When two partial glaciers united, they would create a superficial moraine, such as one of the äsar, whose upper extremity rests against the rock or eminence forming the termination of the chain of heights, which, by separating the two glaciers, has given rise to two deposits of the superficial moraine. That the glacier moved from north to south is shown by the direction of the striæ, etc., whose slight deviations were caused by the slope and inequalities of the surface. This direction is also shown by the phenomena of the so-called stoss side and lee side. "In the Alps when a glacier encounters a rock or an eminence in its passage, the flank turned towards the side whence the glacier proceeds is always more rounded and rubbed than that turned towards the opposite direction." He similarly argues that the fine striæ on the rocks, noticed by M. Robert, which are generally parallel, run in the direction of the valleys and cross all the strata indiscriminately, without regard either to their direction or hardness, and show that they have been traced by hard bodies moving in concert at fixed distances, that is to say, attached to the same solid body, which prevented them yielding to any obstacles they might encounter; and he further argues that the polish on the surfaces left by the ice in the northern regions being better preserved than in other climates is due to the ice having begun to melt there at a later period, and therefore to the weathering having been less.

He urges further that while the advancing ice-sheet would on uneven ground, where its progress was impeded, upturn the soil, and pierce to the solid rock; where it could extend freely it must have stretched over "the true diluvium" without displacing it. In regard to the reason why the great northern ice-sheet moved southward, he argues that the Polar region, from lat. 60° to the Pole, was covered with snow. Between 60° and 70°, this snow was converted into a glacier, which in its dilatation could not extend in any other but a southern direction, because in other directions it had to encounter the resistance arising from snow and ice themselves.[1]

In a memoir entitled "A Period in the History of our Planet," published by Agassiz in the Edinburgh Phil. Journ. for 1843, we find him further developing his views. Thus he says: "The earth had already assumed its present contour, with the exception, perhaps, of the principal range of the Alpine chain, and of the mountains which rose simultaneously with it. . . . Then numbness seized the light 'sailors of the atmosphere,' the clouds and vapours; icy winds drove them

[1] Bibl. Univ. xxxix. 327-346.; Ed. New Phil. Journ. xxxiv. 55-74.

in a solid form to the earth, and like a huge winding-sheet, they enveloped the polar regions, the north of Europe and of Asia. The British Islands, Sweden, Norway and Russia, Germany and France, the mountainous regions of the Tyrol and Switzerland, down to the happy fields of Italy, together with the continent of northern Asia, formed undoubtedly but *one* ice-field, whose southern limits investigation has not yet determined. And as on the eastern hemisphere, so also on the western, over the wide continent of North America, there extended a similar plain of ice, the boundaries of which are in like manner still unascertained. The polar ice which at the present day covers the miserable regions of Spitzbergen, Greenland and Siberia, extended far into the temperate zones of both hemispheres, leaving probably but a broader or narrower belt around the equator, upon which there were constantly developed aqueous vapours, which again condensed at the poles; nay, if Tschudi's observations in the Cordilleras and Newbold's at Seringapatam shall be confirmed, and to these we may subjoin those made by earlier travellers upon the Atlas and Lebanon chains, the whole surface of the earth was, according to all probability, for a time one uninterrupted surface of ice, from which projected only the highest mountain ridges covered with eternal snow. The limits which would seem to be indicated by the various phenomena which we shall afterwards treat of, are very probably referable to a subsequent epoch, when the universal ice crust had already begun here and there to disappear, and particular tracts to emerge like oases in the immense icy desert." Having gone on to postulate that the present North Sea was filled to the bottom with solid ice, and that it reached to at least 8000 feet above the sea level on the Alps, he goes on to say that, "where there prevailed such an uncommonly low temperature as was necessary to envelop the earth to such an extent in a frozen covering, there would be little or no moisture and little or no melting and freezing, so that, in the temperate zones at least, even at the sea level there existed only snow or nevé but no ice, the temperature standing almost constantly below 32° Fahr." Hence, he urges, "there could be no moving of ice-fields in any direction during that period. . . . During the glacial period there was no motion;

not a brook, nor a rill furrowed the surface of the snowy covering, to remind by its purling that all life had not yet become torpid . . . According to our present state of knowledge, there could be no such thing as motion of the glacier ice, until the warmth of the surrounding media had become such as to cause the melting of the superficial strata. . . . Motion demands warmth as the medium and condition of its existence. It was not, therefore, until a return of warmth and the ice began to soften that all those grand phenomena could take place of which the surface of our present earth is witness. What, therefore, the majority of geologists have hitherto believed to be the result of immense floods and currents, and a few to have been caused by the increase and the progressive motion of glaciers, namely the transportation of erratic blocks, of alluvial boulders and the polishing and channelling of rocks—these are to me manifestations of the retreat of the glacial period, phenomena which denote the moment when an alteration in the climate of our earth began to confine the cold within those narrow boundaries which it now occupies—phenomena which denote the places whence torpid winter began her slow retreat towards the rocks of our lofty mountains, and to the ice-bound continents of the north. . . . It was the plains which first freed themselves from their torpid covering. . . . The snowy covering first began to dissolve, and by the trickling in of the water, and its recongelation in the interstices of the looser snowy strata, to change into more compact ice. With this transmutation of snow into ice began the movement of the masses of ice in the direction of the superficial slope, and their effects upon the solid ground on which they rested. . . . clefts opened under the expansive force of the solar heat, and thus furnished the waters which the melting of the superficial strata called forth in copious abundance. . . . The waters thus formed dug their mutable beds out of the extensive surface . . . undermined deep beds in the rolled matters, and thus produced valleys of denudation the direction of which is frequently inconceivable with reference to the present configuration of the surface. . . . Thus began the retreat. The centres towards which its course was directed were, on the one hand, the far north; on the other hand, the lofty mountain ranges of Central Europe, and from these regions descended the rude witnesses of the extension of the icy covering, those huge blocks which lie scattered in the plains of Germany, etc. . . . Torn from the ridges of the Scandinavian chain, these erratic blocks were moved forward on the surface of the ice coverings over the Baltic Sea, whose basin was filled with ice instead of salt water, and deposited on the plains of Northern Germany on the skirts of the ice-field." So also, argues Agassiz, was it the case with the Alps, "from which radiated the erratic blocks skirting them all round. . . . More and more was the ice mass compelled to recoil, more and more was the land denuded to become covered with verdure. The plains of Germany, Russia, France and Italy, were rescued from their frozen state, the Baltic was set free; the North Sea, for the most part, was anew covered with waves. But the more considerable chains of mountains, presenting by their elevation above the sea level a more secure resting-place to the ice-covering, still retained the eternal winter upon their summits and in their valleys, and thus there arose *detached* glacier ranges, which no longer formed a connected whole, like their common ancestress, but separate groups, each of which belonged to a particular mountain chain, etc." [1]

Charles Martins, writing in 1846, also championed the cause of a continental ice in Scandinavia. Arguing against Durocher, he urged that a glacier moves on a much smaller slope than any solid body, that of the Aar glacier, for instance, moving on a slope of 1·30 only. He also argues that a glacier can move along an undulating surface, and cites the glaciers of Alalein, Tschingel and Grünberg, while among ancient glaciers the *roches moutonnées* of the valley of the Aar, such as those of the two Baerenbuhls, those forming the *contrefort* of Naeglisgraelli, the Spitalnollen and the Kirchet, which are covered with undoubted glacial striæ. In regard to certain instances, such as that of Areskutan, cited by Durocher, where the glacier must have mounted a steep of 222 metres, he urges that it is as difficult to see how water could have mounted this as a glacier, and he argues that the levels of the country have altered since the glacial age. Again,

[1] Ed. New Phil. Journ. xxxv. 17-29.

answering Durocher, who doubted the capacity of ice to traverse a depression like the Gulf of Bothnia, he says that he found a greater difficulty in postulating the same movement for a current of water. As a matter of fact, the Swiss glaciers cross depressions and even lakes over which they pass because the specific gravity of ice is less than that of water. Thus the Aletsch glacier crosses the lake of Moirel and the Spitzbergen glaciers bridge over the sea for some distance in the same way, and he concludes that it is not at all improbable that the Swedish glaciers crossed the Gulf of Bothnia as they crossed the Swedish lakes. We can prove, he says, by the lateral moraines left by these northern glaciers that their base did not rest on the floor of the lake, but corresponded to the level of the water in it.[3]

In the year 1850 Mr. Robert Chambers read a paper before the British Association in the abstract of which he asks some pointed questions:—"How could ice," he says, "move over so large a portion of the North American continent, in a direction admitted to be tolerably uniform, allowing for slight deviations easily explicable as owing to irregularities in the original surface, and this without any mountain chain to give it forth? (2) How was this ice capable of ascending slopes and topping mountains of considerable height. (3) How, in such a valley as that of the Forth, could there be an ice-torrent of undeviating flow for many miles and deep enough to envelop hills many hundred feet high."[4]

In 1852 the same acute observer read a remarkable memoir before the Royal Society of Edinburgh on glacial phenomena in Scotland and parts of England, which has been too little noticed. In this he begins by urging, against the diluvial theory that he had seen what he considered an ice-polished surface crossed by a small runnel of water carrying minute gravel, and it was clearly observable where the water crossed, that the surface was changed, became rougher and dimmer, something like the difference between chased and polished goldsmith's work, and that he had examined the beds of many mountain streams accustomed to bring down all sizes of

[3] Bull. Soc. Geol. de France, iv. pt. i. pp. 99-102.
[4] Report, p. 79.

gravel, but never found in any instance those flowing outlines of abrasion which we see in the so-called glacial surfaces. There is also a palpable enough difference between the confused masses of mud and sand mingled with rounded blocks, which are found in connection with polished surfaces, and the sorted materials, gravel, sand, and clay, which are indisputably attributed to watery action. He similarly argues that the wide extent of abraded surface in Norway and Sweden and the uniformity of striation over such wide areas is inconsistent with the light, partial and irregular action of floating masses of ice. Passing on, he accepts the evidence for the former existence of local glaciers in mountain clusters now free from them as overwhelming, and especially commends the researches of Maclaren and C. Martins on the subject. He goes on to urge, however, that local glaciers will not explain all the facts. How Maclaren himself had shown that the eminence between Loch Long and the Gairloch, 600 feet high, is as perfectly smoothed along the top as is the bottom of either of the two valleys. How Professor Fleming had found striæ identical in direction with those of the plains on the summit of the Pentlands, 1400 feet high. How Maclaren, again, had shown that the striæ which cross the valley of Westwater, 800 to 900 feet above the sea, at the western extremity of the Pentlands, and which runs north and south, cross it from west to east, thus persevering in the normal direction in circumstances where, if anywhere, a divergence was to be expected ; and he asks how any group of Highland mountains, 3000 to 3500 feet high, could have sent out glaciers reaching Mid Lothian, 70 miles off, in such volume and depth as to envelop a range of hills to the depth of 1400 feet, and in such unyielding force as to cross a minor valley 800 or 900 feet above the sea without diverging in the least from its course, and he concludes that these results are not consistent with such a cause.

He then proceeds, in great detail and in clear language, to describe how very general the marks of abrasion are in Scotland. "I have, since August, 1850, found them," he says, "along the whole range of the coast north of Argyleshire, namely in Inverness-shire, and the counties of Ross, Sutherland, and Caithness. I have found them in the Isle of Skye, in situations independent of the Cuchullin hills; in the island of Mull; all along Loch Lomond, and Loch Katrine; even the picturesque eminences which constitute the Trosachs being *roches moutonnées* with abraded faces to the west. I have likewise found them in Perthshire, Fife, and Aberdeenshire. They are reported from Ayrshire. I have found a large lateral moraine near Maxwellton House, in Kircudbright, and seen fine smoothings with striation on the surface of Corncockle Muir in Dumfriesshire." And he urges that we should find traces of ice everywhere except on the highest summits, if all the rock surfaces were exposed, and if all those actually exposed had been equally capable of retaining the impressions made by ice. These facts, and especially the direction of the striæ, impelled him to postulate two independent stages of glaciation. "First, a general sweeping of the surface of the district by a deep flow of mobile ice, one great cause, if not the principal, of that enormous denudation which has been referred to, but of which the spoils, from the universality and power of the agent, were in a great measure swept away. Secondly, local and certainly subærial glaciers occupying certain valleys in the higher mountains and producing moraines, composed of brown clay, sand and blocks." And he quotes several instances as settling the question of two systems and epochs of glaciation, where the strong normal streaks athwart the hill from the north-west, a direction in which no local or limited mass of ice could move, are clearly chequered with fainter streaks, produced by simple down-hill movement, which happen to be from W.S.W.; and he points out further that this second glaciation has in certain cases obliterated the traces of the earlier one. And he especially quotes, as utterly destructive of the idea of exclusively local glaciers, the striation ascending obliquely out of Loch Katrine at Stronachlachar, passing over a high jutting hill promontory, reappearing under compact clay, in low ground, at some distance from the loch, and everywhere maintaining precisely one direction and that from N.N.W. To all appearance, the agent which produced these impressions came over a lofty range of hills from Balquidder and passed on to cross a scarcely lower range and descend into the valley of Loch Ard.

Turning to Scandinavia, he mentions how Böhtlinck had noticed that the marks of striation in the intermediate valleys are at an angle of 50 degrees with those on the intervening heights. He himself had observed that on a summit of 4000 feet of elevation at Jerkind and in the immediate neighbourhood of Sneehatte, the southern slope is abraded and polished almost to the top with striæ between north-east and south-west, a line totally irrespective of all the great mountains of the district, such as Sneehatte, which it sweeps laterally There is no higher ground from which the agent could descend to this spot, and it must have *crossed the chain* again on the *col* over which the road passes from Lavanger to Sundsvall in crossing the frontier between Sweden and Norway. "The whole of the space composed of rocks of chlorite schist, is abraded by an agent which has been able to shear sharply through the upturned edges of the strata, leaving clear striæ to mark its course. Surprising to say, that course has not been across from west to east nor from north to south in the line of the little ridges, but it is athwart both of these lines of hollow from north-west to north-east, and thus clearly has been independent of the form of the country."

Chambers explains the immense mass of detrital deposits which exists on the lee side of the Dovrefelds on descending from the bare *cols*, compared with the paucity of similar remains in Sweden, by the chain having been involved in a flow of ice which pressed hard upon and swept bare the hither side, but passed comparatively lightly over what lay beyond, leaving there little of the solid matter with which it was charged.

Again, after referring to the polished surfaces on the shores of the Gulf of Bothnia (in Finland) at Stockholm, and at Gothenburg, he says: "We are many hundreds of miles from the presumed centre of action, and the intervening space presents an infinite number of minor obstructions, all of which, however, have been swept over by the agent, whatever it was. How could any such agent, descending from hills only half as high as the Swiss Alps, travel over twenty times the space, in a condition, too, necessarily attenuated through the wideness of the country over which they must have spread?"

Turning to America he says: "There are proofs all over

Canada and to a point far south in the United States, as well as around Lakes Huron and Superior, of an abrading agent for the most part from the north-west. Mountains of 2000 feet in height bear on their sides and tops striation in that direction, while to the north-westward no mountains of greater elevation to serve as gathering places for glaciers can be pointed out."

Chambers finally concludes that the more general glaciation to which he refers points to a wide extension of the circum-polar ice and a southern movement of that envelope, in the course of which the surface was abraded and the detritus produced."[5]

In 1853 Murchison, who was such a staunch supporter of other views, was momentarily induced to countenance the theory of ice-sheets. He says: "There are many lofty tracts in Scotland, as well as in Norway and other countries, where the striation seems to be quite independent of the outline of the ground, thus indicating a grand and general movement of ice. It is to countries which present such phenomena, that the memoir of Dr. Rink forcibly applies; and it leads us to imagine, that there was a period when Scotland, particularly all the Highlands, was analogous to what Greenland is now, and when an icy mantle extended itself from higher plateaux into the fiords or friths on its sides."[6]

Meanwhile the general adoption of Forbes' conclusions in regard to the motion of ice apparently checked the spread of views about continental ice which were necessarily based on its motion being independent of gravity. It was apparently the researches of Rink in Greenland, and the papers he wrote on its great ice-fields, that caused a reversion to the ice-sheet theory.

Already in 1856 Hampus Von Post argued that local glaciers were not adequate to explain the phenomena of the Swedish surface beds.[7] He was followed a few years later by Kjerulf, *facile princeps* in his knowledge of these beds in Norway.

Kjerulf had been much struck by Dr. Rink's reports about the condition of Greenland, as affording an explanation of the

[5] Ed. Phil. Journal, liv. 229-281.

[6] Murchison, Address to the Geographical Soc., May 1853, 29. Penck, Vergletscherung der Deutschen Alpen, 15, 16.

glacial phenomena of the north. "From his description," he says, "it is obvious that here, as in our own land, we have an enormous stretch of country; that this country is completely covered with ice, which, on its west coast, is ever slowly pressing outwards to the sea, where it 'calves'; that immense loads of these ice 'calves' are every year floated away by currents in the same constant direction; and finally that this ice-cake attains a thickness of no less than 1000 feet.

"Here all the conditions of our problem are provided for. Let us figure to ourselves a universal ice-covering, a complete glaciation instead of mere individual glaciers, and the phenomena of friction takes its place harmoniously with all the other phenomena of the glacial epoch. In the existing state of Greenland is the very analogy we needed to justify our supposition; a gigantic cake of ice slowly moving outwards, and with the might of its tremendous pressure stripping, polishing and striating the rocks over which it glides, just like a common glacier, only in so much greater a proportion as the ice-cake of Greenland surpasses any glacier of Switzerland. . . . From an examination then of the whole upper covering of our country, we learn that the older part of this formation has a markedly arctic character, and that the oldest masses of all are moraines. We are entitled, therefore, I think, to conceive of Norway, at the close of the Tertiary period, as being in a true glacial condition. Its ice-covering had an outward movement like the present inland ice of Greenland. By this ice, 'rolled stones' and débris were borne outwards to the sea shore, where large and small blocks of stone were floated off and carried away on ice-floes. Hence arose those huge moraine masses (far too vast to have been produced by mere separate valley glaciers), which cover the whole flat land near the coast, and which may be traced on both sides of the Christiania Fiord. From this universal covering of moving land ice resulted the first great network of scorings and scratches, as well as the moraine masses extending out to the edge of the sea. Then followed a diminution of the glacial intensity. Instead of a continuous covering of ice, there were separate glaciers, which laboured down from the various mountain slopes into the open valleys at their feet. All the loose material which lay in their way they bore off as moraines, either along their sides or in front, or where two valleys converged as medial moraines. The process of striation thus went on anew in all these valleys, and in this way two different sets of scratches, lying across one another, might sometimes be produced."[8]

Kjerulf's researches in Norway were supplemented by those of his equally famous compatriot Torell in Sweden, and also by those of another Scandinavian geologist, Axel Erdmann, who, writing in 1868, speaks of "a glacial epoch during which the surface of our country was covered with an immense sheet of ice" as being established "by undoubted evidence."[9] He thus describes the state of things: "The Scandinavian peninsula, surrounded by a Polar sea, was covered with ice, except its highest points and its loftiest chains. Glaciers of several hundreds, or rather thousands, of feet thick filled the valleys and the depressions of the upper regions, extended over the low lands and reached the coasts, breaking, triturating, polishing and striating the surface over which they passed, and carried to the sea masses of stones, gravel, sand, and clay."[1]

Professor Geikie has described in felicitous language, which I am tempted to quote, the features of a Norwegian landscape. "No grander display," he says, "of ice action could one wish to see than that which the fiords and fiord valleys of Norway present. The smoothed and mammillated mountain slopes, the rounded islets that peer above the level of the sea like the backs of great whales, the glistening and highly polished faces of rock that sweep right down into the water, the great perched blocks, ranged like sentinels on jutting points and ledges, the huge mounds of morainic débris at the heads of the valleys, and the wild disorder of the crags and boulders scattered over the former paths of the glaciers continue to make a picture which no after amount of sight-seeing is likely to cause a geologist to forget. The whole country has been moulded and rubbed and polished by one immense sheet of ice which could hardly have been less than 6000 or 7000 feet thick." "The Gulf of Bothnia appears to have brimmed with

[8] Ed. New Phil. Journ. xviii., new series, 4-14.
[9] Formations Quaternaires de la Suede, 21.
[1] Id. 19, 20.

ice, which pressed up against and even in some places overflowed the lofty Norwegian frontier, through the valleys of which it found its way into the North Sea. Geikie quotes M. Horbye for the fact that the striæ cross the watershed between the two countries. "Mr. Törnebohm," he further says, "of the geological survey of Sweden, informs me that the glacier-carried erratics of Jemtland clearly show that the ice has passed from east to west—that is, right against the slope of the land; and according to Keilhau, similar blocks, which could only have come from Sweden, are now found in Trondjems fiord. The most remarkable circumstance in connection with some of these blocks consists in the fact, that they occur at a considerably greater height than the rock from which they have been derived. Thus at Äreskutan, Törnebohm found blocks at a height of 4500 feet which could not possibly have come from any place higher than 1800 feet." Mr. Geikie then quotes Nordenskiold for the conclusion that the Scandinavian ice-sheet once filled the Gulf of Bothnia and occupied the whole area of the Baltic sea, overflowing the Äland islands, Gothland, Öland, Bornholm and Denmark, and passing south-east over Finland into Russia across Lake Onega, Lake Ladoga and the Gulf of Finland.[2]

While these views prevailed in Scandinavia, the geologists of Germany had not, apparently, been attracted by them, and we first find them advocated there by the Swedish explorer Torell. In November, 1875, he communicated to the German Geological Society a paper on the results of a journey he made with Berendt and Orth to the Rudersdorf mountains, where he found the rocks marked by striæ, and adduced them as evidence of the former extension thither of the Scandinavian ice-sheet, and argued against the iceberg theory.[3] Similar groovings were met with by Credner at Klein Steinberg in Saxony, and by Penck on the mountain of Deortz near Taucha. Writing in 1879, the Norwegian geologist Helland argued that these facts could best be explained by the theory that the Scandinavian ice-sheet had deposited them. This he argued from the presence of Scandinavian erratics; from the fact that the soft beds of North Germany are precisely

[2] Great Ice Age, 403-405.
[3] Zeitschrift der Deutschen Geol. Gesell. xxvii. 961.

like moraine matter, in that the stones in them are of different sizes, heterogeneously mixed, and are many of them unrolled, etc., etc.; and he argues that since these beds are like a ground moraine in structure, their further limit marks the ice-foot of the ice-sheet.

He then proceeds to give these limits. He says that neither in France nor Belgium have the northern stones occurred. In Holland they abound, and the most southern specimen recorded is one described by Herr Winkler from Oudenbosch in North Brabant, near Breda. They do not occur, apparently, west of the Rhine, but have been found at Wagennigen on that river, and at Maaren, east of Utrecht. Turning to England, M. Helland claims, after microscopic and other examination, as certainly of Norwegian origin, the boulders of zirkon syenite and rhombic porphyry found in Holderness, the former he derives from Fredrikswärn and Laurwig in Norway, and the latter from Törnebohm. He also recognized granite like that of Christiania. These boulders have been found at Withernsea and Hornsea in Holderness, and as far north as Carnelian Bay. It is curious that these stones should have come from south-east Norway, Christiania, and not from its western coast. As these stones have not been found south of the Thames, Helland puts the limit of the Northern ice sheet at a line joining the mouth of the Thames and the Rhine. Turning to the Continent, he draws the frontier at a line passing through Jevennaar in Holland, Rheinberg, Essen, Dortmund, Unna, Werl, Soest, Paderborn, then northward to Lemgo, Bodenwerden, south to Hildesheim, then along the Hartzrand by Harzburg and Blankenburg, west of Hartzgerode, east of Stolberg; then the line makes a great loop by Nordhausen, Muhlhausen, Langensalza, Erfurt as far as Saalfeld, Gera, south of Zwickau, Chemnitz, Perna, Schlukenau, Warnsdorf, Reichenberg in Bohemia; then by the Kiesengebirge and the Sudetes to Teschen in Silesia, Lublin, Kief, Woronesch, Nishni Novgorod and up to within a short distance of the Ural mountains.

He goes on to urge that these blocks were all deposited by one great ice-sheet, which has also left its traces in the striæ of Saxony and of Denmark. That the beds are a ground moraine is proved by their containing striated blocks from

North Germany as well as from Scandinavia, the moving ice having gathered them all on its way. Such striated blocks have been found at Teutschenthal near Halle, and at Rudersdorf and Mischutz in Saxony.

He claims that the ice-sheet flowed over Norway and Sweden, overwhelmed the Danish islands, went over Scania, Bornholm, Öland and Gothland, and spread into North Germany until stopped by the Hartz and the Silesian mountains. On the Hartz the northern blocks are found, according to Lossen, as high as 450 metres, while Credner only puts them at 407 metres. Herr Orth has found them in the Silesian mountains at 439 metres high, and as they occur on the Scandinavian mountains at a height of 1700 metres, this gives a good idea of the mass of ice which must have occupied the intervening district. The mean slope of the Swiss glaciers is about twenty centimetres in twenty-four miles. With this slope the boulders would take 7000 years to travel from Scania to Saxony.[4]

In another memoir, published in the same volume, by Penck, marked by the same learning and knowledge, we find him subscribing to the same views. He urges that the Scandinavian land ice crossed both the Baltic and the North Sea, overflowing North Germany on the one hand, and reaching as far as England and Holland on the other. In Germany it flowed on till it was stopped by the Hartz, the Erz and Riesengebirge mountains, while further east it flowed over the flat lands of Central Russia. Northwards it united with the polar ice. This ice-sheet carried no stones on its back. The stones it carried were underneath it in a huge ground moraine, hence why true medial and side moraines do not exist in Germany. It was only when the ice-sheet had shrunk and retired to Scandinavia that it began to deposit these tokens of true glaciers.[5]

Turning from Germany for a moment to France, we find that a considerable discussion has ensued there as to whether glacial striæ occur in the neighbourhood of Fontainebleau. M. M. Belgrand and Collomb professed to find them there and also at La Padole and at Champcenil. M. Collomb urged that

[4] Zeitschrift der Deutschen Geologischer Gesell. xxxi. 63-106.

[5] Id. 150, 151.

the absence of lateral and frontal moraines there is what might be expected ; they can only exist in mountainous districts, and not when the ice traverses a flat district and has no limiting barriers, and the striæ point towards Orleans and Poitiers, in which direction there are no mountains. He urges again, that the striæ are at an angle with the contour of the country, the valleys, rivers, and denudations of the plateau of La Brie point to the north-west, while the striæ point to the north-east, in a direction almost at right angles, whence it follows that the valleys could not have existed at the time the striæ were made, and he dates the latter at the very beginning of the quarternary if not in the pliocene period.

M. Tardy, addressing himself to the same problem, solved it in another way. He says of these striæ at Fontainebleau, " Si nous cherchons l'origine de ce glacier, nous reconnaitrons que les seuls montagnes dont il pouvait descendre sont les Alpes Scandinaves; ce qui affirme le grand diluvium Scandinave signalé par M. E. de Beaumont, tout en le transformant en un grand glacier continental."[6] He thus brought the Scandinavian ice-sheet into central France. Hogard had, many years before,[7] virtually described the ancient glaciers of the Vosges as forming a covering mantle of ice.

Let us now revert to England. In 1862 Ramsay, in a paper on the glacial origin of lakes, abandoned the theory of icebergs to which he had adhered, and says: "I am constrained to return, at least in part, to the theory, many years ago strongly advocated by Agassiz, that in the period of extremest cold of the glacial epoch, great part of North America, the north of the continent of Europe, great part of Britain, Ireland, and the Western Isles were covered by sheets of true glacier ice in motion, which moulded the whole surface of the country and in favourable places scooped out depressions that subsequently became lakes." He then goes on, as Chambers had done, but without reference to him, to speak of two "periods of glaciation; one, later, of local glaciers and the other earlier, when great original sheets of ice converted the north of

[6] Bulletin Soc. Geol. de France, xxvii. 562.
[7] Coup d'Oeil sur le terrain erratique des Vosges, 1851.

Europe and America into a country like North Greenland . . . the change extended alike over the lowlands of the Northern and the Southern Hemispheres."[8]

Not only did Ramsay explain the origin of many lakes which are apparently basins scooped out of the rock by glacier erosion, but he tried to prove that the hollow occupied by the Menai Straits was scooped out by the great Scotch ice-sheet.[9]

Ramsay's views as to the excavation of lake basins by glaciers were supported by Mortillet in France, Gastaldi in Italy, and J. Geikie and others in England. I shall return to them in a later chapter.

In the same year Jamieson read a paper on the glaciation of Scotand, full of interesting and valuable observations. From these he concludes that "nothing short of ice filling the valleys up to the brim, and covering the country in one great winding-sheet will meet the requirements of the case. . . . Land ice moving in a volume like that seen in Greenland or on the Antarctic Continent, explains these and many other facts better, in my opinion, than any other theory yet proposed, and so far as I can see, the only strong objection against it is the extraordinary climate for this latitude that it requires. . . . We cannot account for such a development of ice in this country without supposing the whole of the atmospheric moisture, or nearly so, to have fallen in a frozen or snowy condition, and to have had to find its way off the land in the shape of solid ice; and such a climate in the latitude of Scotland cannot with any probability be supposed without some great changes in the physical geography of the northern Hemisphere, for it comes to this, that the whole of Scotland must, during the period of greatest glaciation, have been within the snow line.[1]

In a later paper, written in 1865, the same writer says: "It was, therefore, not in the form of narrow glaciers like those of the Alps that the ice existed at this time, but as a thick cake like that of North Greenland, enveloping both hill and dale, and flowing off not so much on account of the

[8] Quart. Journ. Geol. Soc. xviii. 204.
[9] Id. xxxii 116. [1] Id. 178-181.

inclination of the bed on which it rested as owing to the internal pressure exerted by the immense accumulation of snow over the whole interior of the island, somewhat in the way that a heap of grain flows off when poured down on the floor of a granary. The floor is flat and therefore does not conduct the grain in any direction; the outward motion is due to the pressure of the particles of grain on one another; and given a floor of infinite extension, and a pile of grain of sufficient amount, the mass would move outwards to any distance; and with a very slight pitch or slope, it would slide forward along the incline. The want of much inclination in the surface of a country, and the absence of great Alpine heights, are therefore objections of no moment to the movement of land ice, *provided we have snow enough.*" Turning to the glacial markings on Schehallion and Morven, he urges that these two cases are excellently adapted for showing us to what a prodigious extent the snow and ice must have accumulated.[2]

Sir A. Geikie, who, until the year 1861, had been an advocate of the Iceberg theory as explaining the Scotch phenomena, now abandoned that view. In his memoir on the glacial drift of Scotland, published in 1863, he writes: "The fact that the striations and groovings are not confined to the Highland glens, where actual glaciers may well have existed, but extend far away over the Lowlands from sea to sea, seems sufficient proof that, although such markings are undoubtedly traces of the course of a moving body of land ice, this ice could not have been in the form of a mere ordinary glacier. For the central valley is of greatly too large an extent ever to have been filled by any glacier which could descend from the Highlands. Moreover the same markings occur high on the sides and summits of the Highland mountains, which no glacier, in the usual sense of the term, could ever have reached. . . . If these markings are in any one place evidence of the former existence of a stream of ice, their universality must also prove that the ice existed not as mere local glaciers descending the chief valleys, but as one wide sheet covering the whole, or nearly the whole, of the

[2] Journ. Geol. Soc. xxi. 166.

country. . . . The former extension of a vast icy mantle over the surface of our island accounts well for the character of the *roches moutonnées* and for the diversion of the striations from the main ridges of the country. We see at the same time how these markings should not be confined to the valleys, but run along and over the sides and summits of the hills. We see how masses of ice, many hundred feet in thickness, must have accumulated in the Highland glens, overflowing into the adjacent valleys and seeking egress to the sea, by crossing over ranges of high ground, as in Gareloch, Bute and Cantyre, where, it will be remembered, the striations on the rocks run up and over the ridges and are as clearly shown on the hill-tops as in the valleys. . . . Down the whole of the west coast, from Cape Wrath to the Mull of Cantyre, one long expanse of ice filled up the fiords, and stretched out into the Atlantic. From the uplands of Wigton and Galloway, the icy stream swept down into the valley of Solway, and onward for Ireland. From the hills that border the lovely valley of Liddesdale far away into the blue Cheviots, the same universal mantle of ice threw its folds athwart the hills and dales of the north of England. . . . The whole country was thus probably sealed in ice, save perhaps some high cliffs and precipices. . . . The rounded and flowing outlines of the hills, the smoothed contour of the knolls, and the long banks of boulder clay, all tell of the passage of the great ice flow. . . . It appears that the island was covered, as large tracts of Greenland are, with a wide sheet of snow and ice which, constantly augmented by fresh snow falls, pushed its way outward from the mountain chains towards the sea."

In a paper by Mr. S. V. Wood, jun., read before the Geological Society in 1866, on a section at Litcham, he argues that, "when the masses of chalk in the Norfolk cliffs were deposited, the land was occupied by a great glacier, from whose seaward termination they were carried and dropped into the marine sediment. . . . . In speaking of a glacier, I do not refer to any such as those occurring in mountain regions, but to one enveloping the whole surface of the country, as de-

[3] Op. cit. 76-96.

scribed by Dr. Sutherland, and now covering the land around Baffin's Bay. . . . Much of the centre and south-west of Norfolk was, during the Lower Drift period, the seat of this capping glacier, and its principal seaward termination was in the direction of Cromer, where the bulk of its moraine was spread out under the sea."[4]

Writing in 1871, Mr. James Geikie says: "The only theory which seems to satisfy all the requirements of our present knowledge, is that so long ago advanced by Agassiz. Every part of Scotland, with the possible exception of a few peaks or tips of some of the loftier mountains, has certainly been buried under snow and ice. This *mer de glace* must have levelled up the valleys and occupied all the fiords, sounds and shallow seas around our island. Below this deep sea of ice hill slopes were ground and polished, valleys were deepened and smoothed, and the wreck and rubbish resulting from all this work gathered unequally below the ice, according as the direction and pressure of the superincumbent mass determined."[5]

In a paper read before the Geological Society in 1872, on the Ice-sheet in North Lancashire, etc., Mr. Tiddemann says: "The ice producing these effects must have come from the north, and not from the south. It is clear that local glaciers going down hill have not made them; for in this case the ice has not been travelling in the direction of the greatest fall, but considerably across it. Nor can they have been produced by icebergs grounding, as might be suggested; for then there would be some indication of coasting, whereas these markings are all nearly parallel, and going up the hill instead of along it. Nor will an ice-foot explain it; for then we should have scratches radiating from the higher ground. In fact I can conceive of no arrangement but that of land ice pushed up over this hill (i.e. Claughton Fells) by pressure from behind, and obliged to maintain a definite course by ice to the right and left of it."

Again, speaking of the scratches, etc., at Ingleborough, he says: "They certainly show no signs of having been formed by any small local glaciers which had their birth on Ingle-

[4] Quart. Journ. Geol. Soc. xxiii. 84-87.
[5] Geological Magazine, viii. 547.

borough; for they are transverse to the course of any such glaciers." Again, speaking of similar marks at Bowland Knotts: "If these were produced by a local glacier, what were its gathering grounds? Not the Fells immediately west; for then the scratches would be at right angles to their present course. Looking north, the nearest ground equalling this in height is seven and a half miles distant, and between this and that lies a broad valley, excavated to the depth of about 1000 feet. Ingleborough, with its additional 900 feet, could hardly fill up that valley, and make itself felt at that distance. There is no ground equally high nearer than Pendle Hill, twelve miles off, and across two broad valleys. The height at which these scratches occur across this ridge is greater than the height of some of the passes across the Fells to the west of it. There can be little doubt that the ice stream passed entirely over these Fells." Speaking of certain scratches running in directions transverse to each other, he recalls a suggestion of Professor Ramsay, that it is possible that there may have been undercurrents in the ice-sheet caused by the form of the ground. Again, he urges "that such an agent as is needed to explain the facts is to be found in a great ice-sheet pushing on from its northern gathering grounds, recruited by the greater elevations on its course, but overriding the lesser, grinding down and smoothing by its weight and friction rocks presenting but a gentle incline, tearing up and turning over the basset edges confronting its approach. . . . I know not any phenomenon connected with the glacial period which gives so vivid a picture of a resistless force working in an undeviating course over hill and dale, across the ordinary drainage channels of the country as this wreck and ruin of all opposing obstacles." Our author further argues that a barrier of ice filled up the Irish Sea between the Lake district and the Isle of Man, which diverted the drainage of the Lune and Ribble from their natural outflow. This barrier, he says, was but the line of junction of the ice of the Pennine chain with that from the Lake district, and to the eye they must have presented the appearance of one great sea of ice, and this barrier must have been supported and shouldered up by other ice coming from portions of the Lake district still further west. By this ice stream he explains the fact of the striæ on Anglesea being from the N.N.E. and not radiating from the mountains, and that you get foreign blocks from the hills of Cumberland there. Similarly Mr. Close has shown that on the opposite coast of Ireland the glaciation is along the coast and not from the land. Professor Phillips had observed the same thing at Bray Head.[6]

In his "Great Ice Age," first published in 1873, Professor Geikie says: "On every hand we are furnished with abundant proof of the former existence of a great *mer de glace* in Scotland. From the tops of some of the higher mountains down to the edge of the sea, no part of the country has escaped abrasion. . . . Nay, most of the islands which lie off the coasts plainly indicate by striations and other glacial markings that ice has swept over them. They are smoothed not from the centre to the circumference, as would have been the case if they had supported separate glaciers of their own, but the striations go right across them from side to side. It cannot be doubted, therefore, that the ice actually crossed from the mainland over what now forms the bed of the sea. In Lewis I found the northern portion of Long Island glaciated across its whole breadth from S.E to N.W. The land ice that swept over this tract must have come from the mountains of Ross-shire, a distance of not less than thirty miles, and have filled up the whole of the North Minch (sixty fathoms in depth) and overflowed Lewis to a height of 1300 feet at least, for I found Suainabhal, a conspicuous mountain rising to the south of Loch Roag, was merely a huge *roche moutonnée*, strongly glaciated from base to summit by ice which had passed over it from the S.E. . . . In like manner the island of Bute has been scored and smoothed from end to end by a mass of ice which, streaming from the highlands of Argyleshire, filled up the Kyles and then passed southward over the whole island to occupy the bed of the Frith of Clyde between Bute and Arran. Many of the islets scattered along the western coasts tell the same tale." After describing the edge of the Antarctic continent he says: "And such must have been the aspect presented by the margin of the old ice sheet which in the early stages of the glacial epoch mantled Scotland and its

[6] Journ. Geol. Soc. xxviii. 471-491.

numerous isles, filling up the intervening straits and channels of the sea, and terminating far out in the Atlantic Ocean in a flat-topped vertical cliff of blue ice."[7]

Similar views are developed at greater length in the same author's "Prehistoric Europe."

"Thus are we driven to conclude," he says, "that during the climax of the Glacial Period all Scotland was drowned in a widespread *mer de glace*, which coalesced in the north and east with a similar sheet that crept outwards from Scandinavia. To the west the Scottish ice meeting with no impediment to its course overflowed the Outer Hebrides to a height of 1000 feet, and probably continued on its path into the Atlantic as far as the edge of the 100 fathom plateau, where the somewhat sudden deepening of the sea would allow it to break off and send adrift whole argosies of icebergs. The height reached by the upper surface of the ice that overwhelmed the Outer Hebrides enables us to ascertain the angle of slope between these islands and the mainland. This was one in 211, that is to say, the inclination of the surface of the ice-sheet was about twenty-five feet in the mile, an inclination which would appear to the eye almost like a dead level.[8] . . . Glacial striæ and boulder clay have been followed over all Ireland. . . . Our sister island was buried under an ice-sheet hardly less extensive than that which overwhelmed Scotland. . . . The ice flowed off Ireland in all directions save to north-east in Antrim, upon the coast of which it encountered the Scotch *mer de glace*, which forced it to turn away to north-west and south-east; but along the whole western and southern shores, where no obstacle to its passage intervened, it seems to have swept in one broad and continuous stream, out probably as far as that of Scotland, into the Atlantic. The thickness attained by the ice that flowed into the Irish Sea from Scotland, where it coalesced with the *mer de glace* coming from the eastern seaboard of Ireland, and also with that creeping from England and Wales, makes it quite certain that the area now occupied by that sea must at that time have been filled with glacier ice.[9] Despite the pressure exerted by the massive sheet that made its

[7] Op. cit. 77-81.
[8] Prehistoric Europe, 189.
[9] Id. 189, 190.

way outwards from the Pennine chain, the English ice could not escape into the basin of the North Sea, and consequently we find stones from Scotland, Northumberland and Durham plentifully present in boulder clay all along the eastern maritime districts of England. More than this, when we get as far south as the Humber, and follow the spoor of the ice as indicated by the carry of the boulder clay stones, we are led across Lincolnshire into the Midland Counties by Nottinghamshire and Leicestershire. . . . The North Sea was filled with a massive *mer de glace,* continually advancing in a general S.S.W. direction, the presence of which is distinctly traceable in the remarkable deflection of the glaciation all along the sea-board of Scotland, from Stonehaven southwards." Our author urges that the greater elevation and extent of the Scottish mountains alone protected the invasion of Scotland by this alien ice-stream, while in England, where the ground was lower, "it overflowed bodily and crept south-west across the Midland tableland on its way to the valley of thè Somme and the Bristol Channel. This is proved," he says, "by the carry of the local stones and by the appearance in the till at Cornelian Bay and Holderness of boulders of two well-known Norwegian rocks, which were recognized by Mr. Amund Helland, etc."[1]

"The ice," he continues, "which would thus appear to have streamed transversely across England, eventually coalesced with that which overflowed from the basin of the Irish Sea south-east through Cheshire together with that which streamed east from the Welsh uplands, and the united *mer de glace* thereafter made its way into the Bristol Channel. Here it joined the thick ice that flowed out to sea from the high grounds of North Wales. . . . In the south-eastern counties, so far as we know at present, the ice-sheet at the climax of the Glacial Period did not extend farther than the valley of the Thames"[2]

"Thus in time," Mr. Geikie says elsewhere, "all Scotland became enveloped in ice that flowed west to break off in deep water beyond the Hebrides, and east to meet the Scandinavian *mer de glace* which had all the while been creeping outwards

[1] Prehistoric Europe, 192, 193. [2] Id. 193.

into the basin of the North Sea. The North of England was also shrouded in ice, part of its sheet coalesced with the united Scottish and Scandinavian ice-sheet that filled up the German Ocean. How far south in England that ice-sheet flowed, still remains to be more rigorously determined. A broad belt of ice overflowed from the basin of the Irish Sea, and uniting with the glaciers that descended from Wales, spread in the direction of the Severn Valley. In like manner the ice that flowed eastward from the Pennine Chain to coalesce with the *mer de glace* of the North Sea appears to have advanced into Lincolnshire. But in the high grounds of Derby the ice-flow may have been more or less independent, as it appears also to have been in Charnwood Forest. Ireland with its lofty hills and humid climate, was, like Scotland and Wales, more or less buried in snow and ice, and its immense glaciers, uniting with those of Scotland and England, must have filled up the Irish Sea."[3]

We will next turn to Mr. Croll, who, in his well-known work "Climate and Time," says: "At one time not only an arctic condition of climate prevailed in our island, but the greater part of the temperate region down to comparatively low latitudes was buried under ice.[4] Without the conception of continental ice, the known facts connected with glaciation would be perfectly inexplicable. It was only when it was found that the accumulated facts refused to be explained by any other conception, that belief in the very existence of such a thing as continental ice became common. But although most geologists now admit the existence of continental ice, yet nevertheless, adequate conceptions of its real magnitude are by no means so common. Year by year, as the outstanding facts connected with glaciation accumulate, we are compelled to extend our conceptions of the magnitude of land ice. Take the following as an example. It was found that the transport of the Wastdale Crag blocks, the direction of the striæ on the islands of the Baltic, on Caithness, and on the Orkney, Shetland and Faroe Islands, the boulder clay with broken shells in Caithness, Holderness and other places were inexplicable on the theory of land ice. But it was so only in consequence

[3] Id. 351, 352. [4] Climate and Time, i. 7.

of the inadequacy of our conceptions of the magnitude of the ice."[5] Speaking of the glacial period, Mr. Croll says: "All the northern seas, owing to their shallowness, must, at that period, have been blocked up with solid ice, which displaced the water and moved along the sea-bottoms the same as on dry land. In fact, the northern seas, including the German Ocean, being filled at the time with glacier-ice, might be regarded as dry land."[6]

Again he says: "It may now be regarded as an established fact that during the severer part of the glacial period, Scotland was covered with one continuous mantle of ice, so thick as to bury under it the Ochil, Sidlaw, Pentland, Campsie and other moderately high mountain ranges. For example, Mr. J. Geikie and Mr. B. N. Peach found that the great masses of the ice from the N.W. Highlands came straight over the Ochils of Perthshire and the Lomonds of Fife. In fact, these mountain ridges were not sufficiently high to deflect the icy stream either to the right hand or to the left; and the flattened and rounded tops of the Campsie, Pentland, and Lammermoor ranges bear ample testimony to the denuding power of ice. . . . It is therefore evident that the great mass of ice entering the North Sea to the east of Scotland, especially about the Firths of Forth and Tay, could not have been less, and was probably much more, than from 1000 to 2000 feet in thickness.[7] We must consider that the ice from Scotland and England was but a fraction of that which entered the North Sea. . . . The Scandinavian ice, before it could break up, would thus, like the Scottish ice, have to pass the North Sea and into the Atlantic. It could not pass to the north, or to the north-west, for this ocean in these directions would be blocked up by the polar ice." After urging that the deep trough along the coast of Norway would be rapidly filled up, he continues: "Thus the only path for the ice would be by the Orkney and Shetland Islands. . . . In all probability, had Scotland been a low, flat island, instead of being a high and mountainous one, the ice would have passed completely over it. But its mountainous character, and the enormous masses of ice at the time proceeding from its interior, would effectually prevent this, so that the

[5] Id. 385, 386. [6] Id. 438. [7] Id. 442.

ice of Scandinavia would be compelled to move round by the Orkney Islands. Consequently, these two huge masses of moving ice—the one from Scotland and the much greater one from Scandinavia—would meet in the North Sea, probably not far from our shores, and would move side by side northwards into the Atlantic as one gigantic glacier. . . . The median line of the Scandinavian and Scottish ice-sheets would be situated not far from the east coast of Scotland. The Scandinavian ice would press up as near to our coast as the resistance of the ice from this side permitted, the enormous mass of ice from Scotland, pressing out into the North Sea, would compel the Scandinavian ice to move round by the Orkneys and would also keep it at some little distance from Scotland. Where, on the other hand, there was but little resistance offered by ice from the interior of the country, and this might be the case along many parts of the English coast, the Scandinavian ice might reach the shores and even overrun the country for some distance inland."

Turning to the South Swedish ice-sheet, Mr. Croll says: "All this mass of ice must have gone into the shallow Cattegat, and thence passed over Jutland into the North Sea. After entering the North Sea it would be obliged to keep between our shores and the ice coming direct from the western side of Scandinavia. . . . A very large proportion of the Scandinavian ice would pass into the Gulf of Bothnia; after passing down the Baltic, a portion of the ice would probably move south into the flat plains in the north of Germany, but the greater portion would keep in the bed of the Baltic, and of course turn to the right round the south end of Gothland and then cross over Denmark into the North Sea. . . . The Gulf of Bothnia and the entire Baltic were filled with one continuous mass of land ice, derived from the drainage of the greater part of Sweden, Lapland, and Finland. In fact, the whole glacial phenomena of Scandinavia are inexplicable on the hypothesis of local glaciers. . . . The icy stream, after passing Karlskrona, bent round to the west along the present entrance to the Baltic and again invaded the mainland, and crossed over the low headland of Christianstadt, and thence passed westwards in the direction of Zealand. This immense Baltic glacier would in all probability pass over Denmark, and enter the North Sea somewhere to the north of the river Elbe, and would then have to find an outlet to the Atlantic through the English Channel, or pass in between our eastern shores and the mass from Gothland and the N.W. shores of Europe. . . Is it possible that the entire Atlantic, from Scandinavia to Greenland, was filled with land ice? Astounding as this may at first appear, there are several considerations which render such a conclusion probable. The observations of Chambers, Peach, Hibbert, Allan and others show that the rocky face of the Shetland and Faroe Islands has been ground, polished and striated in a most remarkable manner; that this could not have been done by ice belonging to the islands themselves is obvious, for these islands are much too small to have supported glaciers of any size, and the smallest of them is striated as well as the largest. Besides, the uniform direction of the striæ on the rocks shows that it must have been effected by ice passing over the islands. . . . That the whole of these islands have been glaciated by a continuous sheet of ice passing over them was the impression left on the mind of Robert Chambers after visiting them. This is the theory which alone explains all the facts.[8] The glacial phenomena of Scotland are wholly inexplicable upon any other theory than that, during at least a part of the glacial epoch, the entire island from sea to sea was covered with one continuous mass of ice of not less than 2000 feet in thickness."

Such have been the drafts upon space invoked by the more extreme advocates of "omnipotent ice." They have made corresponding drafts upon time. Thus it was calculated, from the average progression of glaciers, that it would require 10,000 years to carry a boulder 100 miles. Now, as the erratics from Scandinavia have been found in Russia, at least 700 miles from Sweden, as Murchison testifies, this means that these erratics were actually in process of transportation for 70,000 years; but this calculation is outdone by others. M. Mortillet, the well-known editor of the "Matériaux pour l'Histoire de l'Homme," says in his recent work on Prehistoric times: "The length of the great Alpine glaciers varies from 110 to 280 kilometres. Eighteen observations made upon the rate of

[8] Op. cit. 451.

progress of modern glaciers gives a mean rate of $62^{m}\cdot66$ per annum. An erratic block, therefore, would take 4468 years to travel from the superior limit of the glacier to its inferior one; but this number is much too small. These eighteen observations were made among the mountains, where the slope of the ground is very great, and as frozen water follows the same law as liquid water, its speed and thrust diminish rapidly with the diminished inclination." M. Mortillet then cites in proof of this the well-known observations of Agassiz, and says: "The slope of the Quaternary glaciers was at least five times less than that of existing glaciers, whence the rapidity of flow was also five times less, which would transform the 4468 years just cited into 22,340 years. As the Quaternary glaciers quitted the narrow mountains and spread themselves out into the plains, it would require a proportional increase in the time required, and we must at least double the number of years here indicated. The quaternary glaciers, again, must have remained a long time at their position of greatest extension, as is proved by the enormous moraines forming series of hills at their termination. Lastly, the retreat of the glaciers, which must have required as long a time as their advance. This advance and retreat, again, was not continuous, but subject to oscillations which would absorb a considerable time. One is therefore below the truth in assigning 100,000 years as the term of existence of the Glacial Period."[9] Applying M. Mortillet's figures to the Russian boulders, and replacing 280 kilometres by 700 miles, we shall have a measure and test of the courage of some prominent champions of the Glacial theory.

[9] Le Préhistorique, 625, 626.

CHAPTER VII.

THE GROWTH AND CULMINATION OF THE GLACIAL NIGHTMARE.

II.—The New World.

"In thrilling regions of thick-ribbed ice."—"Measure for Measure," Act III. sc. 1.

North America: Agassiz, 1846-1850, etc., Dana, Winchell, Newberry, Morris—Ice Cap in the Northern Hemisphere: Agassiz and Dana—Southern Ice Cap: Agassiz and Haast—Tropical America: Hartt, St. John, Belt, A. Wallace.

In the previous chapter I have traced the progress of what I have ventured to call the glacial nightmare in the Old World. I now propose to extend and complete the survey by describing the application of the theory to the New, and I will begin with North America.

In 1846 Agassiz went to America, and he tells us how, on landing in a new continent full of promise for him, he sprang on shore at Halifax and started at a brisk pace for the heights above the landing. On the first undisturbed ground, after leaving the town, he was met, he says, by the familiar signs, the polished surfaces, the furrows and scratches, the *line engraving* of the glacier, so well known in the Old World.[1]

In 1850 Agassiz published a work on Lake Superior. In this he devoted a chapter to the erratic phenomena of that district, in which he says, *inter alia*: "I maintain that the cause which has transported these boulders in the American continent must have acted simultaneously over the whole ground which these boulders cover, as they present throughout the continent an uninterrupted sheet of loose materials, of the

[1] Geol. Sketches, ser. II. 77.

same general nature, connected in the same general manner, and evidently dispersed at the same time." [2]

"It is well known," he says elsewhere, "that the usual characteristic marks of glaciers extend over the face of the land in the eastern half of the continent, from the Atlantic shores to the States west of the Mississippi, and from the Arctic Sea to the latitude of the Ohio, in its middle course, while within the range of the Alleghanies they stretch as far south as Georgia and Alabama. In no other region where these traces have been observed do they extend over such wide tracts of country in unbroken continuity, this being of course owing to the level character of the land itself."[3] After describing the phenomena as they present themselves in America with his usual graphic force, and urging that there are few traces of local and sporadic glaciation, he continues: "In short, the ice of the great glacial period in America moved over the continent as one continuous sheet, overriding nearly all the inequalities of the surface. . . . The polished surfaces stretch continuously over hundreds and hundreds of miles; the rectilinear scratches, grooves, and furrows are unbroken for great distances; the drift spreads in one vast sheet over the whole land, consisting of an indiscriminate medley of clays, sands, gravels, pebbles, boulders of all descriptions, so uniformly mixed together that it presents hardly any difference in its composition, whether we examine it in New England, New York, Pennsylvania, Ohio, Michigan, Indiana, Illinois, Wisconsin, in Iowa beyond the Mississippi, in the more northern territories or in Canada." [4] Speaking of the fact that in America the boulders are for the most part rounded, while there are few erratics like those of Europe which have preserved their angles intact, he says: "It is evident from this that the ice overtopped the rocky inequalities of the land and that the detached fragments remaining beneath the icy covering underwent the same action from friction and pressure to which the whole mass of drift was subjected."[5] The few angular blocks that are found he attributes to the period when on the retreat of the ice some of the higher portions of the rock were laid bare, and he thus accounts for their greater

[2] Silliman's Journal, x. 83, etc.
[3] Geol. Sketches, 78, 79.
[4] Id. 82.
[5] Id. 83, 84.

frequency in New England, where the mountain elevation is greatest.

He urges again that the mineralogical character of the loose materials forming the American drift leaves no doubt that the whole movement, with the exception of a few local modifications easily accounted for by the lay of the land, was from north to south, all the fragments not belonging to the localities where they occur being readily traced to rocks *in situ* to the north of their present resting-places. "The farther one journeys from their origin, the more extraordinary does the presence of these boulders become. It strikes one strangely to find even in New England fragments of rock from the shores of Lake Superior; but it is still more impressive to meet with masses of northern rock on the prairies of Illinois and Iowa. One may follow these boulders to the 40th degree of latitude, beyond which they become more and more rare, while the finer drift alone extends farther south."

Again, in favour of the same conclusion, he cites the *roches moutonnées* of America. "Whenever," he says, "the natural surface of any hill, having a steep southern slope, is exposed, the marks are always found to be very distinct on the northern side and entirely wanting on the southern one, showing that . . . . the mass moved up the northern slope, forcing its way against it, grinding and furrowing the northern face of the hill as it moved over it, but bridging the opposite side in its descent without coming into contact with it. This is true not only of hills, but of much slighter obstacles which presented themselves in the path of the ice. Even pebbles embedded in masses of pudding stone, but rising sometimes above the level of the general surface, often have their northern side polished and scratched, while the southern one remains untouched."[6]

Again, speaking of Maine, he says: "The glacial traces over the State of Maine, and especially between Bangor and the sea-coast, afford the means of estimating approximately the thickness of the ice-sheet which once moved over the whole land, as well as its limitations during a later period, when it had begun to wane. In order to advance across a hilly

[6] Geol. Sketches, ser. II. 85, 86.

country and over mountainous ridges rising to a height of 1200 and 1500 feet in the southern part of the State, and to a much higher level in its northern portion, the ice must have been several times thicker than the height of the inequalities over which it passed; otherwise it would have been encased between these elevations, which would have acted as walls to enclose it. We are therefore justified in supposing that the ice-fields, when they poured from the north over New England to the sea, had a thickness of at least five or six thousand feet."[7]

Elsewhere he gives us another measure of the postulated ice-sheet in depth and extent. He urges that "the slopes of the Alleghany range wherever examined are glacier-worn to the very top with the exception of a few points, but these points are sufficient to give us data for the comparison. Mount Washington, for instance, is over 6000 feet high, and the rough, unpolished surface of its summit, covered with loose fragments, just below the level of which glacier marks come to an end, tells us that it lifted its head alone above the desolate waste of ice and snow. In this region then the thickness of the sheet cannot have been much less than 6000 feet, and this is in keeping with the same kind of evidence in other parts of the country, for wherever the mountains are much below 6000 feet, the ice seems to have passed directly over them, while the few peaks rising to that height are left untouched; and while we can thus sink our plummet from the summit to the base of Mount Washington and measure the thickness of the mass of ice, we have a no less accurate indication of its extension in the undulating line marking the southern termination of the drift. . . . We may trace the southern border of our ancient ice-sheet on this continent by the limit of the boulders; beyond that line it evidently did not advance as a solid mass, since it ceased to transport the heavier materials."[8]

The views here formulated were fully accepted by some of the most prominent American geologists. Thus Professor Dana, whose authority as their mouthpiece none will dispute, says: "The surface of the glacier of America must have been

[7] Id. 151.

[8] Id. 98, 99.

of unblemished whiteness; for from New England west to the Mississippi there was not a peak above its surface, excepting the White Mountains, and these probably had their cap of snow. . . The direction of the scratches and the extent of country they cover appear to show that over New England and Eastern Canada there was a gradual rise in the surface of the glacier towards the *north-west* and over western New York and Western Canada towards the *north-east*, and that the ice-summit—or the region towards which the directions converge and therefore the opposite slopes rise—was over the Canada water-shed nearly north of Montreal. From this, its southern portion, the broad ice-range stretched northward and north-eastward, for this is proved by the *south-west* direction of the drift and scratches over the country from Lake Huron to and beyond Lake Winnipeg. Since evidences of the great southward moving glacier fail over the region west of a line passing from a few degrees west of Winnipeg, south-eastward through Western Minnesota and Iowa, near the meridians of 98° to 100°, and all the way westward to the borders of California and Oregon, if not to the Pacific coast, the ice thinned out towards the interior of the continent, and was mostly absent except about the higher parts of the Rocky Mountains. . . . The lower limit of the glacier of New England may have been over the shallow border of the ocean, perhaps as far out as the fifty fathom or one hundred fathom line of soundings, which on the east passes outside of St. George's Shoal, and off Sable Shoal east of Nova Scotia. Long Island and the other islands off New England have been made by Clarence King and some other writers the course of a part of the terminal moraine, and a range of boulder hills covering New Jersey, west of Perth Ambay, have been pointed out by Professor Smock as another portion of it. In Wisconsin the Green Bay Valley is bordered on the east, west and south by a range of stony hills called the Kettle Range, and this range is described as of moraine origin by Chamberlin and Irving, and as directly connected with the Green Bay ice during part of the era of ice. At the time of its greatest extension the glacier reached to Southern Iowa and the border of Missouri, where occur boulders and even native copper derived from the Lake Superior region, and even into Kansas. A mass of copper found in

Lucas Co., Iowa, travelled 460 miles, if it came from Keneenaw Point, the probable source."[9]

He estimates the ice in the northern part of the glacial districts of America to have had a thickness of from 5000 to 6500 feet, and in the south an average of 2700 feet. The pressure, he says, must have been immense, 6000 feet corresponding to at least 300,000 pounds to the square foot.[1]

Professor Winchell, who is an ardent champion of the views of Agassiz, writes thus in regard to the ice-sheet in North America. "In regions far to the north, the eye probably would not be able to discern any object except that of universal ice. The surface of the ground would be thousands of feet below the traveller, if we may be permitted to presume so hardy a human being. Like Dr. Kane exploring the great Humboldt Glacier of Greenland, he would meet with countless obstacles and dangers. But those obstacles would consist of hummocky ice, or crevassed ice, or perpendicular ice-walls. He would see no soil, no rocks, no vegetation, no animal life. The winds would whistle, storms would rage, snow would be drifted about, and the ineffectual sun would rarely venture to smile on the dreary waste. Farther to the south, the explorer would find isolated spots of bare ground. He would see about them the accumulated débris of boulders, gravel, and dust, from constant winds, spread more or less over the ice-field, staining its painful whiteness, and showing the more grateful aspect of earth and stones. Another hundred miles farther south, and he finds the evidences of the dissolution of the ice-sheet multiplying. Occasional streams of water run on the surface of the ice, or plunge into some of its openings. Deep gorges reveal multitudes of fragments of rock frozen into the ice, and occasional bands of dirt and gravel embraced in the solid ice. The surface is everywhere dirty, or perhaps muddy, from the wasting away of the surface of the glacier. He meets frequent openings, in which generally water may be seen or heard. In these gorges the débris slides down the sloping sides, increasing the insecurity of his footsteps. Still farther south, the general surface is covered with a pulpy earth, mingled with stones and boulders. The ice is evidently much attenuated. The areas of firm, uncovered *terra firma*

[9] Dana, Manual of Geology, 537, 538 [1] Geol. Mag. x. 277

are wonderfully increased in size and frequency. The ice itself is crowded into the valleys, or, if it be in a broad, level tract, like the State of Minnesota, the surface is covered with the débris of the conflict of ice with earth, the ice itself being visible only in those places where crevasses reveal it, or where deep gorges are worn by running streams. Travelling still farther south, the explorer would come upon large areas in which he would not be able to know whether the glacier underlay the superficial drift or not. If he were to stop on one of those wide areas, and make his latitude and longitude certain, by a series of astronomical observations, he might find, to his surprise, after a few years' residence, that his observatory and apparatus had been bodily carried, by an imperceptible motion, some rods to the south. If he were to penetrate the earth on which his foothold seems so steadfast, he might find, equally to his surprise, that he was still riding on the surface of a vast ice-sheet, the earth and soil of which may have furnished him annual crops of potatoes and barley. In other places in the same latitude he would find the ice laid bare over considerable areas, washed clean by the drainage incident to the dissolution of the glacier. The turbid streams would be vastly larger than those which occupy the same beds to-day. They would run with tenfold more violence. The drift-materials would be freed from the clayey portions, and be spread along their channels in curious and varying assortment. In some places the thickness of the whole sheet of drift would be brought under this washing and stratifying process. In others, the ice gently dies out, and lets it down on the rocky surface without any change from the condition in which it lay on the glacier.

"There is still one important point in this discussion that must not be omitted. It is plain to see that, in some parts of the north-west, the advance of the continental glacier would be up gentle slopes, instead of descending an incline. These slopes, of course, present obstructions to the movement of the ice in those directions. It is true also, that the continental glacier would tend to level the country and obliterate such northward slopes. But, in the later part of the ice period, the valleys would be the last relinquished, and would be deeper dug by isolated branches or spurs from the main ice-sheet, which would conform in their direction to the contour of the valleys they might occupy. . . . All the phenomena of the drift in the north-west are, hence, attributable to the approach, long duration, and slow disappearance of the glacier-ice of Professor L. Agassiz."[2] Elsewhere the same geologist says: "The marks of this stupendous glacier are yet visible. The northern limit was chained by eternal frost to its rocky bed. The southern only was free to move, and the whole expansion would be developed along the southern border."[3] "From season to season, and from year to year, the mighty mass marched irresistibly forward, mowing down the forests, crushing tree-trunks, or burying them with the rubbish of the rocks, from ten to sixty feet below the surface. The great glacier moved onward. It bridged Niagara River; it bridged Long Island Sound, and bathed itself in the mild waters of the ocean beyond."[4] "It made a *tabula rasa.* River channels were filled up; in Ohio and Indiana these buried river-beds are of frequent occurrence. The ancient gorge of the Niagara River was filled by the obliterating agency of this continental glacier. It wiped out the trifling furrow, . . making it necessary for the river to begin anew its work."[5] "It bore rocks southward twenty, fifty, and even five hundred miles. These fragments have been transported over lakes, sounds, and seas." The glacial agency is not recognized south of the Ohio River.[6] The south was spared; nay, greatly favoured; for except the chains of mountains, the material ploughed up by the ice was carried by the waters to make the Cretaceous seaboard, the Gulf States, and many more of the Mississippi and Missouri valleys."

Dr. Newberry, in the Ohio Geological Report for 1873, speaks no less emphatically. "We have abundant evidence," he says, "that during what is called the drift period the climate of our continent had changed from the all-pervading warmth of the Tertiary period to an all-pervading arctic cold. While in the former age the climate of our southern states was carried to

[2] Winchell, The Drift Deposits of the North-West, 293-295.
[3] Winchell, Sketches of Creation, 214. [4] Id. 218, 219.
[5] Id. 219, 220. [6] Id. 220, 221.

Greenland, in the latter, *the present climate of Greenland was brought as far south as the Ohio.* The continent of Greenland is now nearly buried under snow and ice. Precisely such must have been the condition of much of the North American continent during the Glacial period, for we find evidence that glaciers covered the greater part of the surface, down to the latitude of 38 or 40 degrees." [7]

"In America," says Mr. Morris, "the ice-sheet was of enormous proportions. Whether it descended from Greenland, or from the central region of British North America, is still a debated question, but its lower limit is somewhat dimly marked out by the line of moraine which it left behind it. This lower limit extends along the Atlantic coast to Long Island. . . . . The limiting line extends through New Jersey in a N.W. direction, and enters Pennsylvania south of the Water Gap. It passes through this State in a somewhat devious course, and leaves it at a point north of Pittsburg. Thence it passes through Ohio, touching the river east of Cincinnati, and extending for a few miles into Kentucky, whence it passes through Middle Indiana and the States farther west as far as Dakota. Beyond this point the moraine disappears. . . . The movement of this great glacier was not due south. Its direction was S.E. over the Atlantic slope. Farther west it was southward and south-westward. Its lines of motion converge northward toward a region N.W. of Montreal, which formed its more immediate centre of motion. . . The thickness of the Atlantic glacier was in due proportion to its extent. It seems to have pushed southward with little conformity to the valley lines, as if every outline of elevation was buried beneath its enormous depth. Indications of its presence are found 5500 feet high on the White mountains, nearly 3000 feet high in southern Massachusetts and the Catskill region, and 1000 feet high in southern Connecticut. These indicate the lowest limit of its depth, which was probably considerably greater, and can scarcely have been less than 6000 or 6500 feet in Upper New England." After describing the phenomena observed all over the district, our author continues :—"The conditions are such as would be pro-

[7] Op. cit. 85, 86.

duced by an ice plane of enormous weight, a mile or more in thickness, with numerous sharp stones frozen into its bottom and pushed onward with irresistible force over the rocks and earth beneath. . . . The lines of scratches observed pursue a nearly regular course over high and low lands alike, without regard to valley slopes, except in the case of the great valleys of the country, thus proving that no mountain range was sufficiently high to seriously distort the course of the ice flow." [8]

When a vast ice-sheet had been postulated for North America, corresponding to that of the Old World, it was a very small step to combine the two, and to make out of them a continuous ice-cap or ice-mound, spreading outwards from the Pole, where it had its focus, in different directions, and this was the next advance that was made. Here also Agassiz was to the front. Thus, he says, *inter alia*, "As to the southward movement of an immense field of ice extending over the whole north, it seems inevitable, the moment we admit that snow may accumulate around the Pole in such quantities as to initiate a pressure radiating in every direction. Snow, alternately thawing and freezing, must, like water, find its level at last. A sheet of snow ten or fifteen thousand feet in thickness, extending all over the northern and southern portions of the globe, must necessarily lead, in the end, to the formation of a northern and southern cap of ice, moving towards the Equator." [9]

Again he urges that Greenland and the Arctic regions have preserved for us the remains of the ice-sheet. "Their shrunken ice-fields, formidable as they seem to us, are, to the frozen masses of that secular winter, but as the patches of snow and ice lingering on the north side of our hills after spring has opened. Let us expand them in imagination till they extend over half a continent, and we shall have a sufficiently vivid picture of this frozen world, and a temperature which would bring the climate of Greenland down to the fortieth degree of latitude would not only render the field of ice far more extensive, but thousands of feet thicker than it is at present. . . . How can we doubt that when, under the same

[8] Encyclopædia Americana, iii. 221, 222. [9] Id. 160.

latitude, Norway, Sweden, Scotland, England and Ireland were covered by sheets of ice many thousand feet in height, the ice-fields of Greenland must have shared in the same climatic influences, and have been much thicker and far more extensive than they are at present?"

Elsewhere he says that in the two hemispheres we have direct reference to the polar areas and the extent of the local boulders from their centre of distribution reaches levels, the mean annual temperature of which corresponds in a surprising manner with the mean annual temperature of the southern limit of the northern erratics, which points to the phenomena of local mountain erratics in Europe, and of northern erratics in Europe and America, having probably been produced by the same cause. The chief difficulty is in conceiving the possibility of the formation of a sheet of ice sufficiently large to carry the northern erratics into their present limits of distribution.[1]

Here also Agassiz is largely supported by Dana, who, speaking of the southern border of the ice-cap in America, says: "The glacier of the Northern Hemisphere was, therefore, not a great enveloping ice-cap extending southward about the sphere. On the contrary, the polar ice had a broad and lofty southward prolongation toward the eastern border of the continent; another narrow and shorter one on the western border; and also a southward extension through local glaciers, along the higher portion of the Rocky Mountains."

Again, Mr. Belt says: "At the present sea-level the ice extended in the northern hemisphere from the Pole to lat. 39° in America, to about the valley of the Thames in England, to lat. 50° in Central Europe, and to lat. 52° in north-western Asia; along the high lands of America it reached to the tropics, and in Central America all the land lying over 2000 feet above the sea supported glaciers. I do not contend that the present low lands of tropical America were ever covered with ice, and it is on the mountain chains of that continent alone that I believe it nearly reached the equator."[2]

Having completed the ice-cap theory in the northern hemisphere, let us now turn to the southern, and especially to

[1] Silliman's Journal, x. 83-101.

[2] Nature, x. 25.

South America, where Darwin had long ago described the boulders and other débris of ice action in Patagonia, attributing them, as we have seen, partially to glaciers in the Andes, and partially to portage by icebergs. The country was examined with great patience and care by Agassiz when he was in charge of the Haslar expedition, and I shall quote from his reports to the head of the United States Survey.

After describing in some detail the phenomena he observed in the Straits of Magellan, he says: "It was not till we rounded Cape Froward that I felt confident that the range of hills immediately in sight along the channel we followed had assumed their present appearance in consequence of abrasion by ice. Now, however, that I have seen the whole length of the Straits of Magellan, have passed through Smyth's Channel, and visited Chiloe, I am prepared to maintain that the whole southern extremity of the American continent has been uniformly moulded by a continuous sheet of ice. Everywhere we saw the rounded undulating forms so well known to the students of glacial phenomena as *roches moutonnées,* combined with the polished surfaces scored by grooves and furrows running in one and the same direction; while rocks of unequal hardness, dykes traversing other rocks, slates on edges were cut to one level. In short, all the surface features of the Straits of Magellan have much the same aspect as the glaciated surfaces of the Northern Hemisphere. Whenever the furrows and scratches were well preserved their trend was northwards."[3]

Our author proceeds to point out in detail the various places where he had observed the phenomena in the Straits, urging that the evidence points to the planing machine having moved south and north, and that the strike-side of the glacier agency was upon the southern slope, and the lee-side upon the northern, while nothing indicates the fitful action of glaciers.[4] In sailing up the western coast of America, he tells us how "the hills in the Guia narrows were everywhere distinctly *moutonnées* to the height of about 2000 feet . . . . that the south side was here also the strike-side was equally apparent from the facts that all these marks were either wanting or less distinct on

[3] Nature, vi. 230. [4] Id.

the north side of the islands. Had any abrading agent advanced from the north, all appearances must have been reversed in these narrow channels, or they must have crossed them at right angles had the action come from east or west. Floating ice is out of the question where the furrows are not horizontal, and here in the narrowest part of the channel, west of Saumarez Island, there is a track where the scratches and furrows are distinctly ascending on the west side of the channel, and horizontal on the eastern side opposite, showing that the pressure of the ice-sheet must have been from S.E. to N.W.[5] Looking south, after passing Saumarez Island, the scenery appears totally different, from the fact that this is the lee side of the glacier action. . . . That local glaciers, however, existed, and extended much beyond their present range, may be plainly seen in many of the inlets crossing the main channels in an east-westerly and west-easterly direction. It is true that general and local glacial phenomena are so interwoven throughout this region that it is at times difficult to appreciate their true connection, but there are also many localities where the difference is quite obvious. . . . as in Switzerland, there seems to be a level above which the ice-sheet has never risen; at least, there is a line above which the mountain ridges remain jagged and abrupt, while below their crests the whole land is *moutonnée*. . . . I know no more interesting locality for the study of glaciers than the vicinity of Saumarez Islands. It shows in the most palpable manner that glaciers only—that is, terrestrial masses of ice moving upon solid ground—can have produced these abrasions; that floating ice cannot have been the cause. Their direction is such that no one could suppose the adjoining cordillera to have been in any way connected with the abrasion or planing of the rock, or with its grooving and furrowing. The country has everywhere a glacier-worn aspect as far as the Gulf of Pennas. . . . The erratic pebbles and boulders from the eastern to the western coast of Patagonia, judging from my observations at Monte Video, in the Bay of St. Matthias, in Possession Bay, at Sandy Point, in all the ports of the Straits of Magellan which we have visited, at Shell Bay, on entering Smyth's Channel,

[5] Id. 270.

throughout Smyth's Channel itself, and upon the shores of Chiloe have the same character. They consist of a mixture of plutonic and metamorphic rocks, among which the hardest siliceous rocks prevail. Their geological identity is further shown by the unfailing presence of a very hard, compact, epidotic rock never absent from these erratic materials, yet never found in place, as far as I know, over the whole extent of country through which I have traced them. . . . This fact is of great significance, inasmuch as it shows that the drift phenomena in this region cannot have been due to the enlargement of the present glaciers, otherwise the drift would consist mainly of the rocks in place, and differ from one locality to the other, and yet their glacial origin is unmistakable, since a considerable proportion of these pebbles and boulders are polished, scratched, grooved and furrowed, like the erratics of the United States and of Northern Europe. It is this uniformity in the character of the drift which has led me from the first to discriminate between the glaciers as they exist, and even as they once existed in their greater extension, in short, between all the phenomena connected with local glaciers and those belonging to what I have called the glacial period, during which the two hemispheres must have been capped with a sheet of ice of enormous thickness and extent. The equatorial limit of this ice-sheet, both in the northern and southern hemispheres, is part of the problem upon which we have thus far fewest facts in our possession. In South America I have now traced the facts from the southernmost point of the continent uninterruptedly to 37° S. lat., on the Atlantic as well as the Pacific coast. Even at Talcahuano, large erratic boulders and *roches moutonnées* exist at the mouth of the Biobio on the hills of Hualpen."[6] Again, he says: "It is certain that the local glaciers of the present time have never had the power in their greatest extension, or lasted long enough to obliterate or even obscure the phenomena of the glacial period. To refer the latter to an enlargement of the present glaciers is simply absurd. . . . After seeing the dispersion of the drift in a south-northerly direction over this part of the South American continent, and observing the relation of

Id. 271.

the local to the general glacial phenomena, I protest anew against the confusion introduced into the subject by those who imagine that what I have called the glacial period was produced by the gradual enlargement and subsequent shrinking of the glaciers now in existence."[7]

In a subsequent letter, describing his journey north from Talcahuano to Santiago, Agassiz describes the great valley of Chillan, extending from the Gulf of Ancud or Port Montt to Santiago and farther north. "This great valley," he says, "extending for more than 25° of latitude, is a continuous glacier bottom, showing plainly that for its whole length the great southern ice-sheet has been moving northwards on it. Nowhere," he adds, "could I find any indication that glaciers descending from the Andes had crossed this valley and reached the shores of the Pacific."[8]

Agassiz was generally fearless in his logic, and he now proceeded to apply the theory by which he had explained the drift phenomena of the northern hemisphere by a polar ice cap to the southern hemisphere. Thus he writes: "After having traced what seems to me palpable evidence of an ice mantle overspreading once the southern part of the continent, the effect of which I have seen from Monte Video on the Atlantic to Talcahuano on the Pacific coast, the question naturally arises how far the southern extremity of Africa as well as New Zealand and Australia, were involved in the extension. I hope I may live to see younger naturalists investigate these regions with the same object. I believe that whenever such an investigation is undertaken by a competent observer, it will be found that over and above all local glacial phenomena, and side by side with them, there is also evidence of a southern circumpolar glacial agency."[9]

Writing in 1865 he says: "During this period two vast caps of ice stretched from the northern pole southward and from the southern pole northward, extending in each case far towards the Equator; and ice-fields such as now spread over the Arctics, covered a great part of the temperate zones."[1]

I have now described how Agassiz applied his ice-sheet

[7] Id. 272. [8] Id. 414. [9] Id. 272.
[1] Geological Sketches, 212.

theory to the southern as well as the northern hemisphere, and must finally turn to the application he and others made of it to the Tropics. In doing so I shall have to revert somewhat in time, since his arguments in this behalf were published some years before his visit to Patagonia.

He first announced his discovery in a letter to Professor Peirce, dated the 27th of May, 1865. In this he says: "Yesterday was one of the happiest days of my life, and I want to share it with you. Here I am at Tijuca, a cluster of hills about 1800 feet high, and some seven or eight miles from Rio, in a charming cottage-like hotel, from the terrace of which you see a drift hill with innumerable erratic boulders as characteristic as any I have ever seen in New England. I had before seen sundry unmistakable traces of drift, but there was everywhere connected with the drift itself such an amount of decomposed rocks of various kinds, that though I could see the drift and distinguish it from the decomposed primary rocks in place on account of my familiarity with that kind of deposit, yet I could probably never have satisfied anybody else that there is here an equivalent of the northern drift, had I not found yesterday, near Bennett's hotel at Tijuca, the most palpable superposition of drift and decomposed rocks with a distinct line of demarcation between them." After describing the decomposed rocks which remain distributed and stratified as they originally were, except in being disintegrated into a soft paste, "a sort of rotten crust of their own substance," he goes on to say that they consist of small particles of primitive rocks, but without a trace of larger or smaller boulders in them, while the superincumbent drift, consisting of a similar paste, does not show the slightest sign of the indistinct stratification characteristic of the decomposed metamorphic rocks below it, nor any of the decomposed dikes, but is full of various kinds of boulders of various dimensions. "I have not yet traced the boulders to their origin, but the majority consist of a kind of green stone, composed of equal amounts of a greenish black hornblende and feldspar." Agassiz says an engineer had told him that the iron mines of Minas Geraes are worked in stone of this kind. He goes on to say that besides this drift he had not seen any trace of glacial action, properly so called, if polished surfaces and scratches and furrows are especially to be considered as such. Later on we read in Madame Agassiz's narrative "that it is hardly possible to hope to find traces of striæ and furrows where the rock is so perishable and disintegrates so fast; but this much is certain, that a sheet of drift covers the country, composed of a homogeneous paste without trace of stratification, containing loose materials of all sorts and sizes embedded in it, without reference to weight, large boulders, smaller stones, pebbles, and the like. This drift is very unevenly distributed; sometimes rising into high hills, owing to the surrounding denudations; sometimes covering the surface as a thin layer; sometimes, and especially on steep slopes, washed away completely, leaving the bare face of the rock; sometimes deeply gullied, so as to produce a succession of depressions and elevations alternating with each other. . . . A number of localities exist where the drift rests immediately above stratified rocks, with the boulders protruding from it, the line of contact being perfectly distinct." Our authoress then goes on to point out several localities where the phenomena can be well studied.[3] Going north from this to the mouth of the Amazon, Agassiz verified the existence of this drift along the whole length of coast, and notably at Bahia, Maceio, Pernambuco, Monteiro, Parahyba do Norte, Cape St. Roque, Ceará, Maranham and Pará. "The sheet of drift," says the graphic reporter of his journey, "which he has thus followed from Rio de Janeiro to the mouth of the Amazons, is everywhere of the same geological constitution. It is always a homogeneous clayey paste of a reddish colour, containing quartz pebbles; and, whatever be the character of the rock in place, whether granite, sandstone, gneiss or lime, the character of the drift never changes or partakes of that of the rocks with which it is in contact. This certainly proves that, whatever be its origin, it cannot be referred to the localities where it is now found, but must have been brought from a distance."[4]

In going up the Amazon, we are told, the drift became more conspicuous, and Agassiz devotes a long chapter to its discussion in that river valley, in which, he says, it differs some-

[2] Agassiz, A Journey to Brazil, 86-88.
[3] Op. cit. 99-101. [4] Id. 146, 147.

what from the deposit farther south, in that the boulders are less frequent and in showing occasional signs of stratification. At one point, namely, at the Serra of Ereré, which rises to about a height of 900 feet on a side arm of the Amazon named the Gurupatuba, Agassiz found some interesting evidence of a glacial character. "On the northern flank of the Serra," he says, "I found the only genuine erratic boulders I have seen in the whole Amazonian valley from Pará to the frontier of Peru, though there are many detached masses of rock; as for instance at Pedreira, near the junction of the Rio Negro and Rio Branco, which might be mistaken for them, but are due to the decomposition of the rocks in place. The boulders of Ererê are entirely distinct from the rock of the Serra, and consist of masses of compact hornblende." Then, again, he makes the very interesting remark that Alpine lichens were growing among the cacti and palms and a crust of Arctic cryptogamous growth covered rocks between which sprang tropical flowers.[5] These erratics of Ereré are sunk in the unstratified mass of clay,[6] and this Rio drift, as he calls it, covers the country everywhere. Agassiz says: "It may truly be said that there does not exist on the surface of the earth a formation known to geologists resembling that of the Amazons. Its extent is stupendous; it stretches from the Atlantic shore, through the whole width of Brazil, into Peru, to the very foot of the Andes. Humboldt speaks of it 'in the vast plains of the Amazons, in the eastern boundary of Jaen de Bracamoros,' and says, 'This prodigious extension of red sandstone in the low grounds stretching along the east of the Andes, is one of the most striking phenomena I observed during my examination of rocks in the equinoctial regions.'[7] When the great natural philosopher wrote these lines, he had no idea

[5] Id. 418. [6] Id. 421.

[7] Id. 422. Bohn's edition of Humboldt's Personal Narrative, p. 134. "Humboldt alludes to these formations repeatedly: it is true that he refers them to the ancient conglomerates of the Devonian age, but his description agrees so perfectly with what I have observed along the banks of the Amazons, that there can be no doubt he speaks of the same thing. He wrote at a time when many of the results of modern geology were unknown, and his explanation of the phenomena was then perfectly natural. The passage from which the few lines in the text are taken shows that these deposits extend even to the Llanos."

how much these deposits extended beyond the field of his observations. Indeed they are not limited to the main bed of the Amazons; they have been followed along the banks of its tributaries to the south and north as far as these have been ascended. They occur on the margins of the Huallaga and the Ucayale, on those of the Iça, the Hyutahy, the Hyurua, the Hyapura and the Purus. On the banks of the Hyapura, where Major Coutinho has traced them, they are found as far as the Cataract of Cupati. I have followed them along the Rio Negro to its junction with the Rio Branco; and Humboldt not only describes them from a higher point on this same river, but also from the valley of the Orinoco. Finally they may be tracked along the banks of the Madeira, the Tapajos, the Xingu, and the Tocantins, as well as on the shores of the Guatuma, the Trombetas and other northern affluents of the Amazons. The observations of Martius, those of Gardner, and the recent survey above alluded to, made by my assistant, Mr. St. John, of the valley of the Rio Guruguea and that of the Rio Paranahyba, show that the great basin of Piauhy is also identical in its geological structure with the lateral valleys of the Amazons. The same is true of the large island of Marajo, lying at the mouth of the Amazons. And yet I believe that even this does not cover the whole ground, and that some future writer may say of my estimate, as I have said of Humboldt's, that it falls short of the truth; for, if my generalizations are correct, the same formation will be found extending over the whole basin of the Paraguay and the Rio de la Plata, and along their tributaries, to the very heart of the Andes.

"Such are the facts. The question now arises, How were these vast deposits found? The easiest answer and the one which most readily suggests itself, is that of a submersion of the continent at successive periods to allow the accumulation of these materials, and its subsequent elevation. I reject this explanation for the simple reason that the deposits show no sign whatever of a marine origin. No sea-shells nor remains of any marine animal have as yet been found throughout their whole extent, over a region several thousand miles in length and from five to seven hundred in width. It is contrary to all our knowledge of geological deposits to suppose that an ocean basin of this size, which must have been submerged during an immensely long period in order to accumulate formations of such thickness, should not contain numerous remains of the animals formerly inhabiting it. . . . It is my belief that all these deposits belong to the ice period in its earlier or later phases, and to this cosmic winter which, judging from all the phenomena connected with it, may have lasted for thousands of centuries, we must look for the key to the geological history of the Amazonian valley. I am aware that this suggestion may appear extravagant. But is it, after all, so improbable that when Central Europe was covered with ice thousands of feet thick; when the glaciers of Great Britain ploughed into the sea, and when those of the Swiss mountains had ten times their present altitude; when every lake in Northern Italy was filled with ice, and these frozen masses extended even into Northern Africa; when a sheet of ice, reaching nearly to the summit of Mount Washington in the White Mountains, that is having a thickness of nearly 6000 feet, moved over the continent of North America—is it so improbable that in this epoch of universal cold, the valley of the Amazons also had its glacier poured down into it from the accumulations of snow in the Cordilleras, and swollen laterally by the tributary glaciers descending from the tablelands of Guiaica? The movement of this immense glacier would be eastward, and determined as well by the vast reservoirs of snow in the Andes as by the direction of the valley itself. It must have ploughed the valley bottom over and over again, grinding all the materials beneath it into a fine powder, or reducing them to small pebbles, and it must have accumulated at its lower end a moraine of proportions as gigantic as its own; thus building a colossal sea wall across the mouth of the valley. I shall be asked at once whether I have found here also the glacial inscriptions, the furrows, striæ and polished surfaces so characteristic of the ground over which glaciers have travelled. I answer, not a trace of them; for the simple reason that there is not a natural rock surface to be found throughout the whole Amazonian valley. The rocks themselves are of so friable a nature, and the decomposition caused by the warm torrential rains and by exposure to the burning sun of the tropics so great and unceasing, that it is hopeless to look for marks which in colder climates and on harder substances are preserved through ages unchanged. With the exception of the rounded surfaces so well known in Switzerland as the *roches moutonnées*, heretofore alluded to, which may be seen in many localities, and the boulders of Ereré, the direct traces of glaciers as seen in other countries are wanting here. I am, indeed, quite willing to admit that, from the nature of the circumstances, I have not here the positive evidence which has guided me in my previous glacial investigations. My conviction in this instance is founded, first, on the materials in the Amazonian valley, which correspond exactly in their character to materials accumulated in glacier bottoms; secondly, on the resemblance of the upper or third Amazonian formation to the Rio drift, of the glacial origin of which there cannot, in my opinion, be any doubt; thirdly, on the fact that this fresh-water basin must have been closed against the sea by some powerful barrier, the removal of which would naturally give an outlet to the waters, and cause the extraordinary denudations the evidences of which meet us everywhere throughout the valley."[8]

In regard to this Rio drift, Agassiz says: "As I stated in the beginning, I am satisfied that the unstratified clay deposit of Rio and its vicinity is genuine glacial drift resulting from the grinding of the loose materials interposed between the glacier and the solid rock in place, and retaining to this day the position in which it was left by the ice. Like all such accumulations it is entirely free from stratification. On the other hand, the ochraceous sandy clay of the Amazon valley, while it was originally ground by glaciers in the upper part of the valley, these materials have subsequently been spread throughout the whole basin and actually deposited under the agency of water."[9]

Under the low temperatures of the ice period, he urges, the climatic conditions necessary for the formation of land ice existed in the valley of the Amazons, and it was actually filled with an immense glacier, whose terminal moraine, he argues, closes the Amazon valley on the east towards the sea. "When these fields of ice yielded to a gradual change of climate and slowly melted away, the whole basin, then closed against

[8] Id. 425, 426. [9] Id. 426, 427, note.

the sea by a huge wall of débris was transformed into a vast fresh-water lake. The first effect of the thawing process must have been to separate the glacier from its foundation, raising it from immediate contact with the valley bottom, and thus giving room for the accumulation of a certain amount of water beneath it, while the valley as a whole would still be occupied by the glacier. . . . It is at this time and under such circumstances that I believe the first formation of the Amazonian valley, with the coarse pebbly sand beneath and the finely laminated clays above, to have been formed, when the greater part of the ice was melted and the basin was filled with water." It was while thus partially or wholly filled with water that Agassiz supposes the boulders of Ereré to have been "carried by floating ice when nothing remained of the ice-fields except such isolated masses—ice-rafts, as it were, or perhaps by icebergs dropped into the basin from glaciers still remaining in the Andes and on the edges of the plateaus of Guiana and Brazil."

He lastly urges that the gigantic moraine which closed the valley was broken through either by some violent movement inside or by the sea outside, whereupon the waters rushed through, denuding and sweeping away a good deal of the deposit and leaving some outliers alone remaining.

Agassiz was not content with visiting the Amazon valley; he went back to the province of Rio to try and find traces of glaciers in its mountain ranges or serras, and also traces of the southern lateral moraine marking the limit of ice which he supposed once filled the Amazonian basin. *Inter alia,* he learnt from a land surveyor, named Dr. Felice, that there is a wall of loose materials, boulders, stones, etc., running from east to west for a distance of some sixty leagues from the Rio Aracaty-Assu to Bom Jesu in the Serra Grande. From his account, says Agassiz, this wall resembles greatly the "Horsebacks" in Maine, those remarkable ridges accumulated by the ancient glaciers, and running sometimes uninterruptedly for thirty or forty miles. The horsebacks are, however, covered with soil and turf, whereas Dr. Felice describes this wall as rough and bare.[1] Agassiz was not able to visit this moraine,

[1] Id. 447, 448.

but he did visit more than one serra in the province of Rio. Thus he says of the Serra of Monguba: "All the valleys have had their glaciers, and these valleys (? glaciers) have brought down from the hillsides into the plains boulders, pebbles, and débris of all sorts."[2] Of the Serra of Aratauha, he says: "The glacial phenomena are as legible as in any of the valleys of Maine or in those of the mountains of Cumberland in England. It had evidently a local glacier, formed by the meeting of two arms, which descended from two depressions spreading right and left on the upper part of the serra, and joining below in the main valley. A large part of the medial moraine formed by the meeting of these two arms can still be traced in the central valley. One of the lateral moraines is perfectly preserved, the village road cutting through it; while the village itself is built just within the terminal moraine, which is thrown up in a long ridge in front of it."[3] Agassiz tells us he specially examined the right lateral moraine of this serra to ascertain its real character. "Everywhere," he says, "in the ridges encircling the depressions on the serra, the loose materials and large boulders are so accumulated and embedded in clay and sand that their morainic character is unmistakable. Occasionally, where a ledge of the underlying rock crops out, in places where the drift has been removed by denudation, the difference between the moraine and the rock decomposed in place is recognized at once. It is equally easy to distinguish the boulders which here and there have rolled down from the mountain and stopped against the moraine. The three things are side by side, and might at first be easily confounded; but a little familiarity makes it easy to distinguish them. Where the lateral moraine turns towards the front of the ancient glacier near the point at which the brook of Patacuba cuts through the former, and a little to the west of the brook, there are colossal boulders leaning against the moraine, from the summit of which they have probably rolled down. Near the cemetery the front moraine consists almost entirely of small quartz pebbles; there are, however, a few larger blocks among them. The medial moraine extends nearly through the

[2] Id. 454. Id. 456, 457.

centre of the village, while the left-hand lateral moraine lies outside of the village, at its eastern end, and is traversed by the road leading to Ceará. It is not impossible that eastwards a third tributary of the serra may have reached the main glacier of Pacatuba. I may say that in the whole valley of Hasli there are no accumulations of moraine materials more characteristic than those I have found here,—not even about the Kirchet; neither are there any remains of the kind more striking about the valleys of Mount Desert in Maine, where the glacial phenomena are so remarkable, nor in the valleys of Lough Fine, Lough Augh, and Lough Long in Scotland, where the traces of ancient glaciers are so distinct. In none of these localities are the glacial phenomena more legible than in the Serra of Aratanha."[4]

Agassiz tells us that two of his subordinates, Mr. F. C. Hartt, whom he sent to the province of Spiritu Santo, the valley of Rio Doce, and that of Mucury, and Mr. O. St. John, who went into the interior in the north and visited the valleys of the Rio San Francisco and the Rio das Velhas, both experienced geologists found similar phenomena.[5] Mr. Hartt, who revisited Brazil the next year, wrote an elaborate work on the "Geology and Physical Geography of Brazil." In this, while differing from Agassiz as to the stratified beds of the Amazon valley, he is careful to explain "that these views do not affect Agassiz's theory of the former existence of glaciers under the tropics, down to the present level of the sea; a theory which," he adds, "I hold as firmly as he." This testimony is the more valuable as he says: "I desire to record the fact that I began my studies of the Brazilian drift with a conviction that Professor Agassiz was wrong, and I feel much gratified that my independent observations have so fully confirmed the results of his own."[6]

Speaking of the valley of Tijuca, he says: "If we follow down the valley we shall find the soil full of boulders, and some of them are many feet in diameter. As we descend the valley still further these boulders are seen lying bare, not only in the brook, where the water has washed away the loose material, but on the hillsides. I think no geologist familiar

[4] Id. 463, 464. [5] Id. 404, 405. [6] Op. cit. 493.

with drift phenomena, who should suddenly find himself in this valley, would have even the slightest suspicion of there being anything else than the most ample testimony of the former prevalence of glacial action over the region. . . . If one descends the valley towards the Cascate Grande, he will see that the valley is heaped with a confusion of immense boulders tumbled one upon another; masses of greenstone, weighing hundreds of tons, piled up with those of gneiss of all qualities. Where these are bare they are all rounded, but I believe this is referable, to a large extent, to a concentric and even decomposition of the surface; but there is no resisting the conclusion that we have here a morainic deposit from a glacier which anciently occupied the valley. . . . Descending to the plain below the Cascate, one traces the drift clays and boulders quite down to the plain, where they end abruptly, and the flat lands are seen to be of alluvial origin resting on sea sands. . . . From Belem to the Parahyba River the same red clay entirely covers the surface, lying even on very high slopes. Nowhere is there the slightest trace of stratification, and it is sharply defined from the alluvial deposits of the river. The same pebble sheet is seen almost everywhere, though in cutting after cutting it may sometimes be wanting. Boulders are rare, and are almost invariably so decomposed as to be seen only in fresh cuttings. The rock on which the drift rests is always smoothly and evenly rounded down."[7] Speaking of the surface deposits of Bahia and the São Francisco, he says: "I must insist upon the fact that the unarranged materials are precisely like our unmodified drift in the north, and that the surface of the rock on which they lie has the moulding of the surfaces on which our northern drift lies; and that if we refer the northern drift to the action of glacier ice, we must do the same thing for the Brazilian surface detritus; contrary as it is to all our preconceived opinions of the distribution of the drift. The fact that neither Professor Agassiz nor myself, nor any one else of our expedition, has been able to discover glacial striæ in Brazil is of very secondary importance. The drift itself exists all over the country, and it cannot be explained away. . . . I can offer but one ex-

[7] Id. 29, 31.

planation of the formation of the pot-holes observed by Mr. Allen, and that is that they were formed by glacial cascades in the same way as the pot-holes seen so often on the surface of ridges in the north have been formed during the drift; for, according to the testimony of Mr. Allen, the pot-holes of the province of Bahia occur on the gneiss plains, far away from any present obstacle over which the water may have flowed. Mr. Allen describes them as being exceedingly well preserved, and having smooth sides. . . . I believe that during the time of the drift the country stood at a much higher level than at present, and that it was covered by a general glacier."[8]

The case for the former extension of a glacial age to the tropics was further pressed by Mr. Belt in his well-known work "The Naturalist in Nicaragua." In it he speaks of *unstratified beds of gravel* being exposed in numerous natural sections on the banks of the river about a mile below Depilto, in Nicaragua. "These beds," he says, "deepen as the river is descended, until at Ocotal they reach a thickness of between two and three hundred feet. . . . *These unstratified* deposits consist mostly of quartz sand with numerous angular and subangular blocks of quartz and talcose schist. Many of the boulders were very large, and in some parts great numbers have been accumulated in the bed of the river by the washing away of the smaller stones and sand. Some of these huge boulders were fifteen feet across, the largest of them lying in the bed of the river two miles below Depilto. Most of them were of the Depilto quartz rock and gneiss, and I saw many in the unstratified gravel near Ocotal fully eight miles from their parent rock. Near Ocotal this unstratified formation is nearly level, excepting where worn into deep gulches by the existing streams. The river has cut through to a depth of over 200 feet, and there are long precipices of it on both sides, similar to those near streams in the north of England, that cut through thick beds of boulder clay."[1]

Elsewhere Mr. Belt tells us that some time after leaving San Rafael, he found a steep ridge fully 1200 feet high, composed entirely of boulder clay. "This clay was of a brown colour,

[8] Hartt, Geology and Physical Geography of Brazil, 570-573.
[1] Op. cit. 259, 260.

and full of angular and subangular blocks of stone of all sizes up to nine feet in diameter. . . . This boulder clay had extended all the way from San Rafael, and ranges of hills appeared to be composed entirely of it. The angular and subangular stones that it contained were an irregular mixture of different varieties of trap, conglomerate and schistose rocks. In the northern States of America it would be unhesitatingly ascribed to the action of ice; but I was at the time unprepared to believe that the glacial period could have left such memorials of its existence within the Tropics, at not greater elevations above the sea than 3000 feet.[2] . . . The evidences of glacial action between Depilto and Ocotal were, with one exception, as clear as in any Welsh or Highland valley. There were the same rounded and smoothed masses of rock, the same moraine-like accumulations of unstratified sand and gravel, the same transported boulders that could be traced to their parent rocks several miles distant. The single exception was, I am convinced, one of observation and not one of fact. . . . I saw no glacial scratches on the rocks; but geologists know how rare these are on natural exposures in districts that have certainly been glaciated, and will not be surprised that, in a hurried visit of only a few hours, I should not have discovered any. Glacial scratches are seldom preserved on rock surfaces exposed to the action of the elements. Even in Nova Scotia, where scratches and grooves are met with wherever the rock surface has been recently laid bare, I do not remember having ever seen any on natural exposures. It is only where protected by a covering of clay or gravel from the action of the elements, that they have been preserved through the ages that have passed since the glacial epoch; and as I did not see any rock surfaces near Depilto that had been recently bared, it is not surprising that, notwithstanding the other proofs of glacial action, I should not have seen any ice scratches or grooves.

"I could no longer withstand the evidence that had been gradually accumulating of the presence of large glaciers in Central America during the glacial period, and these once admitted afforded me a solution of many phenomena that had before been inexplicable. The immense ridges of boulder clay near San Rafael, the long hog-backed hills near

[2] Op. cit. 247, 248.

Toblason, the great transported boulders two leagues beyond Libertad on the Juigalpa road, and the scarcity of alluvial gold in the valleys of Santo Domingo, could all be easily explained on the supposition that the ice of the glacial period was not confined to extra tropical lands, but in Central America covered all the higher ranges, and descended in great glaciers to at least as low as the line of country now standing at 2000 feet above the sea, and probably much lower."[3]

Mr. Belt then goes on to account for the scarcity of alluvial gold in the valleys of Nicaragua as in the glaciated valleys of Nova Scotia and North Wales by the potency of glaciers in scooping out all the contents of the valleys. "When," he says, "the denuding agent was water, the rocks were worn away and the heavier gold left behind at the bottom of the alluvial deposits; but when the denuding agent was glacier ice the stony masses and their metallic contents were carried away, or mingled together in the unassorted moraines.

He then proceeds: "That the transportation of boulders in Nicaragua was due to glaciers, and not to floating icebergs, may be argued on zoological grounds. The transported boulders near Ocotal, are about 3000 feet above the sea, those near Libertad about 2000 feet. The low pass between the Atlantic and the Pacific oceans, through the valley of San Juan and the lake of Nicaragua, is less than 200 feet above the sea, and to allow for the flotation of icebergs at the lower of the two places named, a channel of more than 1800 feet in depth would have connected the two oceans. This supposition is negatived by the fact that the mollusca on the two coasts separated by the narrow Isthmus of Darien, are almost entirely distinct, whilst we know that since the glacial period there has been little change in the molluscan fauna, nearly if not all the shells found in glacial deposits still existing in neighbouring seas. . . . No zoologist of note believes that there has been any submergence of the land lying between the Pacific and the Atlantic since the pliocene period, and icebergs would not have floated without such submergence, so that in the cases I have mentioned the boulders, if ice-borne, have been carried by glaciers and not by floating ice."[4] "While I thus found evidences of the ice

[3] Op. cit. 261, 262. [4] Id. 263, 264.

of the glacial period reaching in the Northern Hemisphere to within the tropics, in the Southern Hemisphere Professor Hartt has found glacial drift extending from Patagonia all through Brazil to Pernambuco, and Agassiz has even announced the discovery of glacial moraines up to the Equator. I have myself seen, near Pernambuco, and in the province of Maranham, in Brazil, a great drift deposit that I believe to be of glacial origin."[5]

In answer to the natural question of what became of the tropical fauna and flora of America at this time, such, for instance, as the Heliconinæ and Morphinæ, groups of butterflies peculiar to tropical America, containing many distinct genera which, on any theory of descent from a common progenitor, must have originated ages before the glacial period—"How is it," says Mr. Belt, "that such peculiar tropical groups were not exterminated by the cold of the glacial period, or, if able to stand the cold, that they did not spread into temperate regions on the retreat of the ice? I believe the answer is that there was much extermination during the glacial period, that many species and some genera, as for instance the American horse, did not survive it, and that some of the great gaps that now exist in natural history were then made, but that a refuge was found for many species on lands now below the ocean, that were uncovered by the lowering of the sea caused by the immense quantity of water locked up in frozen masses on the land."

Elsewhere he says, "The existence of a glacial period, however much derided when first announced, is now a recognized fact. The divergence of opinion respecting it is limited to a question of extent; and after my recent journey in the Amazons, I am led to add a new chapter to the strange history of glacial phenomena, taken from the Southern Hemisphere, and even from the Tropics themselves. I am prepared to find that the statement of this new phase of the glacial period will awaken among my scientific colleagues an opposition even more violent than that by which the first announcement of my views on this subject was met. I am, however, willing to bide my time; feeling sure that, as the theory of the ancient extension of

[5] Id. 264, 265.

glaciers in Europe has gradually come to be accepted by geologists, so will the existence of like phenomena, both in North and South America, during the same epoch, be recognized sooner or later as part of a great series of physical events extending over the whole globe. Indeed, when the ice period is fully understood, it will be seen that the absurdity lies in supposing that climatic conditions so different could be limited to a small portion of the world's surface. If the geological winter existed at all, it must have been cosmic; and it is quite as rational to look for its traces in the Western as in the Eastern hemisphere, to the south of the equator as to the north of it. Impressed by this wider view of the subject, confirmed by a number of unpublished investigations which I have made during the last three or four years in the United States, I came to South America, expecting to find in the tropical regions new evidences of a bygone glacial period, though, of course, under different aspects. Such a result seemed to me the logical sequence of what I had already observed in Europe and in North America."[6]

Such, then, is a sample of the evidence for the existence in the Tropics of phenomena which in the Northern Hemisphere are held to prove the existence of a glacial age. These are not the casual opinions of *dilettanti*, but the matured views of experienced geologists, who have spent months in investigating them. Mr. Alfred Wallace, himself among the foremost authorities upon tropical natural history, thus speaks of the conclusions of Agassiz and Hartt about the Brazilian drift. "This is truly a startling conclusion, and one which has hitherto been received in this country with some incredulity. Professor Agassiz was thought to be glacier-mad; but if we separate his theories from his facts, and if we carefully consider the additional facts and arguments adduced by Mr. Hartt, we shall be bound to conclude that, however startling, the theory of the glaciation of Brazil is supported by a mass of evidence which no unprejudiced man of science will ignore merely because it runs counter to all his preconceived opinions." Again he says, "It can hardly be maintained that the discoverer of glacial phenomena in our own country, and who has since lived in

[6] Id. 398, 399.

such a pre-eminently glaciated district as the Northern United States, is not a competent observer; and if the whole series of phenomena here alluded to have been produced without the aid of ice, we must lose all confidence in the method of reasoning from similar effects to similar causes which is the very foundation of modern geology. . . : The facts as stated by two careful observers, both thoroughly experienced in glacial phenomena, are undoubtedly such as to warrant the main conclusion drawn by them, and it is to be hoped that geologists will not ignore the facts because the conclusions seem improbable, as they so long ignored facts proving the antiquity of man for no other reason."[7]

It is not wonderful that in view of generalizations like these our breath is taken away, and that Agassiz had doubts about his contemporaries and made appeals to the verdict of posterity to justify him. Thus, in a letter already quoted which he wrote to Professor Benjamin Peirce, he says, "You may think that I have given you too many details. I have done so purposely, that no one may accuse me of basing theories on imperfect observations. I am well aware that my results will be questioned, and I shall be thought fanciful by geologists of all schools, as I have been at every step of my glacial researches. But an old hunter does not take the track of a fox for that of a wolf. I am an old hunter of glacial tracks, and I know the footprint whenever I find it."[8]

[7] Nature, ii. 511, 512.

[8] Id. 272.

CHAPTER VIII.

ON THE ALLEGED RECURRENCE OF GLACIAL EPOCHS, AND ON SUPPOSED INTER-GLACIAL BEDS.

"From the Laurentian epoch down to the present day, all the physical events in the history of the earth have varied neither in kind nor intensity from those which we now have experienced."—RAMSAY, Address to the British Association, 1880.

The Champions of former glacial periods—*Pre-Cambrian,* Hicks, Sir A. Geikie—*Cambrian,* James Thompson—*Silurian,* Carrick-Moore, Lyell, J. Geikie, Dawson—*Devonian,* Cumming, Ramsay, Etheridge, Selwyn, J. Geikie—*Carboniferous,* J. Geikie, Haughton, G. Austen, Stirrup, Weiss, Orton, Dawson, W. T. Blandford, T. Hughes, Waagen, Oldham—*Permian,* Ramsay, Hall, Sutherland, Stow—*Trias,* Godwin Austen, Ramsay, Wilkinson—*Oolite,* Ramsay, Judd—*Cretaceous,* Godwin Austen, Escher von der Linth—*Eocene,* Croll, J. Martins, Lydekker—*Miocene,* Gastaldi, Ramsay, Geikie, J. Martins, Stuart Menteath, Tardy—*Pliocene,* Desor and Stopanni—Champions of inter-glacial beds, Venetz, Agassiz, Muhlberg, Bach, Tarancelli, Morlot, Chavenne, Scipion Gras, Heer, Sir A. Geikie, Bennie, Searles Wood and Geikie, Penck, Törnebohm, McGee, Whittlesey, Orton, Winchell.

THE champions of the new views in regard to the Glacial epoch were also, for the most part, champions of Uniformity. Champions, that is to say, of the doctrine embodied in the heading to this chapter, that Nature has always used precisely the same tools and employed them in precisely the same way. It was natural, therefore, that they should begin to look round and try and find traces of former periods when ice was widespread and when the results of its handiwork were similarly widely scattered. It was contrary to much modern teaching to suppose that an event like the glacial epoch was solitary and detached and not a link in a great chain in which the astronomical and other forces which influence the earth produced recurrent phenomena.

Especially in late years has the evidence been collected to prove, first, that evidences of other glacial epochs can be found in the older rocks, and, secondly, that the glacial epoch itself was no uniform period of continuous cold, but was broken by intervals during which temperate conditions existed. I do not propose to collect an exhaustive series of the examples which have been thus adduced, but only to bring together some samples of them, and I think it better to go stratigraphically through the different beds rather than to give a chronological account of the successive discoverers.

In 1876 Hicks published a paper in the *Geological Magazine* on the Northern Palæozoic Rocks. In this, speaking of the Pre-Cambrian land, he says: "This condition of continents of great size in high latitudes with elevated plateaux and high mountains, would lead us to believe that they were covered with ice and snow in their higher parts, and that the plains and valleys had therefore much loose material strewn over them. In the Cambrian rocks in Wales boulders of considerable size, sometimes a foot in diameter, are often found in beds of conglomerate, and the frequent alternations of conglomerate grits, and sandstones, seem to prove that, at least in the earliest epochs, an abundance of loose material must have been near and ready at hand to be easily denuded off as each part became submerged." After arguing against this having been caused by marine currents, he says that some additional force must have been at work,[1] and he concludes that the early Cambrian was a cold epoch, as evidenced from the fact that the Pre-Cambrian continents occupied very extensive areas in the higher latitudes, and that they were traversed by mountainous ranges attaining in some cases to great heights. At no time since, unless in the glacial period, does there seem to have been so much high land in the higher latitudes, and it is, therefore, reasonable to suppose that in the earlier stages at least of the epoch the climate was one of great cold.[2]

In 1880, in a paper on Pre-Cambrian Volcanoes and Glaciers, the same author says: "Evidences showing that glacial phenomena prevailed in pre-Cambrian times were pointed out by me in 1876, and there can, I think, be little doubt that glaciers did, even at that early period, spread over some of the higher lands." As evidence he points to the regular character

[1] Geol. Mag. N.S. iii. 157. [2] Id. 252.

of the masses in the Cambrian rocks which rest immediately on the old land in the Highlands of Scotland. ". . . In Wales also the Cambrian conglomerates, when they rest on the old floor, contain very large angular masses, though I have never recognized blocks as large there as those in the angular breccia which is presumably of Cambrian age, near Gareloch." And he finally concludes that glacial phenomena have prevailed since at least later pre-Cambrian times in proportion as the geographical changes have been favourable or otherwise.[3]

In 1880 Ramsay quotes a letter from Sir A. Geikie, in which he informed him that he had discovered mammillated *moutonnée* surfaces of Laurentian rocks passing underneath the Cambrian sandstones of the north-west of Scotland at intervals all the way from Cape Wrath to Loch Torridon, for a distance of about ninety miles. "The mammillated rocks are as well rounded off as any recent *roche moutonnée*, and in one place these bosses are covered by a huge angular breccia of this old gneiss (Laurentian) with blocks sometimes five or six feet long. . . . In the Gareloch district this breccia is utterly unstratified, the angular fragments standing on end and at all angles, just as they do in many modern morainic mounds wherever large glaciers are found."[4] In a paper in *Nature*, published in the same year,[5] Geikie argues that these mammillated and polished surfaces, like the accompanying breccia, belong distinctly to Pre-Cambrian times; and he concludes thus: "On these far northern shores, then, there still remain fragments of the surface on which our oldest sedimentary accumulations were deposited. These fragments are found to bear on their smoothed hummocky contours a striking resemblance to the surface which geologists now always associate with the action of glacier ice. There can be no doubt at least that they are denuded surfaces. The edges of the vertical and twisted beds of gneiss and schist have been smoothly bevelled off. These rocks, however, would never have assumed such a contour if exposed merely to ordinary sub-aerial disintegration. They would have taken sharp craggy outlines like those which are here and there gradually

[3] Id. N.S. vol. vii. 490, 491.
[4] Ramsay, Address to the British Association.
[5] Nature, vol. xxii. p. 400, etc.

replacing the ice-worn curves of the *roches moutonnées.* They have certainly been ground by an agent that has produced results which, if they were found in a recent formation, would, without hesitation be ascribed to land ice. The breccia, too, is quite comparable to moraine stuff."[6]

In 1870 Mr. James Thompson described certain pebbles and boulders of granite in schistose rocks in Islay in Scotland, which he assigned to Cambrian age. These fragments were some of them angular but mostly rounded, of all sizes, from mere particles to great boulders of granite, resembling the granite of the Isle of Mull. Similar rocks do not occur in Islay, and Mull is at a considerable distance to the north, while there is a deep sea between the two islands, and he suggested the probability of the granite having been transported by the agency of ice. These deposits, he says, resemble the boulder drifts of more recent times, in the following respects: 1. In the absence of stratification in one part of the section, which in another shows signs of regular deposition. 2. In the close proximity of fragments of far-transported rock, varying in size from minute fragments to large boulders. He suggests that the mass had been deposited in a tranquil sea of mud, sand, and blocks from melted drift ice. "The discovery," he says, "of fragments and boulders of granite embedded in these deposits furnish adequate proof that glacial action was not limited to any special geological period."

Turning from the Cambrian beds to the next horizon, we find that as far back as 1848 Carrick-Moore described certain conglomerates occurring in the Silurian beds of Wigtonshire in a way which made subsequent writers, such as Dr. Croll, assign them to a glacial origin. Thus he wrote: "These vertical and highly inclined beds near the Corswall Lighthouse are very remarkable from their containing beds of conglomerate of a coarser nature than any I have ever seen described as occurring in so old a formation, with the exception of those of the Potsdam sandstone, described by Mr. Lyell as Old Silurian. The fragments generally vary from the size of one inch to a foot in diameter, but, in some of the beds,

[6] Nature, p. 403.

[7] Brit. Assoc. Rep. 1870, p. 88.

boulders of three, four, and even five feet diameter occur. They are well rounded, and principally consist of red quartziferous porphyry and a large-grained grey syenite; but serpentine, red jasper and other rocks occur, and I have found one or two instances of large angular fragments of greywacke. There are no rocks in the neighbourhood, as far as I know, from whence any of these rounded fragments could have been derived."[8]

In regard to the boulders in the Potsdam limestone just referred to, Lyell says: "The upper part of the Potsdam limestone at the falls of Montmorenci is remarkable for containing boulders of enormous size, the largest I can remember to have seen in any ancient stratified rock. I measured some of them which were eight feet long."[9]

In his "Great Ice Age," Mr. J. Geikie says: "In the Lower Silurian of the south of Scotland (Glen App and Dalmellington), we find large blocks and boulders (from one foot to five feet in diameter) of gneiss, syenite, granite, etc., none of which belong to the rocks of that neighbourhood. Indeed, no such rocks of older age than the Silurian occur nearer than the Laurentian rocks of the north-western Highlands and islands. Possibly the boulders may have come from some ancient Atlantis; and considering the great size of the blocks and the considerable distance they may have travelled, it is not unreasonable to conjecture that ice may have had something to do with their transport. Similar boulder-beds have been found in Ireland."[1]

Principal Dawson records a somewhat similar instance in the Lower Silurian of Maimanse, Lake Superior, where a conglomerate occurs with boulders two feet in diameter. Again, in the Upper Silurian of Nova Scotia, the same author has detected beds of angular stones and chips, "the materials of which seem precisely similar to that which is at present produced by the disintegrating action of frost on hard, and especially schistose and jointed rocks."[2]

Passing on again to another horizon, namely the Devonian, Mr. Cumming, in his "History of the Isle of Man," pub-

[8] Quart. Journ. Geol. Soc. v. 10. [9] Lyell's Travels, ii. 125.
[1] Great Ice Age, 568.
[2] Great Ice Age, 568; Canadian Naturalist, ii. 6; vi. 416.

lished in 1848, compares the old red sandstone conglomerate there to a consolidated ancient boulder clay formation, and he says, *inter alia:* "Was it so, that those strange trilobite looking fishes of that æra (the Coccosteus, Pterichthys and Cephalaspis), had to endure the buffeting of icy waves and to struggle amidst the wreck of ice-floes and the crush of bergs? These are questions which we may venture perhaps to ask, but which we dare not hope to have solved till we know something more than at present we know of the history of the boulder clay formation itself." [3]

In regard to the Old Red Sandstone of Kirkby Lonsdale and Sedbergh, Ramsay writes: "We have found many stones and blocks distinctly scratched, and on others the ghosts of scratches nearly obliterated by age and chemical action, probably aided by pressure at a time when these rocks were buried under thousands of feet of carboniferous strata. In some cases, however, the markings were probably produced within the body of the rock itself by pressure, accompanied by disturbance of the strata; but in others the longitudinal and cross striations convey the idea of glacial action. The shapes of the stones of these conglomerates, many of which are from two to three feet long, their flattened sides and rectangular edges, together with the confused manner in which they are often arranged (like stones in the drift), have long been enough to convince me of their ice-borne character; and the scratched specimens, when properly investigated, may possibly convince others." [4]

At several places in Victoria, Australia, Messrs. Taylor and Etheridge junior, found a conglomerate consisting of a mixture of granite pebbles and boulders of various colours and textures, porphyries, indurated sandstone, quartz and a peculiar flint-coloured rock in a matrix of bluish-grey very hard mud cement. Rocks similar to the pebbles and blocks composing the conglomerate do not occur in the intermediate neighbourhood, and from the curious mixture of large and small angular and water-worn fragments, it was conjectured that it might possibly be of glacial origin. Scratched stones were not observed, although a careful examination was made.

[3] Op. cit. p. 89.

[4] Reader, Aug. 17, 1865.

From similar mud pebble beds on the Lerderderg River, Victoria, Mr. R. Daintree obtained a few pebbles grooved after the manner of ice-scratched blocks.[5] Mr. Selwyn classed the deposit provisionally as Devonian.

In 1851, Professor Haughton brought forward at the Geological Society of Dublin, a case of angular fragments of granite occurring in the carboniferous limestone of the county of Dublin; and he explained the phenomena by the supposition of the transporting power of ice.[6]

"Great beds of conglomerate," continues Geikie, "occur at the bottom of the carboniferous, in various parts of Scotland, which it is difficult to believe are other than ancient morainic débris. They are frequently quite unstratified and the stones often show that peculiar blunted form which is so characteristic of glacial work. They are confusedly huddled together in a dull, tough arenaceous matrix; and the whole deposit will sometimes continue to exhibit these appearances over a wide area. Ever and anon, however, we detect traces of water action, the stones become more rounded and are spread out in more or less regular layers or beds. The coarse unstratified portions so closely resemble till, that only a practised eye can distinguish the difference at a glance, and many geologists who had not previously tried their hammers on the deposit might well mistake it for a post-tertiary glacial accumulation. It is typically developed in the Lammermuir hills district, and appears also in Ayrshire, Arran and other regions in the west of Scotland. Among these conglomerates my brother has noticed many striated stones."[7]

In 1856, Mr. Godwin Austen, writing about the French carboniferous beds, says: "Whether from local elevation, or from climatal conditions, there are certain appearances over the whole which imply that at one time the temperatures must have been very low, as glacier action can alone account for the presence of the large angular blocks which occur in the lowest detrital beds of many of the southern coal basins."[8]

[5] Croll's Climate and Time, 295.
[6] Climate and Time, 296.
[7] Great Ice Age, 568, 569.
[8] Quart. Journn. Geol. Soc. xii. 58.

In 1888, my friend, Mr. M. Stirrup, read a paper before the Manchester Geological Society on Foreign Boulders in Coal Seams. In this he refers to several such discoveries in Lancashire, the first of which was described by Mr. T. W. Binney in 1851. As a rule these boulders are of a hard siliceous grit or quartzite, are smooth, often polished, with the corners rounded off by abrasion. In form they are roughly quadrangular, irregularly ovoid or elliptical, and occasionally globular, and all have been evidently water-worn, but they are neither scratched nor furrowed. According to Mr. Plant, the miners report them to be infrequent, but when found it is chiefly in groups of five, six, or a dozen. They range in size from two inches in diameter to twenty. In addition to the quartzite boulders, others have occurred of quartz and occasionally of granite. Mr. Stirrup argues that transport by ice, or floating ice-bergs in a summer sea would perhaps better than any other theory explain the position and often isolation of these boulders, in strata which indicate quiet deposition and which are singularly free from any admixture of extraneous matter.[9]

In 1884, C. E. Weiss of Berlin published a paper on Boulders from coal seams in Upper Silesia, in which he refers to a list of such boulders contained in a paper by Herr D. Stur of Vienna. These continental boulders, like the English ones, are rounded and polished, except some of the granite ones, which are somewhat rough, due to the projection of certain crystals of garnet, the result of weathering. They consist of gneiss, granite, quartz, porphyry, granulite, quartzite, etc., and vary in size from one to seven pounds. In one case a boulder of typical granulite, lenticular in shape, from Kattviertz, weighed 121 English pounds. In another there was found a group of ten boulders at the same place, one weighing 176 pounds. Weiss argues that these boulders were derived from the neighbouring mountains of Bohemia.

According to Professor Orton, similar boulders occur in the American coal fields. Thus, in the 5th volume of the "Geology of New York," there is a notice of a boulder of

Trans. Manch. Geol. Soc. xix. 420.

metamorphic sandstone apparently derived from the Cambrian rocks of the Appalachian border. This weighed 200 pounds. "The surface of the boulder," says the report, "appears glaciated, and ice transport would seem almost a necessity for such a block." Boulders of quartzite very similar to those from the Lancashire coal field and similarly worn and smoothed have been described from the Zaleski coal beds in Ohio.[1] A notice of several similar boulders from the Anthracite Mammoth seam in Pennsylvania is given in a later volume of the same journal.[2]

Principal Dawson also mentions the occurrence in the Lower Carboniferous of Nova Scotia, of angular fragments, and chips of various hard rocks cemented together, which he thinks may fairly be regarded as evidence of somewhat intense winter cold, in the same way as the angular débris detected by him in the Upper Silurian of Nova Scotia. He describes also a gigantic esker of carboniferous age, "on the outside of which large travelled boulders were deposited, probably by drift ice; while in the swamps within, the coal flora flourished, and fine mud and coaly matter were accumulated."[3]

In the district of Cuttack, in Peninsular India, there occurs a series of coal-bearing beds known as the Talchir beds, whose age has been much discussed, but which have apparently been proved to be carboniferous.[4] Speaking of these beds, which were examined in 1855-6, Mr. W. T. Blandford says: "The lowest bed which we find resting on the gneiss is most generally 'the boulder bed,' which occasionally assumes the form of a coarse conglomerate. It consists essentially of boulders of granite and gneiss, those of the former comparatively small and the latter of much larger size, frequently from four to five feet in diameter, embedded in a matrix which varies from a coarse sandstone to the very finest shale. The question arises how these enormous blocks of stone, manifestly requiring a great force to abrade and transport them, are found mixed with a sediment so fine that in any except a very sluggish current it must have been swept away, and could not have been deposited? Should any

[1] Trans. Manch. Geol. Soc. xix. 417, etc. [2] Vide vol. xxi. 71.
[3] Great Ice Age, 569. American Naturalist, vi. 416.
[4] See Waagen, Records Geol. Surv. of India, xix. 34.

evidence hereafter accrue allowing the inference that these beds may have been formed in a lake on a high table-land, when the winter temperature was sufficiently low to admit of ice reaching the waters of the lake without melting, then an adequate explanation of the phenomenon may be given, as it resembles exactly the effects of the action of ground ice which, enabling boulders to be carried down by a sluggish current, would undoubtedly produce such an intermixture of large rounded masses of rock and of fine silt, as is seen in the present case."[5]

In a series of blue shales belonging to the same series and occurring in the river Brahmini, there occur rounded pebbles or small boulders of gneiss. Mr. Blandford suggests that their scarcity might induce at first sight their attribution to the portage by roots of trees like the boulders in the chalk, but the total absence of wood or carbonaceous matter seems to preclude this. They rather confirm the view already urged in regard to the boulder bed, that they were due to ground ice.[6] Mr. Williams found a similar boulder bed, composed of enormous boulders of granite, some measuring from three to ten feet in diameter, mixed with large boulders of white quartz chlorite rock, hornblende and mica schist, under the lowest coal beds in the bed of the Boodu Nullah in the Ramghur coal-field in Orissa, and he describes it as being the lowest member of that coal field.[7]

Mr. T. Hughes says of the Talchir boulders that some of them attain an enormous size, and he measured one forty-two feet in circumference.[8] In regard to the boulders in the river south of Ranchi and Partábpür, Mr. Ball says they differ from the underlying rocks and were derived from the granitic gneisses three miles north, one rock of pink porphyritic granite occurring *in situ* north of Tárki seems to have been a prolific source for them. In the country west of Rehr many of the boulders consist of a reddish quartz sandstone, probably of Vindyan age; and Mr. Ball adds that if that supposition be correct, they must have been transported some distance. This could only have been effected through the agency of ice.[9]

[5] Records Geol. Surv. of India, i. 49, 50. [6] Id. 55, 56.
[7] Id. 79. [8] Id. vol. vi. art. iii. p. 7.
[9] Id. vi. 28.

In the Clive group in the Salt range, which was once classed as Cretaceous and now as Carboniferous, Dr. Waagen has described a boulder conglomerate made up of boulders, often of enormous size, true erratic blocks, which are embedded in a dark-coloured shale, intermixed often with gravel or coarse sand. The rocks of which the boulders consist are all of a very hard nature, beautiful granites of different colours, syenites, different porphyries, greenstones, quartz rock, etc., but no rocks of a softer nature, like sandstone, etc. "When I visited the Salt range I observed," says Dr. Waagen, "in these conglomerates, a great number of striated pebbles, and recently Dr. Warth has also sent a large number of them to Europe. Such scratched pebbles and boulders embedded in a soft shale are always indications of the influence of ice action, and thus we must admit, for the formation of these boulder conglomerates, the collaboration of ice on a grand scale."[1] Similar boulder beds, which Dr. Waagen assigns to the same carboniferous age, occur in the western Salt range as well as in the Trans-Indus continuation of the range, and he says: "That ice was really greatly concerned in the formation of these beds appears beyond doubt, partly from the numerous striated and scratched stones that are found in them, partly from the huge erratic blocks which are off and on to be met. . . . If thus all the boulder beds of the Salt range belong to one and the same group, we have all at once a large glacial formation stretching through the whole Salt range, during the time of the coal measures. In speaking of a glacial formation, I do not mean that real glaciers were concerned in the formation of these beds; it was probably floating ice coming out of the mouths of rivers which brought the boulders with it and deposited them along the sea shore." Speaking of the evidence of a large southern continent, including India, South Africa and Australia, in Carboniferous times, Dr. Waagen continues: "This continent had evidently a well-developed river system, and at the Salt range there was probably the mouth of a great river. Down these rivers large masses of ice floated while in other parts of the world the coal measures were being formed; and these ice masses drifted

[1] Records Geol. Surv. of India, xix. 30.

along the shore and, as they melted away, deposited large boulders, gravel and fine silt at different places. . . . The enormous development of boulder beds that have been formed under the influence of ice action on this ancient southern continent makes the supposition of very low temperatures during those times on that continent an absolute certainty. These low temperatures were not of local occurrence only, but spread over the whole continent, thus indicating a true glacial period; a glacial period, however, that was in the beginning restricted to the southern hemisphere and only later on spread also to the northern one." [2]

From a general examination of the fossils, the same author concludes that in the earlier times of the Carboniferous period a rather high mean temperature must have prevailed on the southern continent, as luxuriant forests of carboniferous plants were thriving then of which remains have been preserved to us both in Australia and in South Africa. All of a sudden a considerable lowering of the temperature took place, ice began to be formed in South Africa and in India and all the carboniferous flora was destroyed in these countries, as well as in Australia, by this low temperature. In the meantime, in Australia a new flora began to appear—a flora that was suited to support moderate or low temperatures. Notwithstanding the ice that covered part of the southern continent, the flora spread slowly westward from Australia. This flora was of mesozoic types. During the time this went on in the Southern hemisphere, the coal measures were deposited in the Northern one. . . . This state of things continued for a certain time until, in the Permian period, again a change took place. Now the cold had spread to the Northern hemisphere, where traces of it occur, and the result was the extinction of the carboniferous flora also in Europe, and elsewhere in the north, and the flora of a moderate climate, i.e. the mesozoic flora, succeeded it; and as there were no warm currents which could then reach the Northern hemisphere, the Palæozoic fauna perished there except a few genera.

In the Southern hemisphere the intense cold returned during the Permian times and after. In Australia, in the Hawkes-

[2] Records Geol. Surv. of India, xix. 35, 36.

bury beds and the Bacchus March sandstone, which are probably of Permian age (? *vide infra*), glacial boulder beds occur. The flora endured the low temperature, but the marine fauna was destroyed by the currents of cold water which reached the Salt range towards the end of the Permian period. In conclusion Dr. Waagen says: "The cardinal point that is demonstrated beyond doubt is that towards the end of the Palæozoic times a great glacial period occurred similar in extent and more grave in its effects than the one that took place at the end of the Tertiary era, and that thus the earth has passed already through two severe periods of cold."[3]

Mr. Oldham, writing in 1886, describes the great tract between the Aravalli mountains and the Indus, and known as "the Desert," as containing, near the town of Pokran, an old land surface showing glacial groovings and striæ. In the boulder beds which occur close by, and are, without doubt, he says, "of the same age as the glaciated land surface, there may be found facetted blocks which, like those of the Salt range, would not be ascribed to anything but glacier action. Moreover, this land surface is covered in places by a boulder bed with a hard, intensely tough matrix, differing from the stratified boulder beds of the neighbourhood much in the same manner as 'the till' of Scotland differs from the marine boulder clays of the Midland counties. If the hypothesis that the toughness is due to its being a 'Grund moraine' be accepted, it follows that the same explanation will account for the toughness of the boulder beds of Pokran, and we have yet another proof of the existence of glaciers in this old land surface. The boulder beds of the Desert have been traced for 60 miles to the N.E. of Pokran; in the vicinity of the old land surface the boulders are almost exclusively of porphyry and syenite derived from it, but further north blocks of gneiss of the peninsular type become common; and in N. lat. 27° 30′, E. lat. 72° 30′, there is a block of very coarse-grained granite, of which 10 ft. × 7½ ft. × 3 ft. is exposed above ground. The nearest source from which this block could have been derived is in the Aravallis, fully 150 miles away. The age of these boulder beds appears to be the same as that of the Talchirs.

[3] Records Geol. Surv. of India, xix. 38.

. . . Their extent, combined with the distance from which some of the blocks have been transported, as well as their position on the western margin of the peninsular area, point to the conclusion that they are of marine origin; so that here we have evidence of glaciers having descended to the sea in a district now less than 27 degrees from the equator."[4]

We will now turn to the Permian beds, which form a classic ground for the champions of former glacial periods.

In 1854, Professor Ramsay communicated a paper to the British Association on the former probable existence of Palæozoic glaciers. In this he produced some evidence that glaciers and icebergs existed in our latitudes during a part of the Permian epoch, and that there were also probable traces of the same actions in some of the beds of the Bunter sandstone.

Thus, in speaking of the Permian breccias in South Staffordshire and at Enville in Worcestershire, he says: "The fragments found in the breccia are not derived from the Permian beds lower in the series on which it rests in South Staffordshire and near Enville, nor yet from the Coal measures, Old Red Sandstone, and Silurian strata on which it lies in the Malvern, Abberley and Bewdley districts; nor is it probable, taking the physical structure of the country into account, that they were formed from the waste of rocks now concealed beneath the upper and lower New Red Sandstones of these or of neighbouring areas. In lithological structure they can generally be identified with the Cambrian rocks of the Longmynd and the Lower Silurian slates and igneous rocks, etc., that overlie these between the Stiper stones and Chirbury in Montgomeryshire. They are enclosed in a hardened paste of fine red marl, probably analogous to the boulder clay of Pleistocene deposits. The included fragments and blocks consist of greenstone, felstone, felstone porphyry, felspathic ash, hard ribboned slate such as occurs north of Shelve, quartz rock, purple and green sandstone and slate, black slate and sandstones and limestones containing Silurian fossils. Very few of the stones are well rounded. Many of them are angular, most of them are subangular and have

[4] Geol. Mag. N.S. iii. 303.

those flattened surfaces common in fragments found in moraines. They are frequently well polished and occasionally scratched. They are of all sizes up to four feet in diameter. If they came from the Longmynd district, in places they have travelled at least fifty miles, and there is evidence from the outcrops that even now in their fragmentary state they occupy an area of from twenty-five to thirty-five miles in diameter, while there is perhaps much more concealed beneath the surface. Had they formed sea beaches they would probably have been rounded by the waves, and there is nowhere any sound proof that they were scattered abroad by great tidal marine floods. Blocks two feet and more in diameter are not moved along sea bottoms by tidal currents; and, considering their size, angularity, polish, occasional scratchings in straight and sometimes parallel lines, and also the matrix that encloses them, Ramsay believes they were deposited from icebergs, derived from glaciers that rose in the ancient land of the Longmynds and the overlying Lower Silurian strata.

Traces of a similar glacial action are said to occur in the Bunter sandstone on two horizons, first in a part of the pebble beds near Bewdley and later at the base of the white and brown sandstones between Stourport and Abberleys.

In a paper specially devoted to these beds and their contained boulders, published in the Journal of the Geological Society for 1855, Ramsay enters at greater length into the question. He says, *inter alia*; "Everywhere, in spite of exceptional fragments in the Malvern district, the embedded fragments seem to be derived from one set of rocks; they are all enclosed in the same red marly paste, and they are mostly angular or subangular. A well-rounded water-worn pebble is in places of rare occurrence. The surfaces of a great majority of the pebbles are much flattened, numbers are highly polished, and, when searched for, many of them are observed to be distinctly grooved and finely striated. The striæ in some are clear and sharp, and run parallel to or cross each other at various angles; while in others, though you see their remains, age and surface, decomposition have impaired their sharpness and roughened the original polish of the stone.

[5] British Assoc. Report, 1854, 93, 94.

"If lithological character be any guide, the fragments, with rare exceptional pieces, seem to have been derived from the conglomerates and green, grey and purple Cambrian grits of the Longmynd, and from the Silurian quartz rocks, slates, felstones, felspathic ashes, greenstones and upper Caradoc rocks of the country between the Longmynd and Chirbury. The south end of the Malvern Hills is from forty to fifty, the Abberleys from twenty-five to thirty-five miles, Enville from twenty to thirty miles, and South Staffordshire from thirty-five to forty-five miles distant from that country. The question then arises, by what process were so many angular and subangular fragments transported so far; many of them being a foot, and some two to three, or even four feet in diameter; the whole in places forming a deposit of several hundred feet in thickness? Why, also, are they angular and not well-rounded, like the pebbles of the great conglomerate beds of the Bunter Sandstone; and why have they flattened sides, and often polished and striated surfaces?" After examining in turn the various possibilities of water portage, as by rivers, marine currents, etc., and rejecting them in turn, Ramsay concludes that ice alone can have moved them.

In proof of this he quotes, (i.) the great size of many of the fragments, the largest observed weighing (by a rough estimate) from half to three-quarters of a ton; (ii.) their forms. Rounded pebbles are exceedingly rare. They are angular or subangular, and have those flattened sides so peculiarly characteristic of many glacier fragments in existing moraines, and also of many of the stones of the Pleistocene drifts and the moraine matter of the Welsh, Highland, Irish and Vosges glaciers. (iii.) Many of them are highly polished, and others are grooved and finely striated, like the stones of existing Alpine glaciers, and like those of the ancient glaciers of the Vosges, Wales, Ireland, and the Highlands of Scotland; or like many stones in the Pleistocene drift. He urges that he had never seen stones thus scratched in other breccias, and that an experienced eye would have no difficulty in recognizing the peculiar characteristics of glacial scratching in the Permian fragments. In true water-deposited breccias like the New Red Conglomerate, whose component stones are often from three to nine inches in diameter, these stones are all beautifully rounded, and where they touch in the rock they are not scratched, but indent each other at the points of contact, due to lateral pressure in loose lying pebbles. These marks rarely occur in the Permian breccias because, instead of the pebbles being mingled with sand, they are separated by a hardened cementing mass of red marl.

"I conceive, therefore," he continues, "that the peculiar forms, polish and markings of many of the stones indicate that these characteristics have been produced by the agency of ice of the nature of glaciers, for mere coast ice would have picked up and drifted away numerous rounded pebbles from the beach and not a great majority of angular flattened stones, such as form the breccias wherever they occur. If this conclusion be correct and if the parent rocks whence the stones were derived be properly identified, then it follows that the ancient territory of the Longmynd and the adjacent Lower Silurian rocks, having undergone many mutations, at length gave birth to the glaciers which, flowing down some old system of valleys, reached the level of the sea and, breaking off into bergs, floated away to the east and the south-east and deposited their freight of mud, stones and boulders in the neighbouring Permian seas." He further suggests that "the occasional fragments of granite and syenite may be accounted for by the floating bergs having occasionally stranded on or grazed along some of the higher Malvern hill-tops that, as islands, dotted the Permian sea and thus picked up a few fragments to be mingled and deposited with the foreign material wherever they chanced to melt. . . . There is one point of resemblance between these Permian breccias with their associated strata and the Pleistocene drift deposits worthy of note. In the latter, fossils are much scattered, and in most of the beds of rare occurrence. They are still more scarce in that part of the Permian series with which the breccias are associated. I have thought, that, in like manner, this paucity of life may be connected in these latitudes with the glacial phenomena of the Permian and the Bunter periods; and I no sooner mentioned this to Professor L. Forbes than he suggested that it might also be connected with the great break in life that has taken place between Palæozoic and Secondary times." Ramsay concludes with stating that near Wribbenhall, in part of the pebble beds of Bunter date, there are breccias strikingly like those of Permian strata, and also near Astley, a little S.W. of Stourport, and probably at Alfrick at the base of the White Sandstones; and although it is possible they may have been reconstructed from the waste of the older Permian, yet he felt more disposed, after examining them, to attribute them to direct glacial action, and to explain them as Godwin Austen had explained similar boulders in the same beds.[6]

According to Croll, Mr. James Thompson found a bed of conglomerate which he considered of Permian age, and probably of glacial origin. This conglomerate enclosed angular fragments of various schistose, volcanic and limestone rocks and contained carboniferous fossils.[7]

Writing in 1873, in regard to the Lower Permian beds of Armagh, Professor Hall says of certain boulder beds there: "The pebbles and stones of the boulder bed sometimes reach two feet in diameter, and consist of hard greenish and purple grits, probably of Lower Silurian age; vein-quartz, slate, carboniferous limestone, diorite, etc., some of which have probably been transported from the region lying twenty or thirty miles to the north-west of this locality. Considering the large size of these boulders, it is difficult to conceive that they could have been carried to their present position by any other agency than that of floating ice."[8]

In 1870 Dr. Sutherland communicated a paper to the Geological Society on an ancient boulder clay of Natal. He classes this bed as of Permian age, and says it rests generally on old Silurian sandstones, while upwards it passes first into newer shales and through them into the sandstones and shales which are associated in Natal with Carboniferous deposits. The deposit itself, he says, consists of a greyish-blue argillaceous matrix, containing fragments of granite, gneiss, graphite, quartzite, greenstone and clay slate. These embedded fragments are of various size, from the

[6] Journ. Geol. Soc. xi. 185-205.
[7] Climate and Time, 299.
[8] Explanation of sheet 47 Geol. Surv. of Ireland, p. 12 and 13.

minute dimensions of sand grains, up to vast blocks measuring six feet across and weighing from five to ten tons. They are smoothed as if they had been subject to a certain amount of attrition in a muddy sediment, but they are not rounded like boulders that have been subjected to sea breakers . . . the rocky fragments embedded in the boulder clay formation are all of them of the character of rocks that are contemporaneous with, or inferior to, the Silurian Sandstone. Fragments of a higher series are never found in the matrix. In some instances, very ponderous rock-fragments are found as much as fifty or sixty miles away from the sources of their derivation. . . . Dr. Sutherland inclines to think that the transport of vast massive blocks of several tons weight, the scoring of the subjacent surfaces of sandstone and the simultaneous deposition of minute sand grains and large boulders in the same matrix all point to one agency as the only one which can be rationally admitted to account satisfactorily for the presence of this remarkable formation. He believes that the boulder-bearing clay of Natal is of analogous nature to the great Scandinavian drift, to which it is certainly intimately allied in intrinsic mineralogical character; that it is virtually a vast moraine of olden time, and that ice in some form or other has embedded fragments in the amorphous matrix.[1]

Ramsay, in his address to the British Association in 1880, says: " Mr. G. W. Stow, of the Orange River Free State, has of late years given most elaborate accounts of similar Permian boulder beds in South Africa. These great masses of moraine matter not only contain ice-scratched stones, but on the banks of rivers where the Permian has been removed by aqueous denudation, the underlying rocks, well rounded and mammillated, are covered by deeply incised glacier grooves pointing in a direction which at length leads the observer to the pre-Permian mountains from whence the stones were derived that formed these ancient moraines.

In February, 1850, Mr. R. Godwin Austen speaks of the great blocks of porphyry of the middle beds of the New Red Sandstone in the west of England included in sands and marls indicating no great moving power, as seeming to require some

[1] Quart. Journ. Geol. Soc. xvi. 515, 516.

such agent as that of floating ice to account for their position.[2]

Ramsay, speaking from his own observations of the Rothliegende in the north of Germany, says that though marls and gravels are interstratified with it, these, as in our post-pliocene drift, are exceptional, and the main characteristic of this vast formation (2600 feet thick) in the Thüringerwald, is the flattened and subangular nature of its blocks, some of which are of large size. Similar erratic deposits are now forming in Baffin's Bay and the Western Atlantic.[3]

Mr. C. S. Wilkinson, writing in 1881 about the occurrence of glacial boulders in "the Hawkesbury Series," in New South Wales, which belong to the Mezozoic age, says: "Here we have huge and small angular masses of soft shales embedded in pebble conglomerates and false-bedded sandstones. With these sandstones (Hawkesbury) are interstratified argillaceous shales, and the boulders are of the same material and contain the same fossils. It seems that during the deposition of the Hawkesbury Series, the rapid and changing currents which deposited the false-bedded sandstones, were at intervals succeeded by quiet waters, from which the mud forming the shales settled down, probably during winter time when ground ice formed. Spring time following, the ice broke up, and drifting about broke up some of the newly-formed shales and mingled the shale fragments with pebbles and sand brought by currents from the shores." [4]

Ramsay says: "The Oolitic strata of the east of Scotland contain such numbers of huge angular blocks, that their possibly, though scarcely probably, glacial origin suggested itself to my mind when I noted the facts during a journey with Sir Roderick Murchison in 1859. The local character of the blocks, chiefly but not entirely oolitic, is adverse to the view; but the smashed condition of many of the shells in the interstratified oolitic clays is analogous to the state of the shells in the upper drifts all over Britain." [5]

In describing the breccias of Upper Oolitic age in the north of Scotland, Professor Judd tells us that in certain of

[2] Journ. Geol. Soc. vol. vi. 96, 97. [3] Phil. Mag. 4th ser. xxix. 290.
[4] Geol. Mag. N.S. viii. 287. [5] Phil. Mag. 4th ser. xxix. 290.

their beds they include numerous masses of foreign rock of various sizes. The frequency of these included blocks appears to increase as we go northwards till the upper Oolitic strata are almost wholly made up of fragments of foreign rocks, some of these being of enormous size, crowded together in the greatest confusion and cemented by a sandy or calcareous matrix. The enclosed fragments of rock, as Hugh Miller noticed, are clearly foreign, and contain palæozoic fossils. They are sometimes perfectly angular, and at others present every degree of attrition, and are not unfrequently converted into perfectly well-rounded pebbles. They are of all sizes, from pebbles to masses of such prodigious dimensions as altogether to startle the observer. . . . At Allt-an-aird south of the Green-Table point, is a mass composed of hard light-coloured sandstones, alternating with indurated shales and calcareous flagstones. This mass stands on end, its strata being vertical, the exposed upper edge measures twenty feet by ten feet. Another mass close by is probably forty feet long and at least twenty thick, while in the exposed section on the south of the Dunglass there may be seen several large included blocks, one of which is ten feet long and four feet thick. The blocks are not sorted but heaped together in the wildest confusion, angular and subangular blocks, pebbles, trunks of trees, stems of cycads, masses of coral, shells, shell detritus, sand and mud. The greater number of the included blocks are derived from the Caithness flagstones or middle old red series.[6]

In a paper published in the Quarterly Journal of the Geological Society for 1837, by Mr. Godwin Austen, on a boulder of granite in the chalk near Croydon, he says: "This boulder, together with several large fragments of greenstone, was fairly embedded in the solid chalk." In this paper he quotes the discovery by Major-General Portlock of several blocks of black basalt isolated in the chalk of Ireland;[7] the similar discovery of flattened spheroids of syenite in the chalk of Antrim in 1840 by Mr. Griffith, the mention by Mantel,[8] of pebbles of quartz and some fragments

[6] Quart. Journ. Geol. Soc. xxix. 187, etc.
[7] See Report on Londonderry, p. 93, 94.
[8] Geology of the S.E. of England, p. 78.

of green schist in the chalk, and the following statement of Mr. Dixon: "Small pebbles and large rolled fragments of sandstone and quartz rock are occasionally discovered in the centre of the Upper Chalk. Mr. Coombe found one specimen weighing near fourteen pounds, at Houghton, Sussex; and I have seen others from the same pit of two and three pounds weight; several also have been sent me by Mr. Catt from the pits near Lewes."

Mr. G. Austen says he had an opportunity of examining certain extraneous specimens from the chalk in the possession of the Geological Society, and others belonging to Mr. Harris, which were interesting as an addition to the crystalline rock pebbles; these were some small specimens which had the appearance of being volcanic scoriæ. Mr. Catt's collection contained three specimens of rather soft marly, micaceous sandstone of a greenish-grey colour, one pebble of opaque quartz, two of transparent quartz, and one of coloured quartz; these were all small. There was also a much larger sub-rhomboidal fragment of dark clay slate, with the angles rounded, and lastly a block of fine-grained sandstone (quartzite) weighing thirteen pounds fourteen ounces. All these fragments had been derived from old sedimentary (palæozoic) strata. Mr. Cumming, of Devizes, had further shown him a piece of thin fossil slate (very like common roofing slate) embedded in a chalk pit.

Mr. Godwin Austen refers to the generally isolated character of the finds of extraneous stones in the chalk. In the Croydon find the granite boulder was accompanied by a collection of blocks of smaller dimensions, composed of green stone, which were much decomposed.

Speaking of the various extraneous stones found, he says they are all water-worn, either in the form of shingle or rounded boulders. He discusses Darwin's explanation of the finding of a rounded block of greenstone on the outer coast of a small atoll belonging to the Keeling group, namely that it had become entangled in the roots of a tree and thus transported, a view which commended itself to Sir Charles Lyell as explaining the extraneous stones in the chalk to which his attention had been called by Mr. Catt. He himself would rather invoke the agency of floating ice, and, in answer to

Lyell's objection about the tropical fauna of the chalk, he says: "The presence of a sea is no indication as to whether ice from polar regions may not occasionally be floated into it. All that is requisite is that the area of water should have an extension into the polar regions; and then the liberated ice must at all times have been distributed according to the same laws as influence it now. Polar ice, it must be remembered, is now occasionally floated down, and its load scattered, as low as the Azores and Canaries, where its detritus becomes embedded in deposits containing the forms of the South Lusitanian fauna." He denies the possibility of any conceivable mass of sea-weed having carried boulders of twelve and fourteen pounds weight, and he adds that even if it be assumed that sea plants could at any time have lifted the largest boulders, and that a tangled mass of them could have floated away at the same time all the smaller stones associated with it, the difficulty as to the mass of loose sand still remains. In regard to the kind of ice to which he attributes the portage of the boulders, he says that the detritus borne by icebergs is sub-aerial, and its form angular, and that he had never seen a single angular fragment from the White chalk, nor from any part of the Cretaceous series; and he appeals to the coast ice of the Polar lands, and says, "I know of no other agent but that of coast ice by which a mass of incoherent materials could have been held together as a characteristic portion of a sea beach, and so deposited at a vast distance from the nearest land. Such, I conclude, was the nature of the agency by which the assemblage discovered in the chalk at Purley was kept together and conveyed."[9]

In a second paper, published in 1860 in the same Journal, on the discovery of a mass of coal four feet square and from four to ten inches thick, at a depth of 180 feet in the chalk of Kent, the same author says: "Considering its friability, I do not think that the agency of a floating tree could have been engaged in its transport; but looking at its flat angular form it seems to me that its history may agree with what I have already suggested with reference to the boulder in the chalk at Croydon. We may suppose that during the Cretaceous

[9] Quart. Journ. Geol. Soc. xiv. 252-265.

period some bituminous beds of the preceding Oolitic period lay so as to be covered by water near the sea margin or along some river bank, from which portions could be raised off by ice and so drifted away until the ice was no longer able to support its load."[1]

Ramsay says: "It may not be generally known that Escher von der Linth is aware of boulders in the cretaceous strata of the Alps."[2]

Turning now to Tertiary times.

"The so-called flysch of Switzerland," says Dr. Croll, "has by some geologists been assigned to Eocene times. In it are found certain immense blocks, some of which consist of a kind of granite not known to occur *in situ* in any part of the Alps. Some of the blocks are ten feet and more in length, and one at Halekeren, at the lake of Thun, is 105 feet in length, ninety feet in breadth and forty-five in height. These unmistakably indicate the presence of glaciers or floating ice. This conclusion is further borne out by the fact that 'the flysch' is destitute of organic remains."[3]

In a paper by M. J. Martin, he quotes, in support of his views about inter-glacial climates, the fact that the *flysch* of Switzerland, which is generally barren of life, contains a vast number of pebbles and blocks, notably at Séfey in the Canton of Vaud, where enormous blocks of granite, gneiss, and protogine occur very frequently in this deposit. At Halekerenthal, in the Bernese Oberland, are others of enormous size, as much as 500 cubic metres in size. At Yberg in the Sihlthal great masses of lias and of brownjurassic rocks are mixed with blocks of granite. It is similarly the case at Grabs, and in many other places. The blocks of granite and of limestone are foreign to this district. The same fact has been observed at Bolgen in Bavaria, and at the northern foot of the Apennines in Upper Italy. In view of these facts, he says, we are obliged to admit the work of tertiary glaciers in many places, for no other explanation of the facts is admissible.[4]

In the paper already cited by Mr. Lydekker, he refers to the lower Tertiaries near Leh in Ladakh, and says that in certain

[1] Quart. Journ. Geol. Soc. xvi. 327.
[2] Phil. Mag. 4th ser. xxix. 290. [3] Climate and Time 305, 306.
[4] Bull. Soc. Geol. de France, 3rd ser. vol. ii. 271.

soft and crumbly sandstones there are found embedded great numbers of more or less rounded boulders of gneiss, often attaining a diameter of several feet. Some of the isolated boulders show the beds of the sandstone bending down below them, and the polishing and smoothing of some of the others seem suggestive of ice action. . . . "It is difficult to imagine how these gneissic boulders could have attained their present position in the midst of soft sandy strata without the aid of ice as a means of transport."[5]

In regard to Miocene times we have the observations of Gastaldi and Ramsay about the Italian beds. The latter communicated their results to Dr. Croll, who thus writes of them: "Near Turin there is a series of hills, rising about 500 or 600 feet above the valleys, composed of beds of Miocene sandstone, marl, and gravel and loose conglomerate. These beds have been carefully examined and described by Gastaldi.[6] The hill of the Luperga has been particularly noticed by him. Many of the stones in these beds are striated in a manner similar to those found in the true till or boulder clay of this country. But what is most remarkable is the fact that large erratic blocks of limestone, many of them from ten to fifteen feet in diameter, are found in abundance in these beds. It has been shown by Gastaldi that these blocks have all been derived from the outer ridge of the Alps on the Italian side, namely, from the range extending from Ivrea to the Lago Maggiore, and consequently they must have travelled from twenty to eighty miles. So abundant are they that extensive quarries have been opened in the hill, for the sake of procuring them. These facts prove not only the existence of glaciers on the Alps during the Miocene period, but of glaciers extending to the sea and breaking up into icebergs; the stratification of the beds amongst which the blocks occur sufficiently indicating aqueous action and the former presence of the sea."[7]

Professor Geikie, writing of these North Italian beds and also quoting Gastaldi, says: "The slopes of the Moncaluri-Valenza hills are sprinkled with boulders and large erratics of Alpine

[5] Op. cit. 105.
[6] See Memorie della Reale Acad. delle Scienzi de Torino, 2nd ser. vol. xx.
[7] Climate and Time, 306, 307.

rocks, which were at one time supposed to have been carried across Piedmont by the ice of the Glacial epoch. But subsequent and more detailed observations have led Gastaldi to the opinion that the blocks in question are merely the denuded wreck of certain great beds of conglomerate belonging to the Miocene formation. No one who has visited the ground is likely to dispute this conclusion. One sees embedded in the Miocene conglomerates large erratics of precisely the same character as those that are lying loose on the hill slopes; and the conclusion seems irresistible that these latter are but the relics of those portions of the conglomerate which the denuding forces have carried away."[8]

M. J. Martin, in 1873, in discussing certain beds found in the Chalonais in Burgundy, urged that they were of the same age as others found scattered at Grosmont, Roumont, Magny, etc. Similar beds on the plateaux of the Sénonais and the Gatinais were treated as of Miocene age by MM. Raulin and Leymerie. They resemble again completely the so-called *poudingues* of Nemours, which have been long classed as of the date of the plastic clay. To the same system, again, he urges, belong the clays with flints and with blocks of sandstone called *ladères* in the department of Eure et Loire described by M. Laugel and treated by him as of Miocene age. In this view M. Hebert also concurred, and the section he gave from Souancé to Senonches shows these clays with flint covered with lacustrine beds containing Limnæa longiscata and Planorbis planatus. In addition to these facts he quoted an excavation made in 1870 in the hamlet of Larrey (Dijon) in the Miocene conglomerate, which was formed of coraline and portlandean pebbles; many were rolled, many again polished and striated like glacial boulders. In this excavation masses of half a cubic metre in extent, with one side angular and the other rubbed and striated, rested in the midst of rounded pebbles, having a brilliant and soapy polish, with fine lines scratched on them, in which it was not difficult to find the particle of quartz which acted as a graver. The deposit is considerable, opposite to and of the same age as the conglomerate with *helix ramondi* at the Dijon station. M. Jules

[8] Great Ice Age, 437-439.

Martin concluded definitely that these facts prove there were two glacial epochs, one in Miocene and one in Quaternary times; the first one being characterized by the conglomerate with polished and striated pebbles of Dijon, by the clay with flints and with a puddingstone of flints in the Chalonais plain, and by the blocks of puddingstone and of siliceous sandstone of Grosmont, Roumont, Magny, etc. These Miocene débris, both in the Seine valley and in that of the Saône, came from the Morvan.[9]

In 1866-1868 Mr. P. W. Stuart Menteath published two or three memoirs on the evidence of a Miocene Glacial period in the Pyrenees. In these papers the author classed certain conglomerates (*poudingues*) occurring near Pau, as of Miocene age. These conglomerates he identified with those described by M. Leymerie in the Haute Garonne under the name of Poudingue de Palasson, and with the upper part of the similar beds described by the Abbé Pouech in the department of the Ariège. These beds, containing no fossils but having all the appearance of a glacial formation, form, according to Mr. Menteath, the boundary between the Miocene and Eocene periods. He compares them with the Permian conglomerate described by Ramsay.

M. Tardy, in a communication to the French Geological Society in 1876, calls attention to large boulders of quartz included in the Miocene beds at Bourg, and in the lacustrine marls of Miocene age at Varambon in La Bresse. He concludes that they must have been transported by ice, and compares them with the similar boulders found near Turin and the Grand Chartreuse.[10]

In regard to Pliocene times, the only evidence available is that of the North Italian beds. According to Desor and Stopanni, clays containing unmistakable Pliocene shells, which undoubtedly, according to Desor, lived and died where their remains occur, are covered with morainic débris, including striated stones; and according to Stopanni these striated stones are not confined to the moraines, but appear also in the underlying Pliocene clay. "From their observations," says Mr. Geikie, "Stopanni and Desor infer that during the formation

[9] Bull. Soc. Geol. de France, 3rd ser. i. 390-397.
[10] Id. 3rd ser. iv. 185, 186.

of the marine Pliocene the old Adriatic sea covered all the low grounds of Northern Italy, and extended into the deep valleys of the Alps, down which crawled large glaciers, which eventually entered the sea and shot their moraines upon its bed."[11] Rutimeyer virtually came to the same conclusion. Geikie, in summing up his views, says: "In the present state of our knowledge no hard and fast line can be drawn between these two periods, *i.e.* between Pliocene and Pleistocene times. From early Pliocene times onwards we have proofs of a continuity of terrestrial conditions over a wide region in Europe. And in the long lapse of time represented by the Pliocene and Post-pliocene deposits—a period during which considerable changes in the mammalian fauna were taking place—glaciers now and again descended to the mouths of the great Alpine valleys."[12]

I have quoted a sufficient number of instances from different geological horizons to show how those geologists—and they probably include the majority—who believe in the recurrence of glacial phenomena justify their conclusions. If they appeal to astronomical and secular changes of climate, they are of course bound to prove that the last great glacial age was only one of many; or, as Mr. Croll puts it, if the glacial epoch resulted from these causes then such epochs must have frequently supervened. The case needed more than this, however; it needed also evidence to show that the glacial period itself was broken up into alternate zones of mild and severe climatic conditions. Hence the theory of inter-glacial periods.

Venetz, who, like Charpentier, had not apparently heard of ground-moraines, argued that a glacier cannot at the same time scratch and polish rock surfaces and carpet them with soft débris, and he consequently urged that the former process was due to an earlier glacial action than the latter. He also urged that, inasmuch as when the mountains were covered with ice and snow they would be protected from denudation, it is difficult to understand how, during the maximum of the ice development, there could be angular erratics deposited in the extreme terminal moraines; and he

[11] Great Ice Age, 441, 442. [12] Id. 445.

consequently argued that where there are several successive terminal moraines, they point to several stages of glaciation. He also distinguished between two great moraine zones, an outer and an inner, and attributed them to two distinct glaciations.[1]

In regard to the distinction between the outer moraine and the inner ones, Agassiz had noticed that the true moraines in the Alps are not found at such a distance from the radiating centres as the more evenly distributed débris to which some geologists give the name of ground moraines; and he distinguished between two glacial epochs, one marked by a widespread ice-sheet, evidenced by the ground-moraine, and the other by a later glaciation to which the deposition of the terminal moraines is to be attributed. This distinction between an outer and inner moraine zone was adopted by several later writers on the Alps. Thus Muhlberg, in two memoirs on the erratics of the Aargau, published in 1869 and 1878, adopted this view. So did Bach in his description of the erratic phenomena of Würtemberg, published in 1869 Taramelli, in 1880, applied the same arguments to the Italian phenomena in the Canton Ticino and its neighbourhood. H. Höfer had, some years before, done the same in Carinthia. These various writers had deduced a double glaciation from the geographical distribution of the moraines. Another set of writers came to the same conclusion from examining the internal arrangement of the drift deposits, and professed to find traces of temperate conditions in beds interstratified with the true glacial beds.

The first person who suggested that the glacial deposits instead of being continuous are broken and intermittent, was also Venetz. He pointed out that the depôt of bituminous wood found on the southern flank of the great glacial talus of Thonon, rests on one glacial bed and is covered by another; and, according to Morlot, he prepared a map in which he showed the extent of the Rhone glacier during its second phase.[2]

In 1854 M. Morlot published a paper in the Bulletin de la Société Vaudoise, on the Quaternary deposits of Switzerland.

[1] See Penck, Vergletscherung der Deutschen Alpen, 212, 213.
[2] Bull. de la Société Vaudoise des Sciences Naturelles, iv. 45.

This he followed up by another, published in 1855, on a larger scale, on the subdivision of the Quaternary deposits of the same country. In these papers he urged that the beds called ancient alluvium by M. Necker, in which the traces of the mammoth and its contemporaries occur, and to which he gives the name of diluvium, were deposited by the present rivers and lakes, when the continent was at a much lower level than it is now; that when they were laid down the Alpine country was free from ice to a height of 3000 or 4000 feet above the sea level, and that it is probable the ice zone was even more restricted, since the rivers flowing from the glaciers would otherwise not have had the necessary length and volume to deposit these beds. He went on to urge that the deposit of this diluvium has been generally recognized as anterior to the glacial epoch, since the erratic deposits have been distinctly found in several places superimposed on these beds; as near Geneva, where erratic detritus to a depth of fifty feet overlies them, 100 feet above the level of the lake.

This evidence, however, is not uniform. Thus at Tavel, near Clarens, between Vevey and Villeneuve, he himself had described an instance where a diluvial bed of seven feet thick was imposed on a bed of unstratified blue clay forty-three feet thick, enclosing rounded and striated Alpine boulders ; which clay rested immediately on the *molasse*. M. Ischer afterwards found a similar arrangement of beds near Berne. In other cases the diluvium itself contains Alpine materials which must have been carried by ice and not water, or they would have filled up the lake of Geneva.

From these facts it follows, he urged, that a glacier must have invaded the country before the diluvial epoch. During that epoch it disappeared entirely, to return again afterwards. The first invasion of the glacier was the one in which the advance was the most extended and remarkable, and the greater part of Switzerland must then have been covered with ice. The absence of angular unrolled blocks at this stage, as Mr. Chambers had remarked in Scotland and M. Collomb in the Vosges, shows that few naked mountain summits were then exposed to furnish them. This earlier period, although very intense, was not of long duration, for it left no true moraines. This is especially noteworthy on the Jura.

The second glacial epoch, on the contrary, must have been prolonged, for it has left moraines which often rest on the diluvium. These moraines are numerous, well-marked and immense, but they are not found at the same distance from the mountains as the débris of the first glacier. The Rhone glacier, for instance, did not then extend beyond Geneva. While the first glacier, near Vevey, passed over Mont Folly, 5863 feet above the sea level, and rounded its summit, the second glacier did not rise above Avant and Lalliaz, 3263 feet above the sea level. The latter was therefore 2500 feet below the former. The latter, in retreating, made several halts, leaving moraines at each stage. These intermediate moraines are well marked near Linth, where they have been studied by M. Escher, and may be traced up to the present glaciers, showing that the transition from this condition to the present was gradual.

Following in the footsteps of Mr. Chambers in Scotland, M. Morlot then proceeds to divide the so-called glacial beds into two series, one marked by compact blue clay containing boulders, the till of the Scotch geologists, and the other a yellow or red friable clay like loess, but containing no shells, in which the iron has been oxydized; and he considers that the latter is, to a large extent, the result of the so-called *remaniément* of the former. Speaking generally, he attributes the blue clay to the earlier glacial development, and the yellow or red to the later, but adds that to this there are exceptions.[3]

Near Vevey and Lausanne the clay deposited by the first glacier is blue and compact, while that deposited by the second one is friable and of a red colour. At Soleure, however, which the second glacier did not reach, the clay of the first glacier is also red and friable. He accounts for this by the fact that in this case the deposit was made at the termination of the glacier, where it was thin and accessible to the air, which has oxydized the iron in the clay, thus giving it its red colour.

The occasional occurrence of irregularly stratified rolled

[3] Id. 41, 42.

materials with erratic detritus he explains as the result of ice lakes formed by glacier dams. This kind of deposit was named *diluvium glaciare* by M. Charpentier, and was compared by M. Martin with the material of the asars, which also shows the joint action of water and ice. Its great extent points, he urges, to the second glacier having lasted a long time.

In recapitulating his conclusions, M. Morlot urges that in Switzerland there are evidences of—I. The first glacial epoch, when Switzerland, Scandinavia, Scotland and the Vosges were covered with ice sheets, whose débris contain no fossils and are formed of till or compact blue clay with boulders. It probably lasted no long time. II. The diluvian period, when the glaciers disappeared from the principal valleys of Switzerland, as far as 3000 or 4000 feet above the present sea level. The rivers then flowed at higher levels. It lasted a long time, probably as long as the present period. It was marked largely by a similar fauna and flora as the present period. III. The second glacial epoch, when the glaciers again invaded the Alpine valleys, filled up the lakes, such as Geneva, but did not extend as far as the first. To this period he attributes the loess, and in certain localities the so-called *diluvium glaciare*. IV. Lastly the modern epoch, which was a prolongation of the previous one, was caused by the gradual shrinking of the glaciers, and was marked by the disappearance of the mammoth.[4]

M. Chavanne, in the discussion on M. Morlot's paper, said he had noticed at Morges the glacial mud resting on the molasse and covered in turn by diluvial beds; and Dr. Delaharpe said the same thing occurred at Chamblande below Lausanne. M. Blanchet distinguishes three phases in the glacial epoch: a period of the growth of the great glaciers, a second period in which they were stationary, and a period of their retreat, during which, for the most part, the lateral moraines were formed.[5] M. Guyot, in his researches on the Rhone glacier, also pointed out two phases in its history.

In 1857 M. Scipion Gras read a memoir before the French Geological Society on the Quaternary period in the valley of the Rhone. In this elaborate memoir he distinguishes five

[4] Bull. de la Soc. Vaud. iv. 41, 45; Bibl. Universelle de Geneve Archives xxix. 32-42.
[5] Bull. Soc. Vaudoise, iv. 53.

successive stages in the Quaternary period, while he says the facts point to two distinct glacial epochs in the Alps; one marked by the deposition of what he calls the lower diluvium with its angular blocks and striated pebbles, and the other by the dispersal of erratic blocks which, he says, closes the Quaternary age; and he claims to have arrived at the same conclusions as M. Morlot by another route.[6]

In 1865 O. Heer published his "Urwelt der Schweiz," which was translated into English by Mr. Heywood. In this work he pointed out that the lignite beds at Dürnten in the canton of Zurich, and at Utznach in the canton of St. Gallen, are marked by plants not different from those of our time. These lignites, he shows, are overlaid by great deposits of stratified gravel and sand, upon the surface of which isolated blocks are found. The pebbles contained in these beds have been derived from the Alps. Heer accepted the views of Guyot and Morlot about the double glaciation of Switzerland, and he says, *inter alia*: "It is manifest that the lignites of Dürnten and Utznach were formed before the stratified drift, as they lie below it. . . . The climate was then like that of the present day. This condition of things must have lasted some thousands of years in order to produce the thick deposits of peat which the lignite originated. But gradually the climate became colder, the glaciers advanced from the Alps and changed the whole aspect of the country. Again the temperature became warmer; the glaciers slowly melted, and retreated, after many oscillations, to their present limits. . . . Two glacial epochs are indicated by the mode of dispersion of the boulders in the basin of the Rhone. . . . In Eastern Switzerland, until recently, the upper erratic formation only was known, which overlies the stratified pebble-beds. Next at Wetzikon, Alpine blocks, showing all the signs of glacial transport, have been found under the lignites. This deposit of lignite, however, is of small extent; but it lies horizontally, and presents the same sequence of clay and coal as at Dürnten; so that it cannot be supposed that these erratics have got under the lignite deposit by displacement of the soil." In Mörschweil on the Bodensee also, according to

[6] Bull. Soc. Geol. de France, ser. ii. xiv. 241.

the investigations of Professor Deicke, erratics are overlaid by lignites. Again, Deicke says that on the new road near St. Gallen, erratic blocks occur over and below the diluvium, and in the Lake district of Baden, the erratic blocks lie under the erratics and partly immediately on the molasse as at Stabringen.

Fortified by these facts, the conclusion may be adopted, says Heer, that at two different periods glaciers invaded Switzerland, and that the lignite formation occurred in the interval between them, thus representing an episode of several thousand years in the long glacial epoch, the intercalated period having been sufficiently long to diffuse over the low country a new vegetable covering.

The stratified deposits of rolled pebbles must have taken place at different periods; first, after the epoch of the lignites, which may be called inter-glacial; secondly, during the great extensions of the glaciers—*glacial drift*; and lastly, after the second glacial extension.[7]

A notable advance in the doctrine of interglacial climates was made by Sir A. Geikie, who, in his well-known paper on the Glacial drift of Scotland, read before the Glasgow Geological Society in 1863, drew attention to the fact that in certain places in Scotland, beds of peat and other traces of similar land-surfaces have been found intercalated between beds of boulder clay. He pointed out a notable case at Cowbyres Bridge, about a mile and a quarter from Hawick. "So far," he says, "as it was possible to ascertain the nature of the vegetable remains in the peaty layer, they appeared to be the rootlets of a kind of heath. The boulder clay which rested upon these stratified beds coincided in every respect with that below them, except perhaps in having not quite so many large stones."[8]

He says he was much taken by surprise by the section, but soon obtained other examples of the intercalation of stratified materials in the tumultuous and unstratified till. These for the most part did not contain traces of terrestrial animals or plants. In another notable case, however, at the pit workings

[7] Heer, Urzeit, etc., 531, 532; Trans. ii. 201, 202.
[8] Trans. Geol. Soc. of Glasgow, i. 54, 55.

of the coal field of Airdrie, in digging through the drift beds, layers of peat were from time to time met with in certain laminated clays, likewise decaying twigs and branches. In specimens of these peaty layers which he examined, although much decayed, the vegetable fibres were still distinct, and the substance burnt with a dull lambent flame when put into the fire." These beds were both underlaid and overlaid by till,[1] and Geikie argues from the facts that they possibly indicate that "the boulder clay is not the result of one great catastrophe, but of slow and silent, yet mighty, forces acting sometimes with long pauses throughout a vast cycle of time." [2] He then goes on to call attention to the discovery at Kilmaurs, Ayrshire, in 1817, of two elephant's tusks in the boulder clay itself, while at 17½ feet from the surface other tusks occurred in the same place; and Dr. Scouler, who examined the spot and saw some of the remains, assured Geikie they occurred in a true boulder-bearing till. A second similar discovery was made in 1820, during the excavation of the Union canal, between Edinburgh and Falkirk, where an immense tusk is reported to have occurred 15 or 20 feet from the surface, embedded in the heart of the stiff boulder clay; while a bone of an elephant is also said to have been found at Chapelhall near Airdrie, in laminated sand underlying till. Remains of the elk and reindeer are similarly quoted by Geikie as having been found elsewhere in similar beds. And he concludes from the evidence that "while Scotland was covered during the drift period with a wide sheet of snow and ice. . . . there were periods of varying climate; that sometimes many of the valleys glowed with the vegetation of an Arctic summer, rills of water trickled down their sides or dashed over their rocky ledges, and herds of huge mammals pastured in their solitudes; that again the cold increased, and the ice, once more pushing down the valleys, buried their beauty under one wide pall of cheerless desolation." [3]

In a paper read on April 2, 1868, before the Geological Society of Glasgow, Mr. Bennie gives the result of a large series consisting of 248 borings, of an aggregate length of 21,245 feet, made for water, etc., in the surface deposits near

[1] Id. 61-63. [2] Id. 66. [3] Id. 96.

Glasgow. He points out the numerous instances in which great beds of water-formed materials, as sand and gravel, were laid down upon such glacial débris as boulder clay, indicating a mild interval, and how glacial conditions again prevailed, as shown by clay being superimposed on them again. "The whole series showing," he says, "that the glacial period was not uniform throughout its duration, but was broken up by warmer periods during which water regulated the course of deposition. . . . Between 50 and 60 bores were obtained, having sand, gravel or clay, true water-drifted materials, interstratified with or lying beneath boulder clay. Twenty-five of these had one bed of sand or clay intercalated between two beds of boulder clay, proving that one break occurred. Ten had two beds of sand or clay between three beds of boulder clay. One bore had three and two bores had four beds of sand thus interpolated, while one bore had five such beds, a number sufficient to show that the ice sheet was not continuous throughout the glacial epoch, but disappeared for periods so long that great beds of water-made débris could be deposited in the interval." After quoting some details, Mr. Bennie continues: "Such is the evidence upon which we seek to establish an inter-glacial character for the sands, clays and gravels, which we find interstratified in these bores with what seems to be boulder clay, and the conclusion clearly inferable from them is, that there were several glacial periods; or, if we consider the whole as forming one great epoch, we must conclude that the glacial conditions were not uniform throughout the whole time it lasted, but were interrupted by warmer periods during which sand, gravel and clay, the forms which water-carried drift assumes, were the only sediment possible."[4]

In 1871 Professor J. Geikie published a paper in the *Geological Magazine,* on Changes of Climate during the Glacial epoch. In this he discusses the evidence of the stratified beds included occasionally in the till of Scotland. Among these he especially notes the laminated brick clays resting, he says, in little basin-shaped hollows of the till, and covered by it, in which may sometimes be detected thin lines of peat. From these and the fresh-water beds similarly intercalated, he argues that that

[4] Trans. Geol. Soc. of Glasgow, vol. iii. 133, etc.

section of the glacial epoch represented by the Scotch boulder clay was not one long unbroken age of ice. It was certainly interrupted by several intervening periods of less Arctic conditions, during the prevalence of which the ice sheet must gradually have melted away from the low grounds and given place to streams and lakes and rivers. "At such periods a vegetation like that of cold temperate regions clothed the valleys with grasses and heaths and the hill sides with pines and birch. Reindeer wandered across the country, whole herds of the great ox and the mammoth frequented the grassy vales. . . . Whether during any of our inter-glacial periods the climate was ever mild enough to melt away all the ice and snow from our Highland valleys, the record does not say. What evidence we have, points to the existence of local glaciers in our higher valleys and to moderate summers and severe winters—during such inter-glacial periods as we have any certain records of. . . . We must bear in mind that the inter-glacial deposits are the veriest fragments. They have been preserved only in sheltered hollows from the ravages of the great ice plough—and the interrupted and patchy portions that remain are mere wrecks of what must once have been, in the lower valleys, widespread and continuous deposits. Every renewed descent of the glaciers upon the low ground would tend to effect the removal of these accumulations, and it may well be that of many inter-glacial periods, not a single representative deposit now remains. . . . So far then as the Scottish glacial drifts are concerned, there is no evidence whatever to show that the inter-glacial periods may not have been warm enough to cause all the snow and ice to disappear from the country."[5] Again, Mr. Geikie says, "We shall probably never learn how many great changes of climate took place during the accumulation of the till. This arises from the fact that during every period of intensest cold, when the country was covered with a more or less thick sheet of snow and ice, the loose materials which in the preceding age had gathered in river valleys and in lakes would almost inevitably be subjected to excessive denudation. . . . yet in sheltered hollows we find sometimes two, sometimes three and even four beds of till separated by

Geol. Mag. vol. viii. 551-553.

intervening beds, deposits of gravel, sand and silt. We cannot assume, however, that only four cold periods and three intervening ages of milder conditions were comprised within the first stage of the great glacial epoch. For aught that we know there may have been many revolutions of climate.[6] He sums up the Scotch evidence thus :—

1. The till of Scotland, with its intercalated beds, indicates a vast lapse of time, during which there were several great revolutions of climate; how many we do not know.

2. The beds of till point to intense Arctic conditions having prevailed at the time of their deposition.

3. The beds intercalated with the till, consisting of deposits of silt, clay and gravel, with land plants and mammalian remains, and in some cases with marine shells, clearly show that the intense Arctic cold which covered the country with an ice-sheet, was interrupted not once but several times, by long continuous ages of milder conditions.

4. The direct evidence shows these inter-glacial periods were not characterized by warmer climates than that now prevailing in the forest region of the high latitudes of North America.

5. We are entitled to argue, from the few remaining fossils yielded by them, that the climates of inter-glacial periods were never positively warm.[7]

Turning from Scotland to England, it has long been argued that the drift beds of the Eastern Counties are separable into a series. Mr. S. V. Wood, Jun., urged that there are four distinct boulder clays in this district, namely, (1) the Cromer clay, (2) the great chalky boulder clay, (3) the purple clay of Holderness, (4) the Hessle clay. The Cromer clay is underlaid by the forest bed and overlaid, according to Wood and Geikie, by certain gravels containing marine shells; and here and there, as near Ipswich, says Geikie, associated with beds of brick clay apparently of fresh-water origin, in which fragments of wood, sticks and logs have been found. The same author identifies this layer with the brick earths of Brandon, containing palæolithic

[6] Id. 265-270.

[7] Id. 29, 30.

implements. "Thus," he continues, "we have evidence to show that after the ice-sheet which laid down the Cromer till had melted away, mild conditions of climate ensued. . . . Overlying the Brandon beds, with their flint implements, comes the great chalky boulder clay or bottom moraine of an ice-sheet formed during the climax of glacial cold. It is this boulder clay which has been traced south to the valley of the Thames. Resting upon it occur sands, gravels, and loams of much the same character as those that immediately overlie the Cromer boulder clay. . . . To these deposits succeeds a third boulder clay, that which is known as the purple clay, the presence of which points to a third advance of the ice-sheet. Overlying the purple clay again, we encounter a series of sands and gravels which in the valley of the Humber have yielded remains of Pleistocene mammalia. . . . Lastly, these beds are covered in their turn by a fourth sheet of boulder clay, the Hessle boulder clay. . . . Thus we have evidence in these English sections of no fewer than four glacial epochs separated by intervening epochs of mild climatic conditions."[8]

In the north-west of England and the east of Ireland, the drift series has been divided by Professor Hull and others into three members, an upper and a lower boulder clay, separated by beds of sand and gravel, in some cases containing marine débris, known as the middle sands. Mr. Geikie would identify the lower boulder clay of Lancashire, Cheshire and Ireland with the purple boulder clay of Yorkshire, the middle sands with the Hessle estuarine beds, and the upper boulder clay on both sides of the Irish Channel with the Hessle boulder clay of Holderness; and he proceeds to give in detail the succession of climatic changes which these facts imply, involving, he contends, a succession of glacial and mild periods.

In Germany, Professor Penck has been a notable champion of inter-glacial beds. In his well-known memoir on the erratic formations of Germany, he refers to a section at Rixdorf in Brandenburg, where, he says, a bed containing bones of Pleistocene mammals and shells separates two distinct boulder clays, and he argues that this

[8] Pre-historic Times, 261-265.

proves that the great Scandinavian glacier overran the neighbourhood of Berlin at least twice, and that this cannot be explained by a mere oscillation, since the deposits in question occupy 400 square miles in Brandenburg. The same author, after describing in some detail the various instances of the occurrence of fresh-water deposits in the neighbourhood of Potsdam, Baumgartenbrück, Glindov, etc., urges that they point to an intercallation between two beds of boulder clay, and to being relics of a mild period intervening between two glacial ones.[1] As Penck puts these beds of sand on a different horizon to the one before quoted, it implies that in his view there is evidence of three cold periods separated by two mild ones. In the province of Prussia proper, Penck claims that there are also three boulder clays separated in this case by beds of sand, etc., containing marine shells in some cases mixed with fresh-water ones, pointing to there having been there as in Brandenburg a threefold return of the ice movement; and this could not have been a mere oscillation of the glacier, since it implies that during the intervals the Baltic was free from ice, while it must have been filled with it if the ice carried the Scandinavian boulders found in the several clays.[2] Penck asserts that in Holstein there are also two distinctly marked boulder clays, separated by a bed containing marine shells, and in one case a fresh-water one also.[3] Forchammer urged that there are at least four boulder clays in Denmark, separated by intercalated beds of sand and loam. While in Puggard's famous memoir on Moen, he describes three sharply defined boulder clays, distinguished by their stony contents. Penck has at some length professed to describe the march of the ice on its three successive invasions of that island.[4] At Faxo the same evidence is forthcoming if we are to accept his reading of the section. Turning to Scania in the south of Sweden, Penck quotes several sections, and concludes that here also, as in Denmark and North Germany, there is evidence that the great ice-sheet invaded the country three times, having withdrawn from it completely twice.[5]

[1] Zeitschrift der Deutsches Geol. Gesellschaft, xxxi. 152-157.
[2] Id. 161-167. [3] Id. 168-175. [4] Id. 175-179.
[5] Id. 182.

In summing up his results he repeats emphatically that North Germany was overwhelmed with ice not once, but three times, separated by distinct inter-glacial intervals.[6]

Törnebohm, in a letter to Professor Geikie, writing of the two layers of till existing in Sweden, the lower of which is generally darker in colour and contains fewer stones than the upper, says: "The fact that the lower till exhibits such marked evidence of denudation underneath the upper and overlying mass, seems to point out that during the glacial epoch there was a great interval of comparatively mild climate, during which the ice retreated to the mountains."[7] Geikie also quotes the observations of Nathorst, who, writing in 1873,[8] describes an inter-glacial bed containing leaves of Salix polaris, Dryas octopetala, Limnea limosa, Pisidium, Anodonta, and Cytheridea torosa as intercalated between two boulder clays in Sweden and Denmark. Similar intercalated beds containing similar plant and fresh-water molluscan remains are named as occurring at Klägerup in Scania. Holmström quotes this as evidence of an inter-glacial period between the time when the great *mer de glace* which had covered the whole country had retired and a later period when some local glaciers crept down from the great mass of ice that still lingered in the north, covering with morainic matter the fresh-water clays which, during the interval, had accumulated in pools upon the surface of the older till.[9] Mr. Geikie, to whom I owe this passage, goes on to say that more recently Mr. E. Erdmann, of the Geological Survey of Sweden, has described the occurrence of inter-glacial land and clay in the till or boulder clay of Scania. He believes he has evidence to show that there were more inter-glacial periods than one, for several inter-glacial deposits separated by intervening masses of till occur one above another. In one of the inter-glacial beds fresh-water shells were detected. This bed was covered by seventy feet of boulder-clay.[1]

Leaving the borders of the Baltic and going southwards, Penck claims to have found in Saxony two boulder clays,

[6] Id. 195.
[7] Great Ice Age, 405, 406.
[8] Kongl, Vetenskaps-Akademiens Förhandlingar 1873, No. 6.
[9] Id. 1873, No. 1.
[1] Great Ice Age, 407.

separated by flood and river gravel and sand, pointing to a double glacial period there also.[2]

Penck has especially examined the evidence in upper Bavaria and the Tyrol. He devotes a chapter of his excellent work on the "Glaciation of the German Alps," published in 1882, to the red and white breccias which occur at Höttingen near Innsbruck, near the village of Wallgau on the Isar, and on the way from Weissenbach to the pass of Gaîcht, and contain plant remains.

These breccias, according to Penck, are distinctly overlaid and underlaid by moraine matter, and he thence concludes that there is conclusive evidence there of at least two and perhaps three glacial periods, since it would not be possible to explain the facts, namely the occurrence of these plants, at the high level where they are found by a mere oscillation of a glacier.[3]

He then turns to the thin beds of coal, enclosed in conglomerates, found in the Algau on the banks of the Iller near Sonthofen, and at an elevation of 900 metres above the sea level. Penck affirms of these coal-bearing conglomerates that they also are under and overlaid by morainic débris, and therefore prove the existence of two sharply defined glacial periods. To use his own words: "Im Iller und Innthale sind zwei zeitlich scharf getrennte Vergletscherungen nachweisbar."[4]

He similarly concludes that the false bedded sands and water-worn gravels, to which the name of Nagelfluh is given by the German geologists and which intervene among the morainic beds, similarly point to an intercalation of other conditions between glacial epochs.[5]

He further argues that in Bavaria as in the Aargau and in upper Italy, the occurrence of outer moraines points to at least three glacial periods, not merely local oscillations of glaciers, but affecting the whole Alpine country.[6] In a subsequent page he points out that it is only in the highlands of the Alps of Scotland and of Norway where direct evidence of manifold glaciation is forthcoming, and that when we get into the lowlands this evidence fails us.[7]

[2] Id. 186-193. [3] Penck, op. cit. 228-251. [4] Id. 251-262.
[5] Id. 266-290. [6] Id. 312-313. [7] Id. 324.

Lastly, we may shortly turn to the New World. In 1879 Mr. McGee published a paper in the *Geological Magazine* on the Geology of the Mississippi Valley, in which he separates the surface deposits of that region into six sections, which he contends were formed during three distinct periods, probably separated by considerable intervals. "It is believed also," he says, "that all are due either directly or indirectly to glacial action, and hence that they afford evidence of three separate glaciations." After describing these sections in detail, he concludes that "three different glacial epochs have succeeded each other at long intervals; and from analogy with the present as well as from direct evidence, that a characteristic fauna and flora spread over the land during each of these intervals, whose counterparts are to be sought for in the more recent marine formations. . . . It may also be suggested that earlier glacial eras have probably occurred at long intervals."[8]

Professor Geikie has collected a large number of observations from North America in regard to buried layers of peat and trees, etc., which he claims as evidencing inter-glacial conditions. The most satisfactory instances are from the State of Ohio.

In that State, a number of years ago, Colonel James Whittlesey published a section, obtained in digging an artesian well at Columbia, in which a layer of blue clay enclosing a log of wood was found covered by drift beds with boulders, and underlaid by the stiff glacial clay known as hard pan.[9]

Mr. Orton, in his account of Clermont County in the same State, describes an ancient forest bed formed of carbonaceous clay containing vegetable matter, leaves and wood, with occasional beds of peat as intercalated between the blue boulder clay or hard-pan and yellow clays with boulders.[1]

Mr. Geikie says that the same observer has described as occurring in the valley drifts of Ohio a forest bed seven feet thick between two seams of ochreous gravel which he argues grew *in situ*, during inter-glacial times. He is further of opinion that "this inter-glacial stage must have had an im-

[8] Geol. Mag. N. S. vol. vi. 354 and 418.
[9] Smithsonian Contributions to Knowledge, xv.
[1] Report of the Geol. Survey of Ohio, vol. i. Geology, 440.

mensely long period for the accomplishment of the work which we are obliged to refer to it."[2]

In Minnesota Professor Winchell describes a bed of peat occurring in Fillmore County, as underlying a gravelly clay, "which contains boulders, and which has the appearance of being the same as the glacier deposit known as unmodified drift. The same underlies it. Some portions of the clay above the peat are reported to be blue, while the whole of that which underlies it is of a blue colour. The country is heavily covered with this drift clay, and some very large boulders of granite lie near the wells that have met this peat."[3]

Mr. Geikie says that Professor Winchell had written to him to the effect that in North-Western Ohio, which he had surveyed, he found satisfactory proofs of the existence of two distinct glacier deposits. The buried soil of Ohio, which extends westwards through Indiana and Illinois, enters Iowa and even appears in southern Minnesota, "is distinctly intercalated," he says, "between glacier drift clays. . . . The two deposits cannot be said to differ noticeably wherever I have examined them, the overlying mass as being evidently of *glacier origin* as the drift that underlies the soil-bed."[4] Mr. Geikie, commenting on this, says, "We must agree with Professor Winchell that if the lower hard-pan be the morainic matter of a great ice sheet, the similar deposit which overlies the forest bed in Minnesota, etc., must have had a like origin; in other words, there was a return to glacial conditions after the forest bed period, during which the ice sheet again invaded New England and the North-Western States."[5] He further argues, from the very small traces of erratic deposits in the so-called driftless area, that there were certain regions over which the older ice sheet prevailed, but which the last ice flow did not cover.[6]

The opinion of the best American geologists seems to be very emphatically against a plurality of glacial periods, but most of them claim that there were at least two. Thus, Mr. G. K. Gilbert says, in America, where there is now great activity in the investigation of glacial phenomena, the evidence of a single interglacial period is cumulative and overwhelming, while there is no indication of more than one.

[2] Op. cit. p. 428; Great Ice Age, 453. [3] Great Ice Age, 454.
[4] Id. 462, 463. [5] Id. 463. [6] Id. 465

Dr. Wright, in his most valuable book on the ice age in North America, says, "I presume the majority of authorities of most weight would prefer to speak of two glacial epochs in America, believing that the ice-front withdrew to the Laurentian highlands during the so-called interglacial period, and thence re-advanced to the outer line of moraines which we described in the chapter upon that subject." [7]

This will suffice; I have tried to state the case of those who hold that there has been more than one glacial period as fully and fairly as my space will allow.

This completes my survey of the historical development of the glacial theory as taught by its more ardent advocates. It is time that I should turn to the criticism which I propose to apply to it.

[7] Op. cit. 475.

CHAPTER IX.

APPEALS TO TRANSCENDENTAL PHYSICS AND ASTRONOMY.

"Of late years, however, the opinion has been gaining ground among our hammer-bearers that on this matter of cosmical climate they must, after all, be content to follow the guidance of the astronomer and the physicist, seeing that their own principles refuse, in this particular at least, to yield as much assistance as could be desirable."—J. GEIKIE, "Great Ice Age," 94.

Appeal of Geology to Physics and Astronomy—Decrease in subterranean heat: Lapperent, Laplace, Poisson, Hopkins, Thompson, Phillips, Croll—Different effects of subterranean heat on land and sea: Frankland, Dawson, Croll—Renoir's theory—Variation in the sun's heat: Vicaire, Balfour Stewart, Renoir—Hotter and colder regions of space: Poisson and Hopkins—Variation in the transparency of space: Erman and Babinet—Variation in the transparency of the air—Increased cold alone will not produce a glacial epoch: Lecoq, Tyndall, Frankland, Croll, Woeikof—Shrinkage of the sun's disc: Blandet, D'Archiac, Lapperent, S. Reinach—Shifting of the earth's axis: I. If the earth is rigid: Alessandro degli Alessandri, E. Hill, Jukes, Thompson, Airy, G. Darwin, Croll, J. Evans, Twisden—II. If the earth contains a molten centre: J. Evans, Thompson, G. Darwin, Airy, E. Hill—Geological witness against a shifting axis: Jukes, Haughton, Hutton, J. Geikie, Boucheporn, Falsan—Variation in the eccentricity of the earth's orbit: I. Effect on mean temperature of the earth: Kepler, Croll and Carrick, Herschell, Haughton, Meech and Croll—II. Its effect on the seasons: Kepler, Reclus, Lambert, Meech, Croll, Arago, Geikie—Variation in the obliquity of the ecliptic: Herschell, Laplace, Stockwell, Ball, Drayson, Belt, Meech, Croll, G. Darwin—Effects of precession and nutation: Airy, Burr, Reclus, Meech, Humboldt, Price—General conclusion—Croll, Geikie and Meech.

WE have now traced the growth and culmination of the glacial theory as taught by its most distinguished advocates. We have shown how they claim that the earth not long ago was largely enshrouded in ice even down to the tropics. That similar invasions have occurred over and over again in geological time, and that it is very largely to this agency we must attribute the excavation of valleys and the formation of lakes and fiords.

It is not unnatural that in days when science means, if it means anything, the explanation of natural phenomena by known and not by occult causes, the champions of the Glacial theory as thus taught should have summoned the astronomer and the physicist to their aid. The theory, as developed by Agassiz and his scholars, postulates a portentous change of climate operating over wide areas of the earth's surface; and those who tolerate no appeal involving the tampering with the known laws of Nature must needs find some cause competent and adequate to so great a result. It was natural, also, that the students of astronomy and of physics, some of whom were fascinated by the theory which was so pertinaciously pressed upon them, should have made corresponding efforts to find a basis for the new departure in geology. Nor, indeed, was the appeal entirely a new one. Long before the glacial theory was heard of, geologists had been brought face to face with the fact that in the *temperate* regions of the earth remains of plants and animals are forthcoming whose relatives now live only in the tropics. When bones which belonged unmistakably to elephants and rhinoceroses were found not only in Britain and Germany but under the perpetually frozen soil of Siberia, and when, on the other hand, bones of the reindeer, the glutton, and the polar bear were forthcoming from France and England, it became impossible to doubt that in former times the climate of the earth must either have been different in itself or differently distributed, and the physicist and the astronomer were set to work to find an adequate cause. The problem which now arose only differed from this older problem in involving not only a more revolutionary change but apparently an oscillation to and fro. To those who were champions of Uniformity as well as of the Glacial theory it was natural and necessary, perhaps, that they should demand a cause which was not sporadic, abrupt and aimless, like some single experiment made by nature, but one which acted in a recurrent manner, and thus created a kind of cosmic change of seasons at intervals during the long years that are gone.

I propose to devote the present chapter to the consideration of the various theories which have been forthcoming in response to this demand, and to show how completely they fail in meeting the conditions of the problem. I shall begin with a cause which is terrestrial rather than astronomical, namely, the decrease in the original heat of our planet. This cause has been frequently urged, since men's minds were impressed with the notion that the earth has been gradually cooling from a molten mass, and have in consequence argued that during the earlier geological periods its temperature was very considerably higher than it is now.

M. Lapparent[1] remarks with truth that if we are to measure the effect of subterranean heat by its effect in the case of volcanoes we shall find that it is exceedingly local and limited, since it does not suffice to melt the snow which so often covers the flanks and even summits of volcanoes. This is specially noteworthy in Iceland, which is a cradle of volcanic energy and proves what a bad conductor the rocky fabric of the earth is. If we remember also that where we can meet with fairly complete sections the azoic gneiss and mica schist which form the foundations of the earth's crust and underlie the Cambrian and Silurian strata are found to be thousands of feet thick, we have a measure of the very slight influence the central heat had in constituting the climate of the earliest geological horizons where traces of life occur. Nor is there evidence forthcoming from the geological record either that in palæozoic times the earth was substantially hotter than it is now or that there has been any gradual process of change in the later ages from a warmer to a cooler climate. Nor, again, is it easy to see how, even if this were so, it would in any way account for a glacial climate followed, as the postulated glacial climate has been, by a temperate one, unless it be argued, which it is not, that the subterranean heat of the earth is subject to oscillations.

Again, as Dr. Wright says, if this theory were tenable, the Glacial period should not have culminated in the past, but we should still be looking for its culmination in the future; for both the earth and sun are cooling off.[2]

[1] Traité de Geologie, 1253–1251. [2] Ice Age in America, 407.

Apart from these arguments, there remains a more crucial one. The actual amount of heat which the surface of the earth derives from subterranean sources has been calculated. Laplace arguing from astronomical observation, made in the time of Hipparchus, showed that during the last 2000 years there has been no appreciable contraction of the earth by cooling, for the length of the day has not been sensibly shortened, not even to the amount of $\frac{1}{300}$th of a second, so that the contraction of the earth must have been inappreciably small or none at all, as it could not take place without affecting the length of the day.

Poisson calculated the actual amount of the superficial heat due to the earth's cooling, subject to the conditions that the power of transmitting heat is the same in every part of it, and also that the temperature of the surrounding space remains uniform.

The result is thus stated by Hopkins: "It is very small, amounting only to about $\frac{1}{20}$th of a degree of Fahrenheit. It must have been constantly diminishing for an immense period of time, and has now approximated so near to its ultimate limit, that if the earth's refrigeration should continue under the same external conditions as at present, it would require, as shown by Poisson, the enormous period of a hundred thousand millions of years to reduce this small fraction to half its actual value. Hence the mean superficial temperature of the earth, and consequently the mean temperature of the atmosphere in contact with the earth's surface may be considered as sensibly unalterable for all future time, provided always that the sun, the earth's atmosphere, and the temperature of surrounding space shall remain unchanged. We are here, however, more immediately concerned with past than future changes. We should not be justified in supposing, from what I have stated respecting the slow future variation of the earth's superficial temperature, that it has been equally slow for an equal period of past time; but still it is highly probable that some millions of centuries must have elapsed since the mean superficial temperature could have been greater by a single degree than at present, from the operation of the causes we are now discussing, assuming the permanence of the conditions above-mentioned. However imperfect our geological chronology may be, these enormous periods of time seem almost to preclude the possibility of applying this theory to account for the changes of temperature of which we have evidence in the more recent geological periods."[3]

Sir William Thompson has shown that 10,000 years after the solidification of the earth's crust first began, the amount of heat which would traverse it could hardly affect the surface temperature. He enlarges upon the subject in his usual picturesque way: "Ten, twenty, thirty times the present rate of augmentation of temperature downwards could not raise the surface temperature of the earth and air in contact with it by more than a small fraction of a degree Fahrenheit. The earth might be a globe of white-hot iron covered with a crust of rock 2000 feet thick, or there might be an ice-cold temperature everywhere within thirty feet of the surface, yet the climate could not on that account be sensibly different from what it is or the soil be sensibly more or less genial than it is for the roots of trees or smaller plants. Yet greater underground heat is the hypothesis which has been most complacently dealt with by geologists to account for the warmer climates of ancient times."[4]

Professor Phillips, while accepting the results which Thompson, Poisson, Fourier and others had arrived at, and the direct conclusion which they drew from it, went on to argue that possibly in former times the atmosphere varied in its capacity as a conductor of heat, and that consequently the heat of the earth was retained longer or dissipated sooner as the case might be. "The state of the atmospheric mantle," he says, "which envelops the terraqueous globe, mitigates solar heat and stellar radiation, and, like the clothing of a steam cylinder, prevents excessive waste of the warmth treasured within."

To this argument Mr. Croll replies: "It is quite true, as Professor Phillips suggests, that a diminution in the conductivity or in the diathermancy of the earth's atmosphere, and an increase in its height, would increase the influence of internal heat on the climate. But when we reflect that under the present condition of the atmosphere the internal heat could

[3] Journal Royal Geol. Soc. viii. 58, 59.
[4] Geological Time, Trans. Geol. Soc., Glasgow, 1877, p. 250.

not even sensibly affect the climate after the short period of 10,000 years from the commencement of solidification of the earth's surface, it appears very improbable that our atmosphere could have ever been so far different from what it is at present, that by means of it the internal heat could have produced and maintained that high temperature of climate which is supposed by some to have prevailed during the long Palæozoic ages of the earth's history. And besides, the important fact is overlooked that any change in the condition of the earth's atmosphere which would prevent the dissipation of the earth's internal heat into surrounding space, such as an increase in the quantity of aqueous vapour contained in the atmosphere, would at the same time tend to lower the temperature of the earth's surface by diminishing the quantity of radiant heat reaching the surface from without. Such a state of things would no doubt equalize, to a certain extent, the extremes of summer and winter temperature, but would not very sensibly increase the mean annual temperature of the climate. In fact it would rather have an opposite tendency."[5]

Another modification of the theory that former climatic changes were due to changes in the internal heat of the earth is due to Professor Frankland, who limits his explanation to the Glacial period only. This he published in a paper in the *Philosophical Magazine* for May, 1864, on the "Physical Cause of the Glacial Epoch." "A satisfactory theory," says Professor Frankland, "must take cognizance of the following points in the history of the Glacial period: 1st. That its effects were felt over the entire globe. 2nd. That it occurred, or at least terminated, at a geologically recent period. 3rd. That it was preceded by a period of indefinite duration, in which glacial action was either altogether wanting, or was at least confined to regions of considerable altitude. 4th. That during its continuance atmospheric precipitation was much greater, and at one period the height of the snow-line considerably less than at present. 5th. That it was followed by a period extending to the present time, when glacial action became again insignificant."[6]

[5] Philosophical Mag. 4th ser. xxviii. 122, 123.
[6] Phil. Mag. 4th ser. xxvii. 327.

The main position upon which Frankland based his theory was that, in order to secure a sufficient supply of ice to constitute a glacial epoch, we must, in the first place, have an adequate amount of aqueous vapour in the atmosphere, and this could only arise from the association of a greater amount of heat with the waters of the ocean, and he concludes that the sole cause of the phenomena of the glacial epoch was a higher temperature of the ocean than that which obtains at present.

This hypothesis rests chiefly upon the two following propositions :—

1st. That a higher oceanic temperature would give rise to an increased evaporation, and consequently to an augmented atmospheric precipitation.

2nd. That this increased atmospheric precipitation would augment the average depth of permanent snow upon the ice bearers, and would, within certain limits, depress the snow line.[7]

Having endeavoured to establish these two hypotheses, he next endeavoured to discover a competent cause of a former higher oceanic temperature; and, having put aside all cosmical causes as insufficient, he fell back upon the internal heat of the earth. Poisson showed that an increase of the surface temperature of the earth to the extent of 10° Fahr. would involve a descending rate of increase so rapid as to reach 200° Fahr. at a depth of only sixty feet, a physical condition of the planet incompatible with the conditions of life in the more recent geological epochs. Frankland, admitting this, goes on to say that although it is true of the solid surface of our planet, it is not necessarily so when extended to the floor of the ocean; and this (1) because of the conductivity of water; (2) its convection; (3) its power of penetrating into any chasms formed in the bed of the ocean; (4) its specific heat. These, combined, would continually lead to the abstraction of the heat which the ocean floor acquired by the water above it, which again would part with it by increased evaporation. Such evaporation, as Dalton showed, would increase largely with its heat, and hence would arise the required amount of

[7] Id. 328.

vapour to be condensed into ice and snow, and hence an efficient cause for the Glacial period.

This theory is certainly ingenious.

That the enhanced heat which the solid crust of the earth received should be abstracted from it by the ocean waters, which again parted with it by increased evaporation, the whole effect being to make the sea warmer and the dry earth cooler than at present, is a pretty scheme, but it fails at all points. As Principal Dawson says: "It is contradicted by the fossils of the period, which show that the seas were colder than at present, and if it existed it would not produce the results required, unless a preternatural arrest were at the same time laid on the winds which spread the temperature of the sea over the land. The alleged facts observed in Norway, and stated to support this view, are evidently nothing but the results ordinarily observed in ranges of hills, one side of which fronts cold sea water, and the other land warmed in summer by the sun."[8]

Mr. Croll replied to Frankland in similar terms. "There is no evidence," he says, "that the ocean was warmer during the glacial epoch than at present; but, on the contrary, we have geological evidence to conclude that it must have been much colder than at present. For example, on examining the marine drift of that period, we find that it is decidedly of an arctic character, indicating the low temperature of the seas during that epoch. These beds show, for example, that our British seas during that period contained in abundance numerous species of shells which are now only to be found in more northern latitudes. Croll enumerates eighteen Arctic shells which no longer exist in the British seas, which occur in the glacial drift of Scotland alone.[9]

Besides, if this had been an efficient cause of climatic change extending over the continuous history of the earth, we should have had abundant evidence in the rocks of a gradual and persistent change of temperature from one extreme towards another, which we have not. Lastly, such a theory would in no way account for the intercalation of a glacial period between

[8] Acadian Geology, 66, 67.
[9] Phil. Mag. 4th. ser. xxviii. 124, 125.

the temperate and semi-tropical conditions of Tertiary times and those now subsisting.

The theory ignores the various astronomical and other facts already quoted, and it is based on such postulates as that its effects were felt all over the globe, which is contrary to experience, inasmuch as the glacial phenomena are limited both in time and in space.

These facts are conclusive, and I do not know that Professor Frankland has a single disciple.

M. Renoir, having previously suggested two other theories to account for the glacial climate and discarded them because of their insufficiency, suggests a third, viz., that a very intense glacial epoch coincided with the time when the internal heat of the earth ceased to materially affect its surface, and when at the same time the earth was so far from the sun as to greatly mitigate its heating power. Presently, in order to overcome the resistance of the ether in space, the planets travelled more slowly and approached the sun in a great spiral, which led to the melting of the ice previously accumulated. I do not quote this theory to seriously refute it, but only as an example of the way in which some students of a science we are devoted to, are prepared to throw away every kind of empirical test and of scientific method in the pursuit of wild hypotheses of which M. Falsan says, with this one in view: "n'étaient en définitive que de simples jeux d'esprit."[1]

We may safely put aside a variation in the earth's internal heat as an efficient cause of changes of climate. Let us now turn to a variation in sources of heat outside the earth; and first to the sun.

The researches of astronomers have shown that certain of the stars vary in luminosity, and go through a process of accretion and diminution of light, and in certain instances they have been noticed to break out into immense conflagration and then to subside again into their normal condition. It has been suggested that the sun has passed through similar phases and thus greatly affected terrestrial climate. We can hardly conjecture a cause which would

[1] La Periode Glaciaire, 193.

account for the heat of the sun being immensely enlarged, except either a sudden addition of fuel to its furnace, which would be witnessed by the perturbations which its increased mass would cause among the planets, of which we have no evidence, or a collision with some other mass of matter, and of this we have no evidence either. These rapid and sudden revolutions must be accompanied by the most widespread catastrophes, and if the sun had ever been the victim of one we should have had ample traces of it. The appeals generally made, however, have been to milder changes. Such appeals have been countenanced in England by Professor Balfour Stewart, and in France by M. Renoir and M. Vicaire. M. Vicaire[2] argues in favour of a diminution in the sun's heat, which he attributes to a gradual exhaustion of the combustible elements in the sun's photosphere; but if this were so we ought to have evidence of a gradual refrigeration of the earth, which is not the case, nor can glacial epochs intervening between temperate ones be thus explained. The other inquirers referred to appeal to a change in the amount of the sun's heat occasioned by a prevalence or absence of sun spots. In regard to this cause, which can be directly experimented upon, the evidence is most unsatisfactory. The recurrence of the spots has been discovered to observe a regular periodicity, but their influence upon temperature seems very small, if any. Wolff[3] showed that the succession of sun spots occurs in regular periods of 11·11 years, or nine cycles in a century, and he argues that the years in which spots occur are generally drier and more productive than others. Buys-Ballot was also of opinion that the prevalence of sun spots has a slight effect on climate. On the other hand, Counsellor Schwab, after twenty-six years of observation, does not think the spots exert any influence on the annual temperature. "A writer in the Encyclopædia Britannica, article Astronomy, says that in 1827 the summer was wet and cold, the thermometer at Paris rose only to 23·7 Reaumur, and the sun exhibited no spots; whereas in the summer of 1807 the heat was excessive, and the spots of vast magnitude. Warm

[2] Bull. Soc. Geol. de France, ii. 211.
[3] Comptes Rendus, xxxv. 704.

summers and winters of excessive rigour have happened in the presence or absence of spots." [4]

We have no warrant, therefore, for supposing that an increase or decrease of spots would greatly modify the earth's climate, nor have we the slightest evidence that such a variation on a great scale occurs in them at long intervals. The theory is really without any basis of empirical fact.

Let us now move on.

It has long been known that the sun with its system of planets has a proper motion of its own from one part of space to another. Struvé calculated that if it were seen from the mean distance of stars of the first magnitude the sun would appear to describe an arc of about ⅓rd of a second in a year. If this motion, says Mr. Hopkins, were rectilinear and uniform it would require a period of nearly 700,000 years for the sun to move over a space equal to the distance of the stars of the first magnitude from the earth; and he urges that, whatever mistakes may underlie the hypothesis of Struvé, the sun's true motion would not be materially greater than his estimate. Poisson argued that the result of this motion might be to take our system into regions of space where a large number of heat rays are forming. This view seems to be based on a misapprehension. We know of no source of heat external to the earth except solar and stellar heat. Inasmuch as the fixed stars are situated at enormous distances from each other or they would cause movements which we could verify, this result would mean that the solar system had approached a good deal nearer to some particular star. 'This approximation of position to a single star," says Mr. Hopkins, "could not take place consistently with the preservation not only of the remoter bodies of the solar system, but even of the motion of the earth about the sun, according to its present laws."

Suppose our sun should approach a star within the present distance of Neptune. That planet would no longer remain a member of the solar system, and the motions of the other planets would be disturbed in a degree which no one has ever contemplated as probable since the existence of the solar

[4] Meech, Smithsonian Contributions, ix. 35, 36.

system. But, such a star, supposing it to be no larger than the sun, and to emit the same intensity of heat, would not send to the earth much more than one thousandth part of the heat which she derives from the sun, and would therefore produce only a very small change of terrestrial temperature."[5] The argument just quoted applies to the case of the earth approaching a source of heat and acquiring a correspondingly high temperature, but the same reasoning applies to the reverse problem. In order that it should be cooled and acquire glacial conditions by a similar movement, it must approach some reservoir of cold corresponding to a star as a source of heat, of which we know of none in nature, since it would hardly affect the temperature of the earth at all if it were to pass out of the sphere of radiation of all the fixed stars, and yet retain its present proximity to the sun, which supplies it virtually with all its external heat.

We may, therefore, put aside this cause as inadequate and impossible. It has been urged by Erman and Babinet that the amount of heat received by the earth may be affected by certain parts of space being denser or more occupied by meteoric dust. It is well known that the earth at certain parts of its orbit encounters showers of meteoric dust and so-called shooting stars, and Erman attributes to the effect of these obscuring the sun, the cooler temperature which is said to have been observed between the 7th and 10th of February and the 8th and 13th of May. This view has been extended so as to include cosmic periods, and it has been suggested that a similar cause may have produced changes of climate, but meteoric dust and showers of shooting stars are parts of the solar system, miniature planets in fact, moving in definite and defined orbits, and it is only when the earth traverses their orbit that the phenomenon occurs. There is no evidence that space is sporadically occupied by showers or by areas of this kind unconnected with the solar system.

Others, again, have urged that space itself, even if it does not contain areas filled with meteoric dust, may differ in transparency or density. If this were the case, and the earth had traversed any considerable space of it, it must have retarded its

[5] Journ. Geol. Soc., vol. viii. 61, 62.

movements, or it would certainly, continually or occasionally, affect and obscure the light of the fixed stars, of neither of which facts is there any evidence. So far as observed facts enable us to form an opinion, it is virtually certain that space is uniformly transparent, and does not contain areas marked by special opacity.

A similar view has prevailed in regard to the earth's atmosphere, namely, that it has varied in its capacity for transmitting solar rays. Of this, again, we have no evidence whatever, nor can we suggest a physical cause to account for such a variation which should explain the vicissitudes of geological climate or the necessities of animal and vegetable life, since a cause like this can hardly have been a local one, and must have prevailed over the whole atmospheric envelope. I am not referring to possibilities of greater or less vapour in the air, etc., which we shall consider in the next chapter, but the presence in, or the absence from, the air of elements other than vapour which would render it more or less opaque as the case might be.

In regard to the theories previously discussed, there is a further answer which seems conclusive. It was first suggested, I believe, by M. Lecoq, and I will quote his neat statement of it in his own words. "La principale cause d'alimentation du glacier réside dans l'abondance du névé et dans l'étendue du cirque qui le reçoit. Les ressources du névé n'existent que dans la neige qui tombe ; la neige ne peut se former qu'aux dépens de la vapeur élevée dans l'atmosphère, et celle-ci ne peut être produite que par l'action de la chaleur solaire."[6]

Similar views were afterwards published and clearly stated by Professor Tyndall. "Some eminent men," he says, "have thought, and some still think, that the reduction of temperature, during the glacier epoch, was due to a temporary diminution of solar radiation; others have thought that, in its motion through space, our system may have traversed regions of low temperature, and that during its passage through these regions, the ancient glaciers were produced. If I understand the writings of the eminent men who have propounded and

Des Glaciers et des Climats, 1847, p. 553.

advocated the above hypotheses, all of them seem to have overlooked the fact that the enormous extension of glaciers in bygone ages demonstrates, just as rigidly the operation of heat as well as cold. Cold alone will not produce glaciers. You may have the bitterest north-east winds in London throughout the winter, without a single flake of snow. Cold must have the fitting object to operate upon, and this object the aqueous vapour of the air, is the direct product of heat. The latent heat of aqueous vapour at the temperature of its production in the tropics is about 1000° Fahr.; for the latent heat augments as the temperature of evaporation descends. A pound of water, then, vaporized at the equator, has absorbed 1000 times the quantity of heat which would raise a pound of the liquid one degree in temperature. But the quantity of heat that would raise a pound of water one degree, would raise a pound of cast iron ten degrees: hence, simply to convert a pound of the water of the equatorial ocean into vapour, would require a quantity of heat sufficient to impart to a pound of cast iron 10,000 degrees of temperature. But the fusing point of cast iron is 2000° Fahr., therefore for every pound of vapour produced, a quantity of heat has been expended by the sun, sufficient to raise five pounds of cast iron to its melting point. Imagine, then, every one of those ancient glaciers with its mass of ice quintupled; and imagine the place of the mass so augmented, to be taken by an equal weight of cast iron raised to the white heat of fusion, we shall then have the exact expression of the solar action, involved in the production of the ancient glaciers. Substitute the hot iron for the cold ice—our speculations would instantly be directed to account for the *high* temperature of the glacial epoch, and a complete reversal of some of the hypotheses above quoted would probably ensue.

It is perfectly manifest, that by weakening the sun's action, either through a defect of emission, or by the steeping of the entire solar system in space of a low temperature, we should be cutting off the glaciers at their sources. Vast masses of mountain ice indicate infallibly the existence of commensurate masses of atmospheric vapour, and a proportionately vast action on the part of the sun. In a distilling apparatus, if you required to augment the quantity distilled, you would not surely attempt to obtain the low temperature necessary to condensation by taking the fire from under your boiler; but this, if I understand them aright, is what has been done by those philosophers who have sought to produce the ancient glaciers by diminishing the sun's heat. It is quite manifest that the thing most needed to produce the glaciers is an *improved condenser;* we cannot afford to lose an iota of solar action; we need, if anything, more vapour, but we need a condenser so powerful that this vapour, instead of falling in liquid showers to the earth, shall be so far reduced in temperature as to descend in snow."[7]

Professor Frankland puts Tyndall's view in quite as picturesque and convincing a manner. "The formation of glaciers," he says, "is a true process of distillation, requiring heat as well as cold for its due performance. The produce of a still would be diminished, not increased, by an absolute reduction of temperature. A greater differentiation of temperature is required to stimulate the operation into greater activity. . . . Professor Tyndall proves conclusively that all such theories assume a condition of things which would cut off the glaciers at their source, by diminishing the evaporation upon which their existence essentially depends."[8]

Croll urges that what Professor Tyndall says shows plainly that the glacial epoch was not brought about by our earth passing through a cold part of space. A general reduction of temperature over the whole globe certainly would not produce a glacial epoch. Suppose the sun were extinguished and our globe exposed to the temperature of stellar space (239° F.); this would certainly freeze the ocean solid from its surface to its bottom, but it would not cover the land with ice.[9]

Woeikof says: "A cursory glance at the present condition of our globe shows us that cold alone will not produce permanent snow and glaciers when vapour of water is deficient. There are no permanent snow or glaciers in the Verkhojansk Mountains in North-east Siberia, yet at the foot of them the mean annual temperature is below 4° F. and that of

[7] Heat as a mode of Motion, 188, 189.
[8] Phil. Mag. 4th ser. xxvii. 325.
[9] Climate and Time, 79.

January below —56° F. The reason is that the snow-fall is but small, and thus the snow is easily melted in summer. In New Zealand, on the contrary, owing to the enormous snow-fall in the mountains, glaciers descend to about 700 feet above sea level on the west side (lat. 43° S.). At this height the mean annual temperature must be about 50° F. Snowfall and frost "are of rare occurrence, even in winter." [1]

These arguments seem to me to be conclusive.

In the theories already discussed the supposed influence upon the earth's climate was distributed over its whole surface; we will now turn to a cause which would, if effective, have a differential effect in different areas. The view here criticized was published by M. Blandet,[2] and adopted by D'Archiac, Lapperent and others. According to it the sun is gradually but definitely shrinking in size, and it is by the contraction and concentration of its sphere that the greater part of its heat is generated. Consequently, the apparent diameter of the sun was formerly much greater. In consequence of the great distance separating us from the sun and of its small apparent disc, the rays which reach the earth from it form a virtually parallel bundle which illuminate a hemisphere at a time and no more. If its apparent diameter were increased to 47°, other things remaining alike, the solar rays, instead of forming a cylinder, would form a cone, whose base would be the solar disc. The result would be that if one limit of illumination was one pole, the other limit would be the parallel of 43° on the opposite hemisphere. Thus a considerably larger space than one hemisphere would be continually illuminated, and nowhere on the earth would the nights be of forty-eight hours.

In such a case the sun's density would be very much less and its texture would be largely nebulous and it would give for each unit of surface less heat and light than it does now; but being much larger in size, and the earth being much nearer its periphery, the actual amount of heat received might be the same, while the climate produced would be one of much greater uniformity.[3]

[1] Nature, xxv. 423.
[2] Bull. Soc. Geol. de France, 2nd ser. xxv. 777.
[3] Lapperent, p. 34, 35, and 1257.

Such is the theory in question.

It is quite true if Laplace's nebular theory be sustained, that at some time in the very remote geological history of the earth such a condition of the sun as is here pictured must have ensued. It is also true that the evidence of the earlier geological periods points to a greater uniformity of climate then than now, but it is a long way from these statements to the conclusion that in recent geological times the sun has shrunk in this portentous way; besides such a cause might account for a continuous and regular change of climate in some particular direction; but, as M. S. Reinach says, how are we to explain by it how the so-called glacial epoch came to an end? Any theory which accounts for the beginning of the glacial epoch must also explain its close. Nor do I know of any astronomer who would countenance the notion that the sun has shrunk so rapidly in geological times from normal causes as to justify such a postulate, which is indeed merely an example of the perverse ingenuity of science rather than of its sober logic.

Let us pass on.

It has been a favourite speculation ever since men realized that the remains of tropical animals have been found in temperate and arctic latitudes, while similar remains of animals of an arctic type have occurred far to the south of their present habitat, that the Poles of the earth, and consequently its Equator, have shifted their position, thus shifting the position of the areas marked by extremes of heat and cold on the earth.

The notion is at least as old as the fifteenth century, and may be found in a work by Alessandro degli Alessandri, entitled *Dies Geniales*,[4] and became a favourite theory with the naturalists who had to explain how an elephant could ever have lived in Europe.

The shifting of the poles and the equator, according to these philosophers, was due to a shifting of the earth's axis within the earth itself so that it pierced it through different points and in a different direction. The suggestion that a body like the earth, moving at the rate it does, could, without a very

[4] Lib. v. ch. ix.

serious dislocation, alter the direction of its axis under any circumstances, seems fantastic to the mathematician. As Mr. Hill says: "Mathematicians may seem to geologists almost churlish in their unwillingness to admit a change in the earth's axis. Geologists scarcely know how much is involved in what they ask. They do not seem to realize the vastness of the earth's size, or the enormous quantity of her motion. When a mass of matter is in rotation about an axis, it cannot be made to rotate about a new one except by external force. Internal changes cannot alter the axis, only the distribution of the matter and motion about it. If the mass began to revolve about a new axis, every particle would begin to move in a new direction. What is to cause this? When a cannon-ball strikes an iron plate obliquely, the shock may deflect it into a new direction. The earth's equator is moving faster than a cannon-ball. Where is the force that could deflect every portion of it, and every particle of the earth into new directions of motion?"[5]

This argument applies to every possible case, and notably to the case where the earth is postulated to be a regular sphere, every axis of which is an axis of figure; but the case becomes much more difficult, or rather more impossible to realize, when we consider that the earth is not a perfect sphere. It is, in fact, a spheroid, with an equatorial diameter larger by many miles than its polar diameter, and it thus has a huge solid ring round it. If the earth has cooled down from a liquid mass, and has rotated at its present speed ever since, the oblateness of its shape must have been originally as great as it is now, and it must have had *ab initio*, a protuberant shell bulging out beyond the form of a true sphere, till it reached to the extent of 13¼ miles, or nearly 700,000 feet, about its equator. "It is very difficult," says Professor Jukes, "to see what force, internal or external, could have given to a globe thus weighted and balanced all round, such a permanent tilt as would cause it to spin on any other than its shortest diameter, or could so alter its form as to make any other diameter shorter than its original axis of rotation."[6]

We know of no cause competent to alter the position of

[5] Geol. Mag. 1878, 265.

[6] Athenæum, Sept. 8, 1860, 323.

the axis, except the removal of a large quantity of matter from one part of the earth to the other, so as to alter its centre of gravity, and in this way to cause a gap between its new axis of figure and its old axis of revolution. If this were done on a great scale, and rapidly, it would no doubt cause a change in the latter axis, and, as Professor Haughton says, the earth would begin to wabble, and it would continue to wabble as a top does when going to sleep, until the two axes had again coincided. Meanwhile, however, the crust of the earth would be tremendously shattered and dislocated, and, as Mr. Twisden has shown, two vast tide waves would sweep the earth, submerging the equator every 150 days to a depth of six miles or more. Of this we have no evidence at all. Suppose, on the other hand, that instead of this change having been rapid, in accordance with the recognized theory of uniformity, the change were very slow and the "result of expansion and contraction, upheaval and depression, denudation and deposition." These forces, no doubt, are slowly altering the distribution of solid matter in the earth, and it then becomes a question of calculation what amount of such changes would have an appreciable effect in changing the position of the axis.

Sir Wm. Thompson says: "As to changes of the earth's axis, I need not repeat the statement of dynamical principles which I gave with experimental illustrations three years ago; but may remind you of the chief result, which is that, for steady rotation, the axis round which the earth revolves must be a 'principal axis of inertia,'—that is to say, such an axis that the centrifugal forces called into play by the rotation balance one another. The vast transpositions of matter at the earth's surface, or else distortions of the whole solid mass, which must have taken place to alter the axis sufficiently to produce sensible changes of climate in any region must be considered and shown to be possible or probable before any hypothesis accounting for change of climate by alterations of the axis can be admitted."

Professor Airy analyzed the question very completely. "To begin with the case most favourable for the production of a large effect," he says, "I will suppose the earth to be a perfectly rigid spheroid, which has in distant ages and under proper circumstances assumed the form of equilibrium, and whose axis of rotation coincides with its axis of figure and is a principal axis. And I will suppose a mountain mass to be elevated by something like a gaseous explosion (the hollow being filled up by the influx of neighbouring matter) in latitudes not very different from 45.° It is evident that the principal effect of the elevation would be that the axis of rotation would no longer be a principal axis, that the principal axis of largest moment would now be a little way beyond the present axis of rotation, and that the two other principal axes whose moments formerly were equal would now have moments very slightly different. The effects which follow are scarcely affected by this last modification.[7] It is well known that, under these circumstances, the axis of rotation would wander in the solid earth. But it would not wander indefinitely; its pole would describe on the earth's surface an ellipse not sensibly different from a circle whose centre would be the pole of the new principal axis, and after a certain time it would return to its former position. The greatest change, therefore, in the terrestrial position of the earth's pole would be double the distance of the new principal axis from the former principal axis."[8]

"A very slight investigation," Professor Airy continues, "suffices to show that the angular change of position of the principal axis will depend upon the proportion which the increased moment of inertia resulting from all the changes made by the elevation bears to the excess of the moment of inertia round the polar axis above the moment of inertia round an equatorial axis. This latter excess is about half the moment of inertia of the equatorial protuberance; that is half the moment of inertia of a mass of matter 25,000 miles long, 6000 miles broad, and 13 miles deep. And what mountain mass can compare sensibly with this? Even if a mountain mass contained one-thousandth part of this matter (which I apprehend is very far above the fact) the shift of the earth's pole would be only two or three miles; and this, though it would greatly surprise astronomers and might sensibly affect the depth of water in harbours, would produce no such changes of climate as those which it is desired to explain."

7 Athenæum, Sept. 22, 1860, p. 384. 8 Id.

Professor George Darwin has entered into more detailed calculations. He takes for instance the case of an ocean bed 15,000 feet deep elevated into a continent 1100 feet high, which is a little above the average, and then calculates the amount of deformation that would result. Taking a continent the size of Africa elevated in this way and a corresponding area depressed in the most effective shapes and positions. This would alter the position of the pole less than two degrees. "Moreover," says Mr. Hill, "this is obtained on the supposition that the elevation and depression are obtained by the actual removal of matter from the surface of the one area, and its deposition on the other. If we adopt the easier supposition that elevation is produced by the expansion of subjacent strata; depression by their contraction, the resulting change is diminished to a considerable, possibly to a very great extent. If, for instance, the expansion or contraction take place between the depths of ten and fifty miles, the shift of the pole will be diminished to less than a sixtieth part of its former amount. If the supposed areas occupy areas other than the most effective, then changes must be still further reduced, and may even vanish altogether." [9]

Dr. Croll, writing on the change here postulated, says: "There probably never was an upheaval of such magnitude in the history of our earth. And to produce a deflection of 3° 17′ (a deflection which would hardly sensibly affect climate) no less than one-tenth of the entire surface of the earth would require to be elevated to the height of 10,000 feet. A continent ten times the size of Europe elevated two miles would do little more than bring London to the latitude of Edinburgh, or Edinburgh to the latitude of London. He must be a sanguine geologist indeed who can expect to account for the glaciation of this country, or for the former absence of ice around the poles, by this means. We know perfectly well that since the Glacial epoch there have been no changes in the physical geography of the earth sufficient to deflect the pole half a dozen miles, far less half a dozen degrees. It does not help the matter much to assume a distortion of the whole solid mass of the globe. This, it is true, would give a few degrees additional deflection of the

[9] Geol. Mag. 1878, 264, 265.

pole; but that such a distortion actually took place is more opposed to geology and physics than even the elevation of a continent ten times the size of Europe to a height of two miles."[10]

The fact is, when some geologists approach the realm of physics, they seem as if they were entitled to make unlimited drafts upon time and force. When Dr. J. Evans suggested that a certain elevation of the earth's crust might shift the pole 15° or 20°, Professor Twisden subjected his premises and conclusions to mathematical induction and showed that such a movement instead of moving the pole 15° or 20° would only move it about 10′, and that to obtain a movement of 1° the zone of land in question must be raised five miles, while to move the pole 20° it would require the removal of a sixth part of the equatorial bulge; and that even if a transfer of this enormous amount of matter took place it would not of necessity produce any effect, and might only produce a small effect on the axis of figure.[1]

While these arguments have been generally held to be conclusive against a change in the position of the earth's axis in a solid globe, it has been in recent years suggested, and has been especially urged by Dr. John Evans, that the case might be different if the earth were a shell filled with molten matter. To this argument, it seems to me, there are two objections. First the postulate itself seems to be inadmissible.

Sir Wm. Thompson says: "That the earth cannot, as many geologists suppose, be a liquid mass enclosed in only a thin shell of solidified matter, is demonstrated by the phenomena of precession and nutation. Mr. Hopkins . . . applied mathematical analysis to investigate the rotation of rigid ellipsoidal shells enclosing liquids, and arrived at the conclusion that the solid crust of the earth must be not less than 800 or 1000 miles thick. . . . It has always seemed to me that Mr. Hopkins might have pressed his argument further, and have concluded that no continuous liquid vesicle at all approaching to the dimensions of a spheroid 6000 miles in diameter can possibly exist in the earth's interior without rendering the phenomena of precession and nutation very

[10] Discussions on Climate and Cosmology, p. 5.
[1] Journ. Geol. Soc. xxxiv. 41.

sensibly different from what they are . . . unless the average substance of the earth is more rigid than steel, its figure must yield to the distorting forces of the moon and sun not incomparably less than it would if it were fluid . . . if the rigidity of the earth on the whole were only as much as that of steel and iron, the earth as a whole would yield as much to the tide-producing influences of the sun and moon as it would if it had no rigidity at all; and it would yield by more than three-fourths of the fluid yielding if its rigidity were no more than that of glass. Such a deformation as this would be quite undiscoverable by any direct or astronomical observations; but if it existed, it would largely influence the actual phenomena of the tides and of precession and nutation." In regard to the latter effect, Sir William says that the only ground on which it could be otherwise would be on the assumption that there is an enormous liquid vesicle, or a solid nucleus separated by a liquid layer from the outer crust in the interior, and that the loss of precessional effective moment of inertia, owing to this portion not being carried round in the precessional movement is almost exactly compensated by a diminution of the generating couple in very nearly the same proportion by elastic yielding;" and he urges as very improbable that the defect of moment of inertia due to fluidity in the earth's interior bears approximately the same ratio to the whole moment of inertia as the actual elastic yielding bears to the yielding which would take place if the earth were perfectly fluid.

Again he says: "At the surface and for many miles below the surface, the rigidity is certainly very much less than that of iron; and therefore at great depths the rigidity must be enormously greater than at the surface. That both the rigidity and the resistance to compression should be much greater several hundred miles down than at the surface, seems a natural consequence of the enormous pressure experienced at those great depths by the matter of the earth."[2]

Elsewhere the same distinguished physicist says: "Whatever be its age, we may be quite sure the earth is solid in its interior; not, I admit, throughout its whole volume, for there certainly are spaces in volcanic regions occupied by

[2] Phil. Trans. 153, pp. 573-581.

liquid lava; but whatever portion of the whole mass is liquid, whether the waters of the ocean or melted matter in the interior, these portions are small in comparison with the whole; and we must utterly reject any geological hypothesis which, whether for explaining underground heat, or ancient upheavals and subsidences of the solid crust, or earthquakes, or existing volcanoes, assumes the solid earth to be a shell of 30 or 100, or 500 or 1000 kilometres thickness, resting on an interior liquid mass. This conclusion was first arrived at by Hopkins, who may therefore be called the discoverer of the earth's solidity. He was led to it by a consideration of the phenomena of precession and nutation, and gave it as shown to be highly probable, if not absolutely demonstrated, by his confessedly imperfect and tentative investigation. But a rigorous application of the perfect hydrodynamical equation leads still more decidedly to the same conclusion."

Sir Wm. Thompson has shown that if the earth were a liquid mass covered with a solid crust, "the solid crust would yield so freely to the deforming influence of sun and moon, that it would simply carry the waters of the ocean up and down with it, and there would be no sensible tidal rise and fall of water relatively to it. The state of the case," he continues, "is shortly this: The hypothesis of a perfectly rigid crust containing liquid, violates physics by assuming preternaturally rigid matter, and violates dynamical astronomy in the solar semiannual and lunar fortnightly nutations; but tidal theory has nothing to say against it. On the other hand, the tides decide against any crust flexible enough to perform the nutations correctly with a liquid interior, or as flexible as the crust must be unless of preternaturally rigid matter." [3]

The same position has been maintained with great learning by Mr. George Darwin, who most emphatically supports the views of Thompson, and proves mathematically that far from the earth having a fluid nucleus, its effective rigidity is at least as great as that of steel.[4]

Driven from this position, the advocates of an interior fluidity to the earth have argued that they do not mean that the whole of the interior of the earth is fluid, but that the earth in fact consists of a solid shell enclosing a solid nucleus,

[3] Report Brit. Assoc. 1876, pp. 6 7. [4] Nature, xxvii. 23.

between which and itself there is a fluid zone. In regard to this notion I cannot realize its possibility. Under what physical conditions it is possible for the centre of the earth to be a huge solid sphere floating in a liquid zone which separates it from the outer crust I do not know. Nor would that meet Mr. G. Darwin's objection in regard to rigidity. But granting that in some transcendental manner it is possible to realize that the earth instead of being a rigid solid is really a shell enclosing a fluid mass, it does not seem to me that the problem is any more advanced.

Professor Airy says: "Let us suppose that the earth is not absolutely rigid, but that there is susceptibility to change of form, either from that degree of yielding or fracture to which most solid substances are liable, or from the hydrostatic pressure of internal fluid. This, as I conceive, puts an end to all supposition of change of axis. The first day's whirl would again make the axis of rotation to be a principal axis and the position of the axis is then permanent."[5]

I cannot myself see how the mechanical difficulty is to be overcome. A shell filled with molten matter would behave in regard to the shifting of its axis, so far as I understand mechanical principles, as a solid sphere would, every portion of it would be revolving with the same speed and momentum; and it seems to me that it would be just as difficult to cause a shell to shift round on a nucleus when both are revolving rapidly at the same speed on the same axis and in the same direction, as it would to change the direction of an axis through a solid sphere. Nor can I realize a liquid molten nucleus sharply defined from a solid shell. There would be an intermediate viscous zone, a gradation from one to the other, and this would make the slipping of the shell over the nucleus as difficult as if the earth were a continuously rigid body.

Again it seems to have been forgotten that the earth is not a sphere.

Whether the earth be solid throughout or be a hollow shell, the protuberant belt of thirteen miles thick about its equator which has been there, so far as we know, since its

[5] Athenæum, Sept. 22, 1860, p. 384.

history began, stands completely in the way of our placing the earth's axis of rotation anywhere else than where it is without a portentous dislocation of which we have no evidence. It is no use writing about the way in which a shell can slip over its nucleus unless we can at the same time explain how the belt of solid matter about the old equator can be shifted so as to accommodate itself to a new equator. The argument seems to me to completely ignore the real conditions of the problem.

Mr. E. Hill well says: "The mathematician asks the geologists whether they have realized the earth's size. Its deviation from a sphere is trifling in amount. In what amount? In figure doubtless. Were its section drawn on this page, a microscope would be required to distinguish the curve from a circle. But in quantity the deviation is not trifling. The height of the highest mountain and the depth of the deepest sea together do not equal its extent. All the mountain chains added together would not perceptibly increase the volume of the equatorial protuberance. The mass of all the continents reinforced by that of all the seas would not be the fifth part of it. To change the earth's shape, this vast protuberance must be shifted or masked. Where is the power that can shift it, the elevation that can mask it? What are our puny upheavals and subsidences of an ocean here and a continent there, compared with a girdle of matter thirteen miles in thickness? Remove 10,000 feet of rock from the surface of one half the earth and spread it over the other half. You could not thereby bring the pole halfway to the present arctic circle. Sufficient changes in the earth's surface will undoubtedly shift the pole. But will geologists grant the changes that would be sufficient?" [6]

Apart from all these arguments drawn from astronomical and physical considerations and which make the postulate of a shifting of the earth's axis impossible, we have the best geological reasons against it. Thus Professor Jukes writes: "We have no reason to look to any other spot on the globe than the present north pole as the centre of the cold climate during this Glacial or Pleistocene period. Neither has any

[6] Geol. Mag. 1878, 266.

one yet ventured to point to any other region of the globe as having been possibly its arctic region during any previous geological period, basing his argument on the fossils of that region having a more arctic character than the contemporaneous fossils of surrounding countries." [7]

Again, Dr. Haughton, in a remarkable paper read before the Royal Society, brought together some very interesting evidence, based entirely on geological grounds, to show that the north pole cannot have shifted its place since the Secondary period. He says, *inter alia*, "McClintock found tropical shells of the jurassic period in the Parry Islands, in lat. 76° 20′. Among them were remains of a reptile allied to the Indian gavial. The triassic beds of Spitzbergen, lat. 79°, have afforded species of Nautilus, Ammonites, Ceratites and Halobia, closely allied to, if not identical with those of the St. Cassian beds of South Austria. In the neighbourhood of Cook's Inlet in Alaska, lat. 60° N., shells characteristic of the triassic and jurassic periods have been found; and similar fossils are found along the Pacific Coast of North America.

"It is not possible to explain the occurrence of tropical animals in the above-mentioned localities by any change in the position of the earth's axis, even if so great an amount of change as would be required were possible. This statement can be proved as follows. Let a great circle be drawn joining Spitzbergen with Cook's Inlet, Alaska. This circle will pass nearly through the north pole. In order to explain the tropical climate of these two localities, and also of the Parry Islands, the pole must be displaced at right angles to the great circle joining Spitzbergen and Alaska along the meridian long. 117° E., nearly that of Pekin. The present difference of latitude between New Orleans and Spitzbergen is 45°, so that in order to make the arctic regions tropical we must move the north pole 45° on the meridian of Pekin, bringing it within 300 miles to the north of that city. Hence it follows that during the triassic period Pekin lay under the north pole, covered by the polar ice-cap. Let us now consider what the south pole was doing. It had

[7] Athenæum, Sept. 22, 1860, 333.

moved on the opposite meridian, and had reached the mouth of the Rio Negro, on the east coast of Patagonia, about 1000 miles to the south-south-east of Valparaiso and the Chilian Andes. Jurassic strata have been found in the Chilian Andes at 34° S., containing the tropical, ammonites biplex, which is found also in Alaska, 60° N., and in Europe. This locality lies within 700 miles of the necessary position of the south pole and cannot have enjoyed a tropical climate. The proposed alteration of the north pole is consistent with the occurrence of tropical animals in the Parry Islands, in Spitzbergen, and in Alaska; while the proposed alteration of the south pole would, permit tropical animals to exist in New Zealand and New Caledonia; but the occurrence of jurassic ammonites within 700 miles of the south pole is fatal to the proposed shifting of the axis of rotation, even if that were allowable, to the extent required."

Turning to the Tertiary age, he shows that in Greenland Disco, lat. 70°, Grinnell Land, lat. 81° 44′, Spitzbergen, west coast, lat. 77°, and Alaska and the Mackenzie river, lat. 70° to 6°, have furnished remains of miocene plants evidencing a climate such as is now enjoyed by the northern parts of Italy, and indicate a mean annual temperature of 48°, where the mean annual temperature is now as low as zero. "It can be shown," says Dr. Haughton, " by a similar method to that employed for the triassic and jurassic periods, that the north pole was practically in the same place during the miocene period that it now occupies. If we join the Mackenzie river and Spitzbergen by the arc of a great circle, the north pole must be moved at right angles to this arc, away from Greenland through 30° in order to give all these northern localities a Lombardic climate. The direction in which the pole must be moved is on the meridian of Nagasaki, and it reaches a point close to Yakutsk, within 800 miles of the peninsula of Kamtschatka and the Island of Saghalien off the Amur. Here we meet with a difficulty similar to that offered by the south pole in the triassic period. The Island of Saghalien and the peninsula of Kamtschatka contain miocene coal beds, requiring at least a sub-tropical climate, which would be impossible under the supposed circumstances. Also the islands of Yesso, Nagasaki and Kiusiu, somewhat further off, contain similar coal beds. It is very remarkable that, while there exist so many proofs of a warm climate near the south pole in former geological periods, there is no evidence of cooler climates having ever existed in the tropics." [8]

Elsewhere Professor Haughton says: "My contention in brief is that the tertiary plant remains indicating a climate similar to that of Lombardy are so situated round the north pole that no possible change in the position of that pole (even were such permitted by mechanical considerations) would give them the climatic conditions as to temperature which they require. . . . I claim to have surrounded the north pole with such a network of Lombardic plants, requiring Lombardic heat, but not Lombardic light, as to render the escape of the pole from its present position as difficult as that of 'a rat in a trap surrounded by terriers.'" [9]

To quote evidence from the other hemisphere. Captain Hutton, than whom it would not be easy to quote a better authority, says: "There is no evidence whatever of a glacial period having occurred in New Zealand, although if it had occurred, there is every reason to expect that it would have left sufficiently clear traces behind it. I will also add that as New Zealand is nearly antipodal to Great Britain, any change of climate in one place, caused by a change in the position of the earth's axes of rotation, would also necessitate a similar change in the other place." [1]

Lastly, in regard to the evidence of so-called glacial beds, I shall endeavour to show presently, that while Western Europe and Eastern America were largely loaded with ice, Northern Asia and North-Western America had a sufficiently mild temperature to enable forests to grow and forest animals to thrive there, right up to the arctic circle. This means that the shifting of the north pole, if such shifting is to be available as a cause of the drift phenomena of Europe and Eastern America, must have been towards the region of the North Atlantic rather than towards that of the North Pacific.

The difficulties of the position are further intensified in the

[8] Nature, xviii. 267, 268. [9] Geol. Mag. 1879, 92, 93.
[1] Geol. Mag. 1875, 583.

presence of the demands of the ultra-glacialists. Thus Mr. J. Geikie, in arguing against the theory of a change of axis, says: "The glacial epoch was interrupted by more than one warm interglacial period. It is not therefore only one change in the position of the poles, but several that we should require, and all these stupendous changes so improbable in themselves, must be supposed to have taken place at quite a recent date, and within a comparatively short space of time."[2] It is not every glacialist who has been so modest. M. Boucheporn[3] would explain the European phenomena by shifting the pole to some point in the Baltic north of Prussia and Poland. The American drift he would explain by another shift, and he does not scruple to invoke other changes of the same kind to meet the various exigencies of the problem. "Mais," says M. Falsan, "il se heurta à tant de difficultés qu'il ne put faire accepter son système."[4]

I will now turn to those recurrent astronomical causes to which appeals have so continually been made in modern times to explain the glacial period; and first let us consider a cause which, if subject to variation, may affect climate in two ways, first, in increasing or diminishing the total heat received during a year by the earth; and, secondly, in altering its distribution between the seasons, namely, a variation in the eccentricity of the earth's orbit. The ancients believed that the earth moves round the sun in a circular orbit. If it moved in a circular orbit, and its axis were always at right angles to the plane of its orbit, and if it always moved at the same rate, there would be no seasons, day and night would be equal everywhere, and every place on the earth would have the same climate which it has at the two equinoxes all the year round. It was Kepler who first proved that the orbit is elliptical and that the sun occupies one of the foci, so that at one portion of the year the earth is nearer to the sun than at another. When at its greatest distance, or, to use astronomical language, when it is at its *aphēlion* it is about 93,750,000 miles distant from the sun, when nearest, that is at its *perihēlion*, it is about 90,259,000.

[2] Great Ice Age, 85, note. [3] Recherches geologiques, i. 188.
[4] Falsan, op cit. 195.

The mean distance, as estimated by astronomers since the corrections of Encke, Hansen, Foucault and Hind, is 91,839,000 miles,[5] and the amount of eccentricity of the orbit is now about a seventeen-thousandth of its major axis.

It has long been known, however, that the earth's orbit is not constant, but that it is altered in position and extent by the attraction of the planets, which cause it at some times to be more eccentric than at others. The variation in the eccentricity of the earth's orbit, dependent as it is on a very complicated series of planetary and other attractions, involves a difficult problem and proves to be very irregular. On the first day of this century the eccentricity was ·016792, a century later this will have diminished by ·0000416. It has been thus diminishing for 13,000 years, and will go on diminishing for 24,000 years more, when it will reach its minimum of ·0033. It will then begin to increase, and go on doing so for 70,000 years, and then attain to ·0211. In another 100,000 years it will reach ·046.

Having established that the earth's orbit varies in eccentricity, let us now see how this affects climate. First, in regard to the quantity of heat received by the earth as a whole, it has been proved, from the known laws of heat and the properties of the ellipse, that the total amount of heat received by the earth in a year varies in inverse proportion to the length of the minor axis of its orbit; and inasmuch as the minor axis is longest when the orbit is nearest a circle and shortest when it is most eccentric, it follows that the more eccentric the earth's orbit is the more heat does it receive from the sun in a year.

Herschell calculated the reciprocal proportions of the two elements, and concluded that if the eccentricity were as great as a fourth of the major axis, it would not raise the mean temperature of the earth more than three per cent. The amount of this variation takes in the whole range of planetary eccentricities from that of Pallas and Juno downwards.[6]

Professor Haughton thus states the case: "If H denotes the general amount of sun heat received by the earth when her orbit is circular, the annual amount of sun-heat received when her orbit is most eccentric will be H (1—0·00333); that

[5] Reclus, The Earth, 6. [6] Trans. Geol. Soc. iii.

is to say, the total variation of annual sun-heat received by the earth due to variation of eccentricity of orbit is only $\frac{1}{333}$rd part of the total heat received, or in other words, if the sun at present be sufficient in one year to melt 100 feet of ice at the Equator, it would melt $3\frac{1}{2}$ inches less when the eccentricity of the orbit was at its maximum. This variation of sun-heat received by the earth is quite inadequate to affect geological climates sensibly." [7]

Mr. Meech says that 100,000 years ago, according to Leverrier's tables, the eccentricity was ·0473 as compared with ·0168 now. The fraction of 365·24 days, counting the days at twelve hours each in respect to solar illumination, amounts to between four and five hours of sunshine in a year, and by so small a quantity only has the sun's annual intensity, during 100,000 years past, ever exceeded the yearly value at the present time. Nor can it depart from its present annual value by more than the equivalent of five hours of average sunshine a year, for 100,000 years to come.

Leverrier's superior limit of variation ·0777 may have been reached at some very remote but unknown period. At that epoch the annual intensity is computed, as before, to have exceeded the intensity of the present by *thirteen hours* of sunshine in a year. On the other hand, the inferior limit of eccentricity being near to zero, indicates only four minutes of average sunshine in a year less than the present annual amount. Between these two extreme limits, all annual variations of the solar intensity, whether past or future, must be included.[8] In this view Croll, who has examined the problem with some care, also concurs. We may take it, therefore, that increasing the eccentricity of the earth's orbit will only very slightly affect the earth's mean climate, and that whatever effect is produced is in the direction of increasing and not decreasing its temperature; and, as a matter of fact, inasmuch as the earth's orbit has been becoming more circular, its mean temperature has also for a long time been slowly declining.

Let us now see if any more can be made of the second

[7] Trans. Roy. Ir. Acad. Oct. 1881, pp. 50, 51.
Smith. Contr. ix. 36.

possible result of altering the eccentricity, namely in so far as it affects the seasons.

Kepler not only showed that the earth's orbit is elliptical, but that the earth moves faster as it gets nearer to the sun and slower as it recedes from it. And this according to a beautiful and symmetrical law which may be thus stated:—If a line be supposed to connect the earth in motion with the sun at rest, which is called in astronomy a radius vector, this line will describe or sweep over that portion of the whole area or surface of the ellipse which is included between its consecutive positions; and the motion of the earth will be such that *equal areas* are thus swept over by the revolving radius vector in *equal times*, in whatever part of the circumference of the ellipse the sun may be moving.[9]

This translated into ordinary language means that the earth does not move round the sun at the same rate, but moves faster in that part of its orbit where it is nearer than it does in that which is farther from the sun. In fact, says Reclus, from the 20th of March to the 22nd of September (i.e. between the two equinoxes), during the spring and summer of the northern hemisphere, the earth takes 186 days to travel over the first and largest half of its orbit; while during the winter period, from the 22nd of September to the 20th of March, only 179 days are required to accomplish the second half of its journey.[1]

It follows from this that if we excluded the question of the obliquity of the earth's axis, and limited ourselves strictly to the effects of eccentricity, there would be no distinction between the two hemispheres, but summer and winter would take place all over the world at the same time. And if the eccentricity were to remain as it is now, the world's summer would always be shorter than its winter by about seven days. When the earth's orbit was most nearly circular the difference would almost disappear; when, on the other hand, its eccentricity was greatest, there would be a difference of about thirty-six days between the lengths of the seasons.

It was Lambert who, in his "Pyrometrie," published in 1779,

[9] Herschell, Outlines, 209. [1] Reclus, 11.

first combined Kepler's law last quoted and the known qualities of heat into a law of much value in the discussion of climate, namely that the amount of heat received by the earth from the sun is proportional to the area described by its radius vector, or, as Sir John Herschell states the case, the momentary supply of heat received by the earth varies in exact proportion to the angular velocity; and from this it follows that equal amounts of heat are received from the sun in passing over equal angles of it, wherever in the ellipse those angles may be. Thus, says Lambert, it may come to pass in the case of comets, that in traversing the 180 degrees of longitude while near the sun, they may receive as much light and heat in a few days or weeks as they shall in traversing the remaining part of their orbits during many centuries of years. Of all the planets, he adds, Mercury has the greatest eccentricity, and this planet receives from the sun, during the 32¾ days while it is nearest him, as much light and heat as it receives during the 53⅕ days while it is farthest from him.[2]

Mr. Meech after discussing the mathematical issues involved in this position concludes thus: "During each of the four astronomic seasons of spring, summer, autumn and winter, the intensities received from the sun are precisely equal. For in each season the earth passes over three signs of the zodiac, or a quadrant of longitude. In the year 1855, for example, spring in the northern hemisphere, commencing at the vernal equinox, March 20th, lasted eighty-nine days; summer, beginning at the summer solstice, June 21st, continued ninety-three days; autumn, commencing at the equinox, September 23rd, continued ninety-three days; and winter, beginning at the winter solstice, December 22nd, lasted ninety days; yet, notwithstanding their unequal lengths, the amounts of heat and light which the whole earth received were equal in the several periods.[3] Not only so, but the amount of heat received by each hemisphere during the year is precisely the same. Croll has described this result lucidly: "Whatever may be the eccentricity of the earth's orbit, the two hemispheres must receive equal quantities of heat per annum;

[2] Haughton, Trans. Roy. Ir. Acad. xxviii. 49.
[3] Smithsonian Contributions, ix. 12.

for proximity to the sun is exactly compensated by the effect of swifter motion. For example, whatever extra heat the southern hemisphere may at present receive from the sun during its summer months, owing to its greater proximity, is exactly compensated by a corresponding loss arising from the shortness of the season; and, on the other hand, whatever deficiency of heat we in the northern hemisphere may at present have during our summer half-year, in consequence of the earth's distance from the sun, is also exactly compensated by a corresponding gain, owing to the length of the season.

Arago puts it in another way. The sun is not always equally distant from the earth; at present its least distance is observed in the early days of January, and the greatest in the early days of July; but a time will come when the minimum will occur in July and the maximum in January. Would the two summers under these circumstances differ sensibly? At first sight everyone would probably answer in the affirmative, since there is a difference in the distance of the earth amounting to a thirtieth of the whole. "Let, however, the consideration of the velocities be introduced into the problem, and the result will be on the side opposite to that we originally imagined; the part of the orbit where the sun is found nearest the earth is at the same time the point where the luminary moves most rapidly along. . . . The hypothesis we have adopted would give on account of the lesser distance, a springtide and summer hotter than they are in our days; but on account of the greater rapidity, the sum of the two seasons would be shorter by about seven days. Thus then, all things considered, the compensation is mathematically exact."[4] We have got to this point, then, that the earth receives precisely the same amount of heat during each of the four astronomical seasons, and that each hemisphere also receives a precisely equal amount of heat during the year, whatever the variation in eccentricity of the orbit may be. The only substantial effect upon climate, as measured by the seasons, caused by such a variation, is to make the seasons longer or shorter, that is to say, to distribute the heat over a greater or less number or days, and thus to make the seasons on the one

[4] Op. cit. 221-224.

hand more contrasted, or on the other more equable; but inasmuch as precisely the same amount of heat in each season is received, whatever be the eccentricity, it will be seen that the total direct effect, as measured by the number of heat-rays received by the earth, and by each hemisphere annually, is *nil.*

Under any circumstances this change could not produce such a climatic effect as is required by the Glacial theory. The tempering of the relative severities of summer heat and winter cold might be a great element in affecting the distribution of vegetable and animal life, but not in affecting the amount of cold or heat on the earth as indexed by such a measure as ice. It takes the same number of thermal degrees to melt the same quantity of ice; and if we apply the same amount it does not matter how we apply it, whether quickly or slowly, the result must be the same. The alternation of very hot summers and very cold winters will have precisely the same effect as that of moderate summers and moderate winters, if the amount of actual heat received and the amount of ice to be melted is the same in both cases, and if it only differs in the length of time over which it is distributed.

In regard to the daily average of heat received, it must, on the other hand, be remembered that while the increase of the eccentricity of the earth's orbit would increase the disparity between summer and winter, it, on the other hand, would, as we have seen, tend to increase the total mean heat of the earth. As the eccentricity enlarged, therefore, there would be a larger quantity of heat to distribute, which would tend to raise the temperature of each day in winter, and thus partly compensate for the sun's heat being distributed over a longer time.

Treating the issue apart from meteorology, and simply as a question of the effective heating-power of the sun as applied to the earth, it seems plain that, except the slight increase in the mean annual temperature already referred to, the alteration in the eccentricity of the earth's orbit can have little or no effect upon climate. Arago says: "The changes of form and of the position of the terrestrial orbit are mathematically inoperative, or at most their influence is so minute that it is not indicated by the most delicate instruments." [5]

Croll speaks equally emphatically. "All the conclusions arrived at by astronomers," he says, (i.e. the conclusions adverse to an appeal to increased eccentricity as a cause of glacial climate) "are perfectly legitimate so far as the direct effects of eccentricity are concerned; and it was quite natural, and in fact proper, to conclude that there was nothing in the mere increase of eccentricity that could produce a glacial epoch. How unnatural would it have been to have concluded that an increase in the quantity of heat received from the sun should lower the temperature and cover the country with snow and ice!"

Mr. J. Geikie speaks exactly in the same way. He approves of the argument of the astronomers that the earth receives the same total amount of heat in the aphelion as in the perihelion part of its orbit, and says that this is true however much we may vary the eccentricity, and that the only effect of increased eccentricity is to make the earth's climate as a whole warmer.

We will now examine another astronomical cause to which a good deal has been assigned in the shape of climatic change, namely, the change in the obliquity of the earth's axis.

The degree of obliquity of the earth's axis determines, of course, the area over which the heat received from the sun is distributed.

If the earth revolved at right angles to the plane of the ecliptic, the two hemispheres would have the same seasons at the same time; each would have its summer when the earth was nearest the sun, and its winter when it was farthest away; and all places on the same parallel of latitude, either north or south of the equator, would receive an equal quantity of light and heat in each twenty-four hours. The portion of the globe lighted by the sun would invariably extend from one pole to the other, and in both hemispheres the days and nights would always be twelve hours long.

[5] Arago, op. cit. 221-224.

Lastly, and most important for our purpose, inasmuch as the sun would in this case be always over the equator, the equatorial region would receive the greatest quantity of heat possible while the two poles would be in a condition of most wintry cold.

If the earth's axis, instead of being at right angles to the plane of the ecliptic, were to be moved so as to incline somewhat to it, it is obvious that the sun would not be always over the equator, but would travel slightly into each hemisphere, and *pro tanto* mitigate the climate of that hemisphere, and cause the other to be more rigorous, which effect would be reversed every six months; and the more the axis was inclined, the more would this effect ensue, causing a diminution of equatorial heat and an enhancement of the heat of each pole during the summer, and the reversal of this process during the winter.

As a matter of fact the earth's axis and the plane of the ecliptic are now inclined to each other at an angle of about 23 degrees. To the observer of the heavens the result is that the sun appears to cross the equator twice a year, in March and September, at the equinoxes, when alone there is equal day and night everywhere. After crossing the equator in March it seems to move northwards until it reaches its farthest limit in June, where it remains stationary a short time, whence the point is known as the solstice (*solis statio*), or rather the summer solstice. It then moves southwards, crosses the equator, and in December reaches its farthest limit southward, which is known as the winter solstice.

This apparent movement of the sun makes it pass in turn directly over every part of the zone bounded on either side by the so-called Tropics of Cancer and Capricorn, and the distance of these tropics from the equator, measures, in fact, the precise degree of obliquity of the ecliptic, and the contrast, other things being equal, between summer and winter in each hemisphere. If this obliquity be subject to variation, it will vary the distance to which the sun will travel every half-year, either north or south of the equator. The greater the inclination the greater the distance it will travel, that is to say, the greater will be the distance between the equator and the two tropics and *vice versa*.

As a matter of fact, the obliquity of the ecliptic is subject to variation, due to the manifold action of the planets.

"The amount of this variation in the plane of the ecliptic," says Herschell, "is about 48″ per century, and has long been recognized by astronomers, by an increase of the latitudes of all stars in certain situations and their diminution in the opposite regions. Its effect is to bring the ecliptic by so much per annum nearer to coincidence with the equator, to cause the two tropics to approach the equator, and the arctic and antarctic circles to approach their respective poles at the rate of forty-five feet annually; the result being to make terrestial climates more uniform. This is the present effect, but the movement is not continuous but cyclical, and the result is an oscillatory movement, by which, after many years, the obliquity returns to its former position, thus causing a certain cyclical variation of climate. Laplace was the first to measure *the amount* of the change, and he calculated thatthe oscillation was about 1° 22′ 34″ on each side of 23° 28′, which was the obliquity in the year 1801. Mr. Stockwell[6] has recalculated the position and arrived at virtually the same result as Laplace, and shown that the limits of the variation of obliquity are from 24° 35′ 58″ to 21° 58′ 36″, whence, he says, it follows that the greatest and least declinations of the sun at the solstices can never differ from each other to any greater extent than 2° 37′ 22″.[7]

[6] Smithsonian Contributions to Knowledge, vol. xvii.

[7] It is true that Colonel Drayson, in a paper entitled "The Last Glacial Epoch," claims to have shown that astronomers like Laplace, Leverrier, Herschell, Airy, Stockwell, and others are mistaken, and that the obliquity has once been much greater, so great in fact as to bring the arctic circle down to lat. 54° 34′ 13″ N.; but it is quite plain that he has been mistaken, and Mr. Belt explains how. His main argument is that observations extending over the last 400 years show that the pole of the earth, instead of describing a small circle round the pole of the ecliptic (*vide infra,* precession), makes a much wider circle round a point 6 degrees distant from the pole of the ecliptic, and 29° 25′ 47″ from the pole of the heavens. If precession acted alone, the curve performed by the earth's axis would be a circle; but inasmuch as this curve is affected by nutation, and its outline is accordingly sinuous, Mr. Belt says that Colonel Drayson has mistaken for the segment of a larger circle the motion of the terrestrial pole down the trough caused by one of the sinuosities in question as it dips below the mean distance from the centre. (Belt, Quart. Journ. of Science, 2nd ser. iv. 444). And we may rest assured that the long roll of distinguished astronomers who have examined the problem have not in fact been mistaken in their induction.

Mr. Belt, who relies upon increased obliquity to explain his glacial epoch, urges that while the astronomical forces as measured by Laplace and others undoubtedly limit the range of variation in the inclination of the earth's axis to the amount already referred to, yet that the conditions would be greatly modified if the earth's crust were sensibly augmented in certain places, and he calls on astronomers to reconsider the problem in view of such a change having taken place. Additional data, he urges, respecting the exact figure of the earth have accumulated since the problem was last treated, and this has gone to show that the earth's contour is more irregular than was supposed.[8]

The challenge has not been left unanswered. "Professor George Darwin" has, in fact, shown that "supposing the whole equatorial regions up to latitude 45° north and south were sea, and the water to the depth of 2000 feet were placed on the polar regions in the form of ice, and this is the most favourable redistribution of weight possible for producing a change of obliquity, it would not shift the arctic circle by so much as an inch."[9] We may take it, therefore, that the calculation of Laplace fixes the extreme possible variation in the obliquity of the ecliptic very closely, and we will proceed to examine the question on that hypothesis.

We will first consider it as an isolated one, in so far as it is affected by varying obliquity only, and apart altogether from other elements.

There are two ways in which the obliquity of the earth's orbit affects climate: namely, first, in regulating the relative quantities of heat received by either hemisphere in summer and in winter; and secondly, in regulating the distribution of that heat over the surface of the hemisphere.

Sir Robert Ball, in his recent publication, entitled "The Cause of the Ice Age," claims to be the first tc have shown that, with the present obliquity, "of the total amount of heat received from the sun on a hemisphere of the earth in the course of a year 63 per cent. is received during the summer, and 37 per cent. is received during the winter."[1] Sir

[8] Belt, op. cit. 463.

[9] Croll, Discussions on Climatology, 4.

[1] Op. cit. 90.

Robert Ball accepts the rigid limitation in the extent of the variation of the obliquity proved by Laplace, and he allows that within any reasonable period of geological time this limit has not been approached. He accordingly treats the present obliquity as virtually constant, and says of the numbers above quoted, that "they are both functions of a single element, namely, the obliquity of the ecliptic. As this fluctuates but little, at least within the periods that are required for recent ice ages, the numbers we have given are regarded as sensibly constant throughout every phase through which the earth's orbit has passed within glacial times."[2] Being a constant, it may, it seems to me, be completely disregarded in this discussion, nor can I see how Sir Robert Ball connects it with an ice age. Whatever potency it has is being exerted now just as much as it would be then. If it were an efficient cause of an ice age, we ought to be passing through one now. We may therefore disregard this law, as an element which remains the same, and which was as much present in the supposed interglacial mild periods as in the glacial severe ones, and which may be discarded when we are considering the effects of change and flux of climate.

Let us pass on, therefore, to the second effect of the inclination of the earth's axis; namely, the variation in the area over which the heat received in each hemisphere is distributed. It is clear that as the sun leaves the equator and approaches one pole, it must mitigate its climate for the time being, while it increases the severity of the other from which it moves away; and that the farther we make the sun travel north and south, that is, the more we increase the obliquity, the more pronounced will be the effect in question, until, when the utmost limit of obliquity is reached and the earth's axis lies in the same plane as that of the earth's orbit, each hemisphere, with its pole for a focus, will have six months of continual sunshine and six months of continual darkness, and thus have a very hot summer alternating with a very cold winter. This seems plain enough, although Mr. Belt, in his well-known paper, seems to have entirely gone astray on the point, and argued as if when

[2] Id. 121.

there was no obliquity, perpetual spring would prevail at the poles, while he attributes the present severity in the Arctic and Antarctic regions to the obliquity of the axis; overlooking the fact that the poles could not suffer more deprivation of sunshine than when the sun was farthest removed from them, namely, when it was continuously over the equator.

Colonel Drayson, who admitted that the effect of altering the inclination of the axis would be to intensify the contrast between summer and winter in each hemisphere, nevertheless seemed to suppose that, if we could so change the angle of obliquity as to bring the Arctic circle into the Temperate Zone, we should cause an arctic climate to prevail in the temperate regions, and thus do much to solve the question of the cause of the glacial epoch.

Croll points out, with his usual perspicuity, the fallacy of this notion. "The polar regions," he says, "owe their cold not to the obliquity of the ecliptic, but to their distance from the equator. Increased obliquity in part tempers their climate; it diminishes the heat received in the equatorial regions, and increases that about the poles. With the present obliquity the proportion is 100 to 42·47. If the obliquity were increased to 35, the amount at the equator would be reduced to 94·93 parts, and that at the poles increased to 59·81, being an increase at the poles of nearly one half. At latitude 60°, the present quantity is 57 parts, but this would increase to 63 under similar conditions. If the Arctic circle were brought down to England, the temperature of these islands would be enhanced and not diminished. The winters would be colder, but the summers would be hotter in a much greater degree, and the greater the obliquity the more would the increase of summer heat exceed its decrease in winter. If the obliquity were increased until it reached its absolute limit of 90°, when the earth's axis would coincide with the plane of the ecliptic, the Arctic circle would then extend to the equator. Would this," says Croll, "produce a glacial period? A square foot at the pole would receive as much heat as a square foot at the equator if the sun remained on the equator the entire year, and the total amount received would be as the ratio of half the circumference of a circle to the diameter, *i.e.*, 1·5708 to 1, or half as much heat again; and all this heat received by the poles would be concentrated into six months, so that they would be receiving twice as much heat as is received by the equator at present in the same time, and more than three times what it received then. There would be continuous sun there for six months without the ground having the opportunity of cooling for a single hour. Nothing living on the earth could exist in the polar regions under so fearful a temperature as would then prevail in the summer months. How absurd to suppose that this would tend to produce a glacial epoch. Not only would every particle of ice in the polar regions be dissipated, but the very seas around the pole would be at boiling-point. Colonel Drayson, in fact, accepts this conclusion, and his glacial period is an alternation of great cold in winter and great heat in summer; but alternations of this kind would not produce a glacial epoch nor accumulate land-ice and icebergs. Great cold in a single winter would involve a deep snowfall, which would disappear in the following summer; land-ice of any thickness requires the accumulated snows of centuries for its production." This reasoning of Croll's seems conclusive.

In the previous discussion, I have endeavoured to isolate the direct effect of increased obliquity.

In calculating this effect as a practical issue, we must remember that the question cannot in practice be entirely detached, but that the result of changed obliquity is complicated with the result of changed eccentricity in the orbit and changed velocity in the earth's motion, or, as Mr. Meech puts it, the variation in the intensity of the sun in different latitudes is a function of the eccentricity and the obliquity.

Taking this into consideration, the problem of measuring the actual result can be approximately solved.

Mr. Meech has calculated and given tables of the sun's present annual intensity in different latitudes. The general result he arrived at is, that if we assign twelve thermal months to the equator, at the Tropic of Cancer or Capricorn there will be but eleven months; in the latitude of New Orleans there will be ten and a half, in the latitude of Philadelphia nine and a half, at London eight, at the polar circle six, and at the north or south pole five.[3] Taking these figures as a basis—·0187 as the eccentricity of the earth's orbit 10,000 years ago, and taking the formula for obliquity from Struve and Peters' tables, he arrived at the result that the annual intensity within the torrid zone then averaged one thermal day and a half less than now. From latitude 35° to lat. 50° it was virtually the same as at present. Above 50° the annnal intensity was then greater in an increasing rate towards the pole at which point it was between seven and eight thermal days greater than at the present time; in other words, the poles, both north and south, 10,000 years ago received twenty rays of solar heat in a year where they now receive but nineteen. Owing to change of obliquity of the ecliptic the sun may be compared to a swinging lamp; at the former period it apparently moved farther to the north and to the south, passing more rapidly over the intermediate space.

In this calculation the difference of obliquity allowed for is 1° 15′, which is very near Laplace's maximum variation of 1° 22′ 34″ above or below the present obliquity. It thence follows that since the sun and earth were put in their present relation to each other, the annual intensity upon the temperate zone has never varied; between the tropics it has not varied from its present annual amount by more than about $\frac{1}{240}$th part, and is now very slightly increasing. The most marked variation has been at the poles, where the variation amounts to four times that at the equator. In its annual amount, the polar cold is now very slowly increasing from century to century, which effect must continue so long as the obliquity of the ecliptic is diminishing.[4] Croll has followed up Mr. Meech's calculations somewhat farther. He says, according to Mr. Meech, if 365·24 thermal days represent the present total annual quantity of heat received at the equator from the sun, 151·59 thermal days would represent the quantity received at the poles. Adopting the same mode of calculation, it follows that when the obliquity was at its maximum, the quantity of heat received at the equator would be 363·51 thermal days, and at the poles 160·04 thermal days. The equator would therefore receive 1·73

[3] Op. cit. 30.

[4] Meech, op. cit. 38.

thermal days less heat and the poles 8·45 thermal days more heat than at present.[5] When the obliquity was at its maximum, therefore, the Poles would be receiving nineteen rays for every eighteen they are receiving at present, or nearly as much heat as latitude 76° is now getting. This, all other things being equal, would mean a rise in the mean annual temperature equal to 14° or 15°, *provided*, of course, says Dr. Croll, that this extra heat was employed wholly in raising the temperature, which would be the case if the polar regions were free from ice and snow, but as they are covered with ice and snow, the extra heat would simply melt it, and the ice-covered surface could never rise above 32° ; the actual result would be the melting of an eighteenth more snow and ice than at present.

We seem driven by these results to the conclusion that, even if it were possible to alter the obliquity to any amount, no increase or decrease, however great, could possibly account for a glacial climate or a warm period. The change in the obliquity would no doubt increase the contrast between summer and winter in certain latitudes, but the increased cold of one season would be compensated by the increased warmth of the next, and the ice made in the winter would be melted in the summer. We see an example of this in the planet Mars, in whose case the obliquity of the ecliptic reaches an angle of 28° 42′. The result is that in Mars there are very hot summers and very cold winters, and we can actually watch how in the summer the snows which have covered almost a third of the planet's disc in the preceding winter disappear, this taking place in each hemisphere alternately.[6] In order to obtain glacial climates or the reverse, we must have some cause operating over several seasons, and not merely a succession of severely contrasted seasons, in which one exactly undoes the work of the other.

Dr. Croll sums up the case against the appeal to alterations in the obliquity of the ecliptic thus:—This theory appears to be beset by a twofold objection: (1) it can be shown from celestial mechanics that the variation in the obliquity of the ecliptic must always have been so small that

[5] Climate and Time, 399. [6] See Le Hon, L'homme fossile 340.

it could not materially affect the climatic conditions of the globe; and (2) even admitting that the obliquity could change to an indefinite extent, it can be shown that no increase or decrease, however great, could possibly account for either the glacial epoch or a warm temperate condition of climate in polar regions.[7]

Elsewhere he speaks of the absurdity of the supposition that an increase of obliquity can possibly account for a Glacial period.[8]

Let us, in conclusion, turn to another element in the earth's motion which is also liable to variation, namely, the well-known motion of the earth's axis due to precession of the equinoxes and nutation. In explaining this difficult question I shall avail myself of Professor Airy's lectures and of a paper by Mr. Burr in vol. i. of the "Intellectual Observer."

Hipparchus when observing, as was the fashion among ancient astronomers, the passage of certain stars across the mouth of a long shaft or well at the spring equinox, found on comparing his observations with those of Aristillus and Timocharis, who lived 150 years before, that the stars did not cross the field of vision at the same time, but a little earlier every year. After closely observing for thirty years he found that they moved westward about a degree in every seventy-five years, with the result that the solar year is 20 minutes 22·9 seconds shorter than the sidereal year. As the result observed was to carry the equinox to meet the sun, the movement was called the precession or acceleration of the equinoxes. It was reserved for Newton to explain this motion by means of his gravitation theory: "He arrived at the theoretical conclusion that the earth is not a perfect sphere, but flattened at the poles, which subsequent accurate measures showed to be correct. The equatorial diameter is twenty-six miles greater than the polar, producing a ring or belt of matter thirteen miles high at the equator. As, according to the law of gravitation, every particle of matter in the universe attracts every other particle of matter wherever situated, with a force directly proportioned to their mass, and varying inversely as the square of the distance, the sun

[7] Climate and Time, 8.

[8] Id. 416.

and moon, both large masses of matter and comparatively near the earth, act powerfully on it; but as there is more matter at the equator than elsewhere, and this projecting matter is nearer to the sun and moon than the matter at the centre of the earth, the projecting portion is more strongly attracted than the centre, and by its endeavour, in conformity with this attraction to approach the sun and moon, tilts the equator a little out of its position, so as to bring the earth's axis more nearly perpendicular to the equator. . . . This part of the earth cannot move alone. The earth's axis—which may be imagined as passing through it like a bar of iron from pole to pole, and stretching to its vanishing points in the heavens—must move too; and this effect also was early noticed, viz., that the place of the pole in the sky did move, and that it was clearly the direction of the pole that changed and not the stars themselves."

"If the path of the pole be traced out among the stars, it will be found that this extremity of the earth's axis moves in an arc of 23½° radius round the pole of the ecliptic."[9] The immediate effect is best seen, says Sir John Herschell, in the apparent position of the Pole-star. At the time of the construction of the earliest catalogues it was 12° from the pole; it is now only 1° 24′, and will approach yet nearer to within half a degree; after which it will again recede and slowly give place to others, which will succeed in its companionship to the pole. After a lapse of about 12,000 years, the star A Lyræ, the brightest in the northern hemisphere, will occupy the remarkable position of a Pole-star, approaching to within about 5° of the pole.[1] It will be seen that the result of this motion is that the earth's axis, in moving about its central point (which is fixed) as a pivot, generates a double cone about the axis of the ecliptic, and thus each half of it imitates the gyrations of a child's top. From this cause there is no variation in the angle at which the earth's axis is inclined to the ecliptic; the angle remains the same, but the axis itself points in a different direction.

The rate of motion is very slow. It has taken 2000 years to move 33°, and if it were not interfered with by other

[9] Op. cit. 358. [1] Outlines, etc., 191.

motions, to which I shall presently turn, it would be about 25,868 in going completely round.

The motion of precession here described is complicated by another motion called nutation. This was discovered by Bradley, who, while engaged in trying to find the parallax of a star in Draco, noticed that it had a proper motion which was not explainable by precession only, as previously understood. It was then seen that, inasmuch as precession is caused by the efforts of the sun and moon to pull the equatorial projection towards themselves, and that as the sun is not constantly vertical over the equator itself but moves from one tropic to the other, thus causing the seasons—the effect of this attraction in the way of disturbing the position of the equator itself must be greatest at the two solstices, when the sun is farthest from the equator and nothing at all at the two equinoctial points, when it is immediately vertical over the equator, where its pulling will have no tendency to tilt the axis.

Precession, *as affected by the sun,* is therefore greater in summer and in winter than at other times, and this irregularity constitutes Solar Nutation. A similar irregularity is caused by the moon every month, when it is north or south of the equator, and this forms a part of Lunar Nutation.

Another part of it, which is more important, is also more complicated. The ecliptic or orbit of the moon is always at an angle of 5 degrees to that of the earth. The two points where the orbits intersect, called nodes, are not fixed, but in consequence of the sun's attraction they are constantly moving, and go completely round the circuit of the orbit in about nineteen years. On the other hand, since the earth's equator is at an angle of 23½°. to that of its ecliptic or orbit, it follows that the inclination of the moon's orbit to the earth's equator is constantly changing. For half the nineteen years it is very little inclined to the earth's equator, and for the remaining half is much inclined to it. In the former case, precession goes on very slowly, and in the latter very rapidly; and this irregularity in the rate constitutes the great part of Lunar Nutation.

The tilting of the axis from side to side of its normal position is modified by the earth's rotation into motion in a small ellipse, having a major axis of 18″, which is completed in about nineteen years. The effect, if the rotation existed alone, would be the motion of the earth's pole in this small ellipse; but as precession is going on at the same time, and carrying the pole of the earth around the pole of the ecliptic in a circle of $23\frac{1}{2}°$ radius, the two motions become compounded, and the result is an undulation of the precession curve, the small ellipses of nutation never being completed, but the motion of the pole being turned into an undulatory circle instead of a uniform sweep round the ecliptic pole.

The effect of the combination of precession and nutation with the alteration in the obliquity of the earth's axis is, according to careful calculations, to increase the annual advance of the equinox from 50″·2 to 62″, and thus to reduce the cycle of 25,800 years due to precession to one of 21,000 years.

The result of the three combined motions has been tersely stated by Mr. Meech in another way. He says: "The perigee advances in longitude 11·8″ annually; by which the instant when the earth is nearest the sun will date about five minutes in time later every year. The time of perihelion, which now falls in January, will at length occur in February, and ultimately return to the southern hemisphere the advantage which we now possess."[2]

The effect of the two motions was also clearly stated by Reclus. "Every year the exact moment of the March equinox anticipates by about 20 minutes the time at which the corresponding equinox fell in the year preceding. Each revolution of the earth round the sun brings a fresh advance of 20 minutes in the determination of the equinox; and as during the long course of ages the axis of the earth does not intermit in this swaying motion, the time must come when the conditions of the seasons will be entirely changed. The relative position of the two hemispheres towards the sun will be reversed, and the long winter and short summer which characterized one will be transferred to the other."[3]

In the year 4000 B.C. the perihelion, or point of nearest approach of the earth to the sun, coincided with the autumnal

[2] Smiths. Contrib. ix. 38.   [3] "The Earth," 12.

equinox of our hemisphere or with the 22nd of September, while in the year 1250 it coincided with the winter solstice.[4]

The effect and the only effect therefore of precession and nutation, when combined with obliquity of the axis, is to reverse the position of the two hemispheres towards the sun every 10,500 years, and thus to reverse the conditions which prevail in them.

The ultimate result will be of course that sometimes our summer will take place when the earth is nearest to the sun, and sometimes when it is farthest from the sun.

Humboldt, it seems to me, stated the concrete result quite conclusively long ago. "It might be supposed at the first glance," he says, "that the occurrence of the perihelion at an opposite time of the year (instead of the winter, as is now the case, in summer) must necessarily produce great climatic variations, but on the above supposition, the sun will no longer remain seven days longer in the northern hemisphere; no longer, as is now the case, traverse that part of the ecliptic from the autumnal to the vernal equinox in a space of time which is one week shorter than that in which it traverses the other half of its orbit, from the vernal to the autumnal equinox. The difference of temperature, which is considered as the consequence to be apprehended from the turning of the major axis, *will on the whole disappear*, principally from the circumstance that the point of our planet's orbit in which it is nearest to the sun is at the same time always that over which it passes with the greatest velocity."[5] This seems conclusive.

If the combination here mentioned were a really effective factor in altering climate, we ought assuredly to find more evidence of it in the southern hemisphere. As Mr. E. L. Price says: "At the present time an aphelion winter and a perihelion summer both occur in the southern hemisphere, and we hear of no abnormal cold or increase of ice. Certainly South America is not invaded by ice, and our ships are going round Cape Horn as usual. It is not known that glaciers on the Andes have grown higher or longer, while it is known

[4] Lapperent, 33.
[5] Humboldt's Cosmos, Bohn's edition, iv. 458-9.

that by the rise of Norway of but three feet in a century, the mountain glaciers have grown larger by three hundred feet."[6]

So much for the combination of precession and nutation with an obliquity of the ecliptic.

If, however, we add the element of a varying eccentricity the case becomes a little different. As we have seen, the effect of precession is to reverse the position of the two hemispheres every 10,500 years. This cycle is much shorter than the grand cycles in which the periods of greatest and least eccentricity occur, so that several of the former may occur during one of the latter. The result of this is that sometimes our summer in the northern hemisphere will coincide in time with the greatest possible distance which the earth can get away from the sun, and sometimes our winter will similarly occur. To use astronomical language, either our summer or our winter may occur when the earth is in perihelion, and at the time of the earth's greatest eccentricity; the difference between this condition of things and their occurring when the earth is in aphelion with the same eccentricity has been supposed to offer a potent cause of climatic change.

In order to test the problem Mr. Meech has taken the most extreme care. He says: "Let us then look back to that primeval epoch when the earth was in aphelion at midsummer and the eccentricity at its maximum value assigned by Leverrier near to ·0777." He then calculates that this would diminish the midsummer intensity by about 9°, and increase the midwinter intensity by 3° or 4°, the temperature of spring and autumn being nearly unchanged, which he says does not seem adequate to the geological effects required.[7]

This exhausts the possible astronomical changes which can affect climate, and it seems to follow that under no circumstances known to us could astronomical causes produce such a result as the glacial period. Croll, a very acute person and a great champion of ultra-glacial views, says it would be absurd to assert that they could.[8] Mr. J. Geikie, the other great English champion of similar views, agrees with him. "Each of these arguments," he says, "is strictly consonant with well-

[6] The Glacial Epochs, 8. [7] Smiths. Contrib. ix. 40.
[8] Climate and Time, 13.

ascertained fact, and cannot possibly be gainsaid; and therefore it may be at once admitted that purely astronomical causes alone will not account for that wonderful alternation of extreme climates to which the geological record bears witness."[9]

A more neutral person, an astronomer and mathematician, Mr. Meech, sums up his lucid and admirable examination of the problem in these words:—"The preceding discussion seems to prove rigidly that, under the present system of physical astronomy, the sun's intensity could never have been materially different from what is manifested upon the earth at the present day. *The causes of notable geological changes must be other than the relative position of the sun and earth under their present laws of motion.*"[10]

[9] Great Ice Age, 103. [10] Op. cit. 41.

Printed in Dunstable, United Kingdom

91219924R00231